Seth G. Pritchard, PhD
Jeffrey S. Amthor, PhD

Crops and Environmental Change

An Introduction to Effects of Global Warming, Increasing Atmospheric CO_2 and O_3 Concentrations, and Soil Salinization on Crop Physiology and Yield

Pre-publication
REVIEWS, COMMENTARIES, EVALUATIONS . . .

"In this book the authors provide an overview of results on the effects of global environmental changes on crop productivity. This is a bold and difficult undertaking considering the enormity of the problem and the rapid expansion of this field. They have consciously focused on four environmental factors: global warming, soil salinization, and increases in atmospheric CO_2 and O_3 concentrations. These four changes in environment are largely a direct consequence of human activities. Each of these environmental factors has been previously examined in isolation, but, up until now, their combined effects and interactions on the physiology of plants have received limited attention. This book brings together these changes in the environment and its impact on plant growth with respect to physiology and ecology.

This book should be of interest to nonscientists concerned with production agriculture and to students and professionals with interests in the area of plant stress. The book focuses on such areas as cellular response, water relations, metabolism, photosynthate partitioning, plant nutrition, vegetative growth, reproductive growth, and biotic environment. Each chapter deals with the effects of the four individual factors and discusses the impacts of these factors on crop physiology."

Narendra K. Singh, PhD
Professor
Department of Biological Sciences,
Auburn University, Alabama

More pre-publication
REVIEWS, COMMENTARIES, EVALUATIONS . . .

"One of the problems with conducting research within the expanding realm of climate change is distilling useful information from the burgeoning well of data. Pritchard and Amthor have synthesized an excellent compendium of plant responses to the major facets of climate change. Their holistic approach to plant physiological responses to climatological stimuli provides an excellent springboard for diving into the often murky waters of climate change literature.

This text is as much ecological as it is agricultural. Many of the responses observed in agricultural systems are also applicable to many natural systems that receive regular disturbances. In addition, the background that each chapter provides on physiology makes this text a suitable choice as a general plant physiology text. Every scientist conducting research regarding environmental responses to climate change needs this book (so do their students!).

Micheal Davis, PhD
Assistant Professor of Botany
Biology Department,
University of Southern Mississippi

"*Crops and Environmental Change* is a valuable, carefully worked, and conceptually wide-ranging analysis of the response of plants, particularly crops and their yields, to environmental conditions. This well-written and -illustrated and excellently referenced text focuses on the rapid increase in CO_2 and the associated temperature rise. In addition, the consequences of the related increases in the concentration of the gaseous pollutant ozone and of soil degradation by salinity are effectively discussed. These topics are central to our understanding of what is happening to the biosphere as a consequence of human activity and the possibilities of ameliorating the potential damage to human food supply in a period of massive increase in the world's human population.

The book is timely and worthy of the task it set out to tackle—to understand the phenomena, process and mechanisms, and possible solutions. This careful and balanced analysis of the literature, with its sensible explanations, well- drawn conclusions, and identification of unresolved issues and possible ways forward is based on concise, clear summaries of metabolic and physiological processes. This should help readers with different backgrounds to understand the problems and solutions. Such a well-judged condensation of information and analysis will be relevant to the needs of a readership in a wide range of disciplines and different stages of training, from advanced undergraduate courses in plant sciences (and for some environmental sciences) through to professionals in agriculture, plant ecology, plant breeding, and plant molecular genetics researching the topics. Indeed it should be required reading, as it shows the difficulties likely to be experienced in achieving crop adaptation to rapid environmental change while showing how the task may be tackled. This will be an important addition to the current literature, useful for some time to come."

David W. Lawlor, PhD
Senior Principal Research Scientist (retired)
Crop Performance & Improvement Division,
Rothamsted Research,
Hertfordshire, UK

More pre-publication
REVIEWS, COMMENTARIES, EVALUATIONS . . .

"With *Crops and Environmental Change,* Seth Pritchard and Jeffrey Amthor have addressed perhaps *the* key topic to be addressed by science and societies facing large-scale environmental changes. What could be more crucial than consideration of how environmental changes are affecting, and will continue to affect, the world's abilities to produce crops? After all, when people don't eat, they don't worry about much else, and hungry people are unlikely to muster the energy to tackle the other impacts of environmental change around the globe.

Pritchard and Amthor are well-qualified to speak on this subject. Both are in professional positions that give them credibility on the topic of environment and crops (Pritchard in the U.S. Department of Agriculture, Amthor in the U.S. Department of Energy). Both are widely considered to be firsthand experts in this field of science by virtue of their own notable research contributions and collaborations in the scientific community. If you're going to dive into the wide-ranging topic of crops and environmental change you need a reliable source. You'll be hard-pressed to find any better than this book.

The authors write in a style that is easy to read, thankfully much more so than the typical research paper. Accessibility notwithstanding, the book is no lightweight. The text is supported by tables and figures, lists of key definitions, and an extensive number of literature citations. This book will be of great value to students, decision makers, and leading scientific experts as a key entry point to the wide array of topics in the vast, ever-growing, and often overwhelming body of knowledge relating environmental changes to crop production. The more people who make a serious choice to dive into this crucial topic, the more likely we will be to have a healthy population and economy in the future."

Steven R. Shafer, PhD
National Program Leader, Global Change,
USDA Agricultural Research Service,
George Washington Carver Center,
Maryland

Food Products Press®
An Imprint of The Haworth Press, Inc.
New York • London • Oxford

NOTES FOR PROFESSIONAL LIBRARIANS AND LIBRARY USERS

This is an original book title published by Food Products Press® an imprint of The Haworth Press, Inc. Unless otherwise noted in specific chapters with attribution, materials in this book have not been previously published elsewhere in any format or language.

CONSERVATION AND PRESERVATION NOTES

All books published by The Haworth Press, Inc. and its imprints are printed on certified pH neutral, acid-free book grade paper. This paper meets the minimum requirements of American National Standard for Information Sciences-Permanence of Paper for Printed Material, ANSI Z39.48-1984.

Crops and Environmental Change

An Introduction to Effects of Global Warming, Increasing Atmospheric CO_2 and O_3 Concentrations, and Soil Salinization on Crop Physiology and Yield

FOOD PRODUCTS PRESS®

Crop Science

Amarjit S. Basra, PhD
Senior Editor

Bacterial Disease Resistance in Plants: Molecular Biology and Biotechnological Applications by P. Vidhyasekaran

Handbook of Formulas and Software for Plant Geneticists and Breeders edited by Manjit S. Kang

Postharvest Oxidative Stress in Horticultural Crops edited by D. M. Hodges

Encyclopedic Dictionary of Plant Breeding and Related Subjects by Rolf H. G. Schlegel

Handbook of Processes and Modeling in the Soil-Plant System edited by D. K. Benbi and R. Nieder

The Lowland Maya Area: Three Millennia at the Human-Wildland Interface edited by A. Gómez-Pompa, M. F. Allen, S. Fedick, and J. J. Jiménez-Osornio

Biodiversity and Pest Management in Agroecosystems, Second Edition by Miguel A. Altieri and Clara I. Nicholls

Plant-Derived Antimycotics: Current Trends and Future Prospects edited by Mahendra Rai and Donatella Mares

Concise Encyclopedia of Temperate Tree Fruit edited by Tara Auxt Baugher and Suman Singha

Landscape Agroecology by Paul A. Wojkowski

Concise Encyclopedia of Plant Pathology by P. Vidhyasekaran

Molecular Genetics and Breeding of Forest Trees edited by Sandeep Kumar and Matthias Fladung

Testing of Genetically Modified Organisms in Foods edited by Farid E. Ahmed

Agrometeorology: Principles and Applications of Climate Studies in Agriculture by Harpal S. Mavi and Graeme J. Tupper

Concise Encyclopedia of Bioresource Technology edited by Ashok Pandey

Genetically Modified Crops: Their Development, Uses, and Risks edited by G. H. Liang and D. Z. Skinner

Plant Functional Genomics edited by Dario Leister

Immunology in Plant Health and Its Impact on Food Safety by P. Narayanasamy

Abiotic Stresses: Plant Resistance Through Breeding and Molecular Approaches edited by Muhammad Ashraf and Philip John Charles Harris

Multinational Agribusinesses edited by Ruth Rama

Crops and Environmental Change: An Introduction to Effects of Global Warming, Increasing Atmospheric CO_2 and O_3 Concentrations, and Soil Salinization on Crop Physiology and Yield by Seth G. Pritchard and Jeffrey S. Amthor

Teaching in the Sciences: Learner-Centered Approaches edited by Catherine McLoughlin and Acram Taji

Crops and Environmental Change

An Introduction to Effects of Global Warming, Increasing Atmospheric CO_2 and O_3 Concentrations, and Soil Salinization on Crop Physiology and Yield

Seth G. Pritchard, PhD
Jeffrey S. Amthor, PhD

Food Products Press®
An Imprint of The Haworth Press, Inc.
New York • London • Oxford

Published by

Food Products Press®, an imprint of The Haworth Press, Inc., 10 Alice Street, Binghamton, NY 13904-1580.

Cover design by Jennifer M. Gaska.

Library of Congress Cataloging-in-Publication Data

Pritchard, Seth G.

Crops and environmental change : an introduction to effects of global warming, increasing atmospheric CO2 and O3 concentrations, and soil salinization on crop physiology and yield / Seth G. Pritchard, Jeffrey S. Amthor.

p. cm.

Includes bibliographical references and index.

ISBN 1-56022-912-8 (hard : alk. paper) — ISBN 1-56022-913-6 (soft : alk. paper)

1. Crops—Physiology. 2. Crops—Effect of atmospheric carbon dioxide on 3. Plants—Effect of global warming on 4. Soil salinization. I. Amthor, Jeffrey S. II. Title.

SB112.5.P75 2005
632'.1—dc22

2004016359

CONTENTS

Preface and Acknowledgments

A disadvantage of writing a book that touches on so many topics is that our expertise, or lack thereof, was often stretched to its limits (and perhaps beyond). The advantage, and the reason we undertook this project, is that we might integrate and interpret effects of environmental change on crops across the wide range of topics covered. The intent was to provide an *introduction;* much longer books could be (and in some cases have been) written on any of the individual topics touched on here.

There is perhaps another significant disadvantage of writing a book of this nature. Herein we have concerned ourselves with the influence of four key global environmental changes on crop physiology and yield: global warming, increasing atmospheric CO_2 and O_3 concentrations, and soil salinization. It might be argued that other environmental changes could be even more important for the future of food production. For instance, increasing UV radiation resulting from the thinning of stratospheric ozone in addition to pollution of air with oxides of both sulfur and nitrogen could prove to be important global environmental changes. Degradation of soil through erosion and desertification are also pressing problems that must be addressed by the agricultural community. Unfortunately, the list of environmental changes that crops might face in the future goes on and on. We therefore chose to discuss the four environmental changes we feel are of greatest consequence based on our current understanding. Another reason we chose these four changes was because a significant knowledge base has been established in these areas and we felt that a synthesis of these topics might be especially valuable for young scientists trying to find their way through a complex literature.

A significant issue that becomes apparent from even a cursory summary of existing knowledge is that from the crop's perspective the important point is the net effect of all the environmental changes that occur, or might occur, at any given place and time. Most research focuses on only one, or a very few, environmental changes instead of the concomitant multiple changes occurring in our world. For example, increasing CO_2, O_3, and global warming are all related (i.e., one is unlikely to occur without the others) through energy production from fossil fuels and all presumably will change together. Unfortunately, however, the bulk of past research has focused on one factor at a time. As a result, much is known about the effects of a change in CO_2 concentration on yield of some crops, but much less is known about the effects of a change in CO_2 *in combination with* a change in temperature, O_3, and soil salinization on the yield of those same crops. It is obviously the effects of all the environmental changes occurring, rather than just CO_2 (or just temperature or just O_3 or just soil salinization), that is of importance to any crop. Although much of this book focuses on single factors, it was our intent to emphasize that the net effect of multiple environmental changes on a crop is far more important than the effect of a single factor on that crop.

We are grateful to many people who supplied data, photographs, Internet URLs, and reprints and preprints, as well as helping in other ways with this book. An incomplete list of names is John Angus, John Brennan, David Connor, Dave Easterling, Tony Fischer, Jürg Fuhrer, Margaret Goodbody, Catherine Grieve, Mary Beth Kirkham, Cathy Lester, Matt Jenks, Bob Loomis, Mike Miller, Steve Prior, Jim Reynolds, Hugo Rogers, Michael Scanlon, Jim Specht, Dave Tingey, Dave Tremmel, and Lew Ziska.

Most of all, I (SGP) would like to thank Heather and Zoey for their support during the course of this project and my parents, Bruce and Donna, for coming to all my ball games.

ABBREVIATIONS

ABA	Abscisic acid
ADP	Adenosine diphosphate
AOS	Active oxygen species
APOD	Ascorbate peroxidase
ATP	Adenosine triphosphate
C_i	Intercellular [CO_2]
CAM	Crassulacean acid metabolism
CAT	Catalase
[CO_2]	CO_2 concentration
CFC	Chlorofluorocarbon
CTC	Closed-top chamber
DHAR	Dehydroascorbate reductase
DGDG	Digalactosyl diacylglycerol
ET	Evapotranspiration
FACE	Free-air CO_2 enrichment
FAO	Food and Agriculture Organization of the United Nations
G_1	Gap 1
G_2	Gap 2
GC	Growth chamber
GCM	General circulation model
GR	Glutathione reductase
HI	Harvest index
ha	Hectare
HSP	Heat shock protein
IPCC	Intergovernmental Panel on Climate Change
LAI	Leaf area index
LGM	Last glacial maximum
MDHAR	Monodehydroascorbate reductase
MGDG	Monogalactosyl diacylglycerol
$NADP^+$	Nicotinamide adenine dinucleotide phosphate (oxidized form)

NADPH	Nicotinamide adenine dinucleotide phosphate (reduced form)
NASS	National Agricultural Statistics Service (USDA)
OPPP	oxidative pentose phosphate pathway
OTC	Open-top chamber
PEP	Phosphoenolpyruvate
POD	Peroxidase
P_i	Inorganic orthophosphate
PA	Polyamine
PCR	Photosynthetic carbon reduction cycle
ppb	Parts per billion
ppm	Parts per million
PAR	Photosythetically active radiation
PSI	Photosystem 1
PSII	Photosystem 2
Put	Putrescine
RGR	Relative growth rate
RPPP	Reductive pentose phosphate pathway
rubisco	ribulose 1,5-bisphosphate carboxylase-oxygenase
rubP	ribulose 1,5-bisphosphate
SFD	Seed filling duration
SFR	Seed filling rate
SLA	Specific leaf area
SOD	Superoxide dismutase
SORG	United Kingdom Stratospheric Ozone Review Group
SOS	Salt overly sensitive pathway
SPAC	Soil-plant-atmosphere continuum
SPAR	Soil-plant-atmosphere research
Spd	Spermidine
Spm	Spermine
SQDG	Sulphoquinovosyl diacylglycerol
TCA	Tricarboxylic acid cycle
USDA	United States Department of Agriculture
USSR	Union of Soviet Socialist Republics (Soviet Union, former)
WUE	Water use efficiency

Chapter 1

Introduction

> The person who does not worry about the future will shortly have worries about the present.
>
> Chinese proverb

Our environment—the one we live in and grow our crops in—is changing. Although environmental change is older than agriculture, and farmers have always had to contend with some type of environmental change, several notable, ongoing environmental changes have especially important implications for crop yield, production, and quality (see definitions in Box 1.1):

1. increasing global atmospheric carbon dioxide (CO_2) concentration;
2. climatic changes associated with increasing atmospheric concentrations of CO_2 and other "greenhouse gases," most importantly global warming;
3. increasing ozone (O_3) concentrations in the lower atmosphere across large crop-growing regions; and
4. soil salinization in areas of irrigated crops.

These environmental changes are largely caused by human activities. This book is about the effects of these four environmental changes on the physiology, growth, and yield of major field crops. Because of great uncertainty about future hydrologic cycle changes, changing patterns of amounts, timing, and geographic distribution of precipitation will be considered mainly in passing, though effects on crops of water stress in combination with the environmental changes being considered is included. The goal is to provide an introduction to the ramifications, both positive and negative, of these ongoing environ-

BOX 1.1. Definitions of Crop Quality and Yield

Production: Amount of desired crop product (e.g., wheat grain or potato tubers) harvested from a specific geographic area (e.g., a nation). The time scale is generally annual.

Productivity: Amount of desired crop product (e.g., wheat grain or potato tubers) harvested per unit land area of crop harvested. Amount is expressed in many units (e.g., bushels or kg) and so is land area (e.g., acres or ha). We use $kg \cdot ha^{-1}$. The time scale is usually annual, but for land cropped more than once during a year (e.g., irrigated annual double- or triple-crop continuous rice in tropical lowland Asia or irrigated annual double-crop rice-wheat systems in southern Asia) productivity may be expressed per crop cycle. Note that biological productivity may exceed harvested productivity because some fraction of grain (or tubers, etc.) grown may be lost (left on field) during harvest. Harvested productivity is the value reported for a crop. Productivity values usually include moisture in the seed, tuber, or other plant part harvested.

Quality: Chemical, nutritional, and/or physical properties of the harvested part of a crop plant. For example, protein concentration of grain, mineral or vitamin concentration of tubers, digestibility of forage, or amino acid mixture of grain protein.

Yield: Same as productivity.

mental changes for present and future crop production and food supply. An attempt is made to "explain" effects of environmental change on yield, at least to the extent that physiological understanding imparts mechanistic understanding of yield. In this regard, past environmental changes are considered when they help explain the effects of present and future changes on crops or can be used to test proposed explanations. In many cases, however, knowledge is limited to empirical experimental observations rather than mechanistic understanding of the effects of environmental conditions and changes on yield. In those cases, only "descriptions" of experimental results and correlations between environmental conditions and yield are possible, though some speculation about cause-and-effect relationships is included. In short, the objective is to explain the effects of environmental change on crop yield, but when explanations elude us, the goal is to describe the empirical database. We begin by noting that through several links, the environmental changes considered in this book are related to human population growth and increases in standards of living.

HUMAN POPULATION GROWTH

Although estimates of past population are imprecise, it is a fact that the human population increased significantly during recent millennia. For example, the population increased from perhaps 50 million about 3,000 years ago to about 425 million by 500 years ago (Figure 1.1), a 750 percent increase over 2,500 years. Population then increased 1300 percent during the past 500 years, with about 80 percent of that increase occurring since 1900 (Figure 1.1). During the period from 1995 to 2000, the population increased about 395 million (more than 216,000 per day), but that increase was about 12 million less than the increase from 1990 to 1995, showing that the rate of recent growth declined. Nonetheless, population growth remains exponential.

Many projections of future population exist, though they all entail some uncertainty. The U.N. Population Division gave low, middle, and high projections of 7.8, 9.3, and 10.9 billion people, respectively,

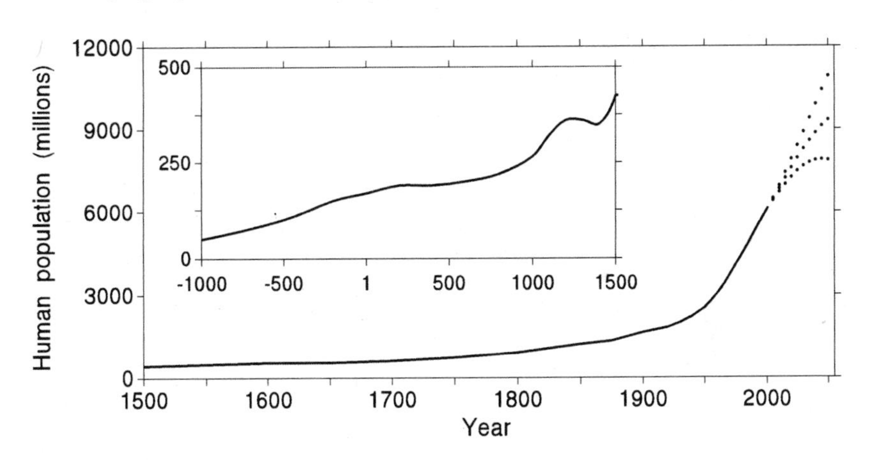

FIGURE 1.1. Human population history and projections to year 2050. Population was assumed to be unchanged from 1600 to 1650 due to losses in Europe (Thirty Year's War) and China (fall of Ming dynasty) with growth elsewhere. Inset shows spline fit to estimates from Kremer (1993) for the 2,500-year period ending in 1500, including population decline in the period 1200-1400 due to Mongol invasions and the black death. *Source:* Values for 1500-1940 are from Kremer (1993), values for 1950-2001 are from FAO (http://apps.fao.org), and multiple projections to 2050 (dotted lines) are based on low, medium, and high fertility models of the U.N. Population Division (http://esa.un.org/unpp).

for the year 2050 (Figure 1.1). The U.S. Bureau of the Census (www.census.gov, accessed October 24, 2003) projected a global population of 9.1 billion by 2050. These and other projections indicate that a population increase on the order of 30 to 60 percent should be expected during the next fifty years.

ENVIRONMENTAL CHANGES

Past population growth and increases in standards of living were accompanied by (and caused) many environmental changes. These changes occurred on local to global spatial scales. Some changes persisted (or will persist) for only short periods, but others will remain for centuries to come. The magnitude of present population increase is evidence that environmental changes will continue (and some changes may accelerate) in the future because the rates of most processes driving the changes increase with population size. Many environmental changes are innocuous with respect to crop yield, production, and quality, but others are either beneficial or harmful to crops. As mentioned previously, this book focuses on four environmental changes: increasing atmospheric CO_2 concentration, global warming, increasing O_3 concentration in the lower atmosphere, and soil salinization.

Atmospheric Carbon Dioxide Concentration

Past Changes

Measurements at Mauna Loa Observatory, Hawaii, show a clear atmospheric [CO_2] increase since 1958 (Figure 1.2). (Square brackets denote concentration, so "[CO_2]" is read "CO_2 concentration." The concentration units of parts per million on a volume basis, ppm, are used to describe atmospheric [CO_2].) From March 1958 (the first year of data collection) to March 2003 (the most recent year of data available at the time of writing), monthly mean atmospheric [CO_2] at Mauna Loa increased over 19 percent from 315.7 to 376.1 ppm (on average, over 1.3 ppm per year). Because the atmosphere is relatively well mixed, the Mauna Loa record reflects global changes on the time scale of years, though there are discernable geographic features with

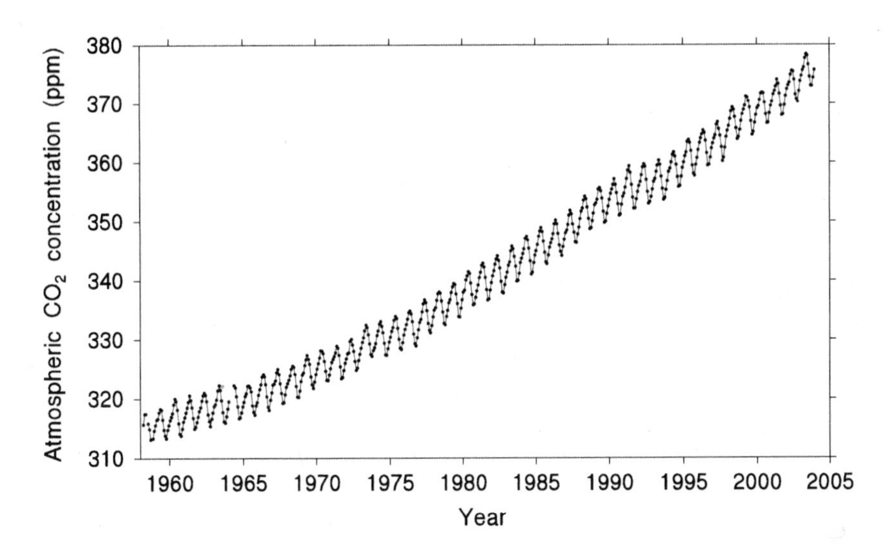

FIGURE 1.2. Atmospheric [CO_2] record from Mauna Loa Observatory, Hawaii, from March 1958 through December 2003. Air samples were collected continuously at the top of four 7-m towers and one 27-m tower on a barren lava field of an active volcano (19°32'N, 155°35'W, 3397 m above mean sea level). This is the longest continuous record of atmospheric [CO_2]. The site is favorable; air chemistry is only minimally affected by local vegetation and human activities, and influences from volcanic vents are excluded from the data. The record is a reliable indicator of regional atmospheric [CO_2] in the middle troposphere. (The atmosphere below about 11,000-16,000 m elevation is called the troposphere.) The sawtooth pattern corresponds roughly to net CO_2 uptake by the northern hemisphere terrestrial biosphere during summer and net CO_2 release during winter. Fossil fuel emissions are mostly aseasonal. Breaks in the line in 1958 and 1964 indicate missing data. *Source:* Data from Keeling and Whorf (2001).

respect to seasonal cycles of atmospheric [CO_2] (Conway and Tans, 1996). Also, local atmospheric [CO_2] often exceeds the global atmospheric mean value for locations near significant CO_2 sources, such as urban areas and/or fossil fuel-based energy production plants.

A longer view on past global atmospheric [CO_2] change is obtained from analyses of gases trapped in polar ice. At locations where ice accumulates over time, the [CO_2] in ice at different depths is an indicator of atmospheric [CO_2] at different times in the past. Based on ice cores collected at Law Dome, Antarctica, global atmospheric

[CO_2] was relatively stable (over decades to centuries) during the period 1000-1550, declined between 1550 and 1600, remained relatively stable until about 1750, and then increased from about 1750 to the 1950s (Figure 1.3a). The dip in atmospheric [CO_2] around 1600 was probably natural, and perhaps related to the Little Ice Age (Etheridge et al., 1996). Law Dome ice-core data are in good agreement with measurements at Mauna Loa Observatory after 1958 (not shown).

An even longer atmospheric [CO_2] record was obtained from ice collected at Vostok Station, Antarctica (Figure 1.3b). Those data show that atmospheric [CO_2] declined from about 230 ppm around 100,000 years ago to less than 200 ppm by 20,000 years ago, the time of the Last Glacial Maximum (LGM; a period of low global temperature). Then, between about 20,000 and 10,000 years ago, [CO_2] increased to about 260 ppm. This roughly corresponded with the beginning of crop domestication, and it is possible that there was a causal relationship between postglacial [CO_2] increase and crop domestication (Sage, 1995), but the topic is unresolved. Climatic, social, and/or other changes following the LGM may also have been critical to early crop domestication (Smith, 1995).

Historic atmospheric [CO_2] data are clear; large increases occurred between about 20,000 and 10,000 years ago, and large increases occurred during the past 200 years. The recent changes continue. What caused the increase between 20,000 and 10,000 years ago is incompletely understood, but the cause of most of the increase since 1750 to 1800 is known. From about 1750 to the mid-1950s, CO_2 released from the terrestrial biosphere was the main driver of atmospheric [CO_2] increase. Forest clearing (including burning and accelerated decomposition of wood, the forest floor, and mineral soil organic matter) and plowing of grasslands to establish crops and pastures (which oxidized a significant fraction of soil organic matter in the plow layer) released perhaps 8 to 15 Pmol CO_2 to the atmosphere by 1950 (Wilson, 1978; Houghton et al., 1983; Richards, Olson, and Rotty, 1983). (A 1.0 Pmol increase in atmospheric CO_2 causes a global [CO_2] increase of about 5.65 ppm.) Rapid expansion of global agriculture between the mid-1800s and the early 1900s was particularly important to CO_2 releases from the terrestrial biosphere. Since the early to mid-1900s, fossil-fuel burning released large amounts of CO_2 into the atmosphere, and today fossil burning is the major driver

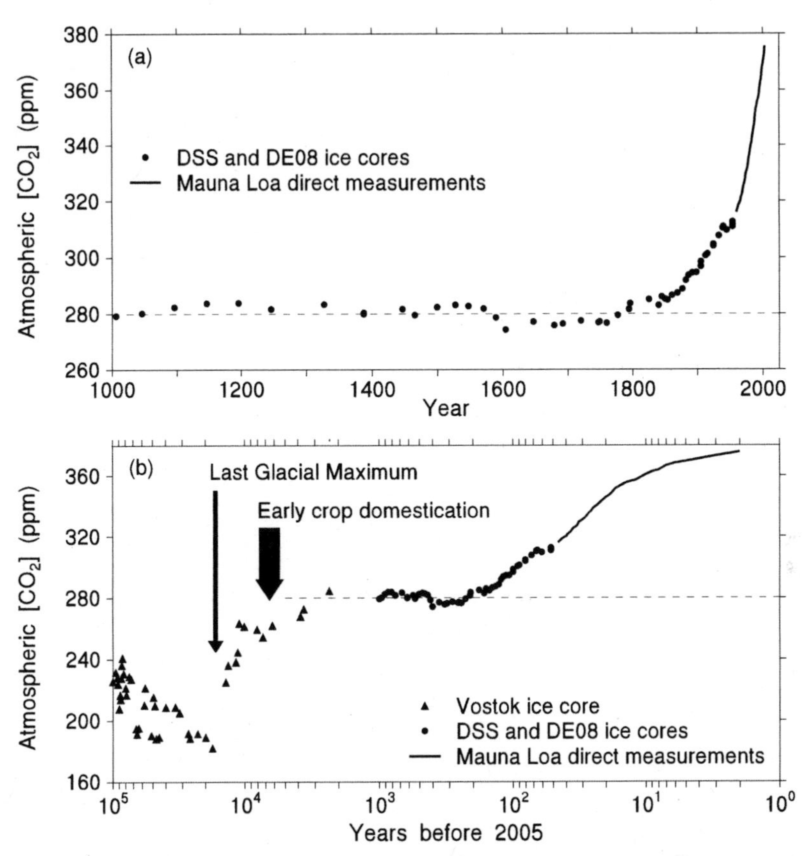

FIGURE 1.3. (a) Atmospheric [CO_2] inferred from measurements of gases trapped in the DSS and DE08 ice cores from Law Dome, East Antarctica (Etheridge et al., 1996). Direct atmospheric measurements at Mauna Loa Observatory, Hawaii, are also shown (annual means from Figure 1.2). The dashed line (= 280 ppm) approximates "preindustrial atmospheric [CO_2]." (b) Atmospheric [CO_2] during the past 100,000 years inferred from measurements of gases trapped in ice at Vostok Station, East Antarctica (Barnola et al., 1999). [CO_2]s in air in Vostok ice corresponding to the atmosphere 100,000 to 414,000 years ago were all in the range 184-299 ppm (not shown). Law Dome and Mauna Loa Observatory data from (a) are included to extend the record to the present. Arrows mark approximate times of the Last Glacial Maximum (Yokoyama et al., 2000) and early crop domestication (Evans, 1993; Smith, 2001). Width of arrows indicates approximate time range. Note the logarithmic *x*-axis in (b). The *y* axes differ in (a) and (b).

of atmospheric CO_2 increase. Through the year 2000, about 24 Pmol of CO_2 had been released to the atmosphere from fossil fuels (Marland, Boden, and Andres, 2000).

Some of the CO_2 released from land-use changes and energy production was absorbed by the ocean (Broecker and Peng, 1982), so the net increase of atmospheric CO_2 was less than the release of CO_2 during the past 200 years. Moreover, a large area of northern temperate forest is now regrowing, and therefore storing carbon, following agricultural abandonment and/or extensive wood harvesting in the late 1800s and early 1900s (Bormann and Likens, 1979; Casperson et al., 2000). Also, some cropland soils may now be recovering some of the carbon lost during earlier cultivation (Buyanovsky and Wagner, 1998). That recovery of soil carbon can be accelerated through the use of reduced-tillage crop management.

Future Changes

Future atmospheric [CO_2]s are uncertain, though [CO_2] will probably continue to rise for at least several decades, if not centuries. Some global carbon-cycle model predictions put global atmospheric [CO_2] in the range 600 to 1,000 ppm by the year 2100 (e.g., Cox et al., 2000; Prentice et al., 2001). How rapidly and to what extent actual atmospheric [CO_2] will increase as this century develops will depend in part on the amounts of energy used by humans and the sources of that energy (e.g., fossil fuels and biomass versus nuclear power, direct solar energy conversion, and hydrogen). These are impossible to predict with certainty.

Much experimental research addressing effects of elevated [CO_2] on crops includes [CO_2] treatments in the range of 550 to 750 ppm (in addition to present ambient values). Such experiments may apply to the period 50 to 100 years from now. With respect to effects of elevated [CO_2] on crops, this book largely mirrors that research.

Crop Canopy Carbon Dioxide Concentration

The atmospheric [CO_2] most directly affecting crops is that surrounding the crop, i.e., the [CO_2] in the crop canopy airspace. Canopy airspace [CO_2] is related to the mean global atmospheric [CO_2] on the annual to centennial time scale, but it is also affected by other factors on the hourly to seasonal time scale. Local CO_2 emissions from metropolitan areas, power plants, and other intense sources increase local ground-level atmospheric [CO_2] above the global mean atmospheric [CO_2]. At night, crop and soil respiration can greatly increase local [CO_2], whereas during the day, photosynthesis in and around a crop can reduce local [CO_2] (Figure 1.4). Thus, strong diel cycles of atmospheric [CO_2] often occur in a crop as a result of local photosynthesis and respiration. Nonetheless, the long-term, ongoing increase in global atmospheric [CO_2] does significantly affect [CO_2]s inside crop canopies, and future canopy [CO_2]s should increase about in proportion to the global atmospheric [CO_2] increase.

The continuing increase in atmospheric [CO_2] per se may affect (probably already has affected) crops for two reasons: (1) elevated [CO_2] inhibits photorespiration and stimulates photosynthesis in C_3

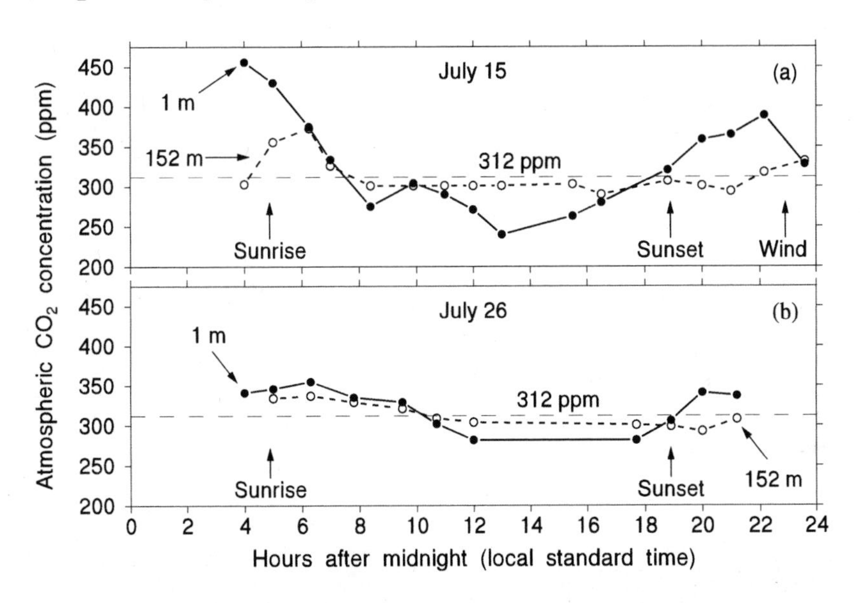

FIGURE 1.4. Atmospheric [CO_2] in and above a maize crop during two days in 1952 in Iowa. Measurements were made 1 m above the ground (•—•) (i.e., within the crop) and 152 m above the ground (◦--◦) on a still day (a) and a day with steady breeze (b). The points represent average [CO_2]s during the about 0.8-1.5 h period before each point. The dashed horizontal line at 312 ppm represents the approximate global mean atmospheric [CO_2] during 1952 (from Figure 1.3). A period of wind at the end of the still day eliminated the [CO_2] difference between heights. *Source:* Data from Chapman, Gleason, and Loomis (1954).

crops such as wheat, rice, potato, soybean, barley, and cotton, and (2) elevated [CO_2] increases water-use efficiency in both C_3 and C_4 crops, adding maize, sugarcane, and sorghum to the list of crops that could benefit from increasing [CO_2]. In general, these effects of elevated [CO_2], which are discussed in more detail in later chapters, enhance yield. Effects of elevated [CO_2] on crop quality are less clear or are inconsistent.

Global Warming and Related Climatic Changes

Increasing atmospheric concentrations of CO_2 and other "greenhouse gases" such as CH_4, N_2O, and chlorofluorocarbons (CFCs, used extensively in the 1960s and 1970s as refrigerants and aerosol propellants) may cause global warming (Watson, Houghton, and Yihui, 2001). Related climatic changes, such as altered patterns of cloud cover, precipitation, and frequency and intensity of extreme weather events may also occur (Easterling, Meehl, et al., 2000; Watson, Houghton, and Yihui, 2001). These present and likely future climatic changes have important implications, some negative and some positive, for crops.

Two key questions concerning global (and regional) climatic changes are: (1) Have they already begun (and if yes, when)? and (2) What will be the rate and extent of future changes?

Temperature Changes from 25,000 to 1,000 Years Before Present

The isotopic composition of ice at different depths at Vostok Station, Antarctica, provides a record of Antarctic temperature changes during the past several hundred thousand years (Jouzel et al., 1987; Petit et al., 2000). Presumably, changes in Antarctic temperature corresponded with larger, global climatic changes. The Vostok record indicates that Antarctic temperature increased 8 to 9°C during the past 25,000 years, with two-thirds of that increase occurring between about 16,000 and 12,000 years ago (Figure 1.5). The increase in temperature to modern values by about 9,000 years ago corresponded approximately with many estimates of early crop domestication (Figure 1.5). A causal relationship between postglacial global warming and crop domestication is possible.

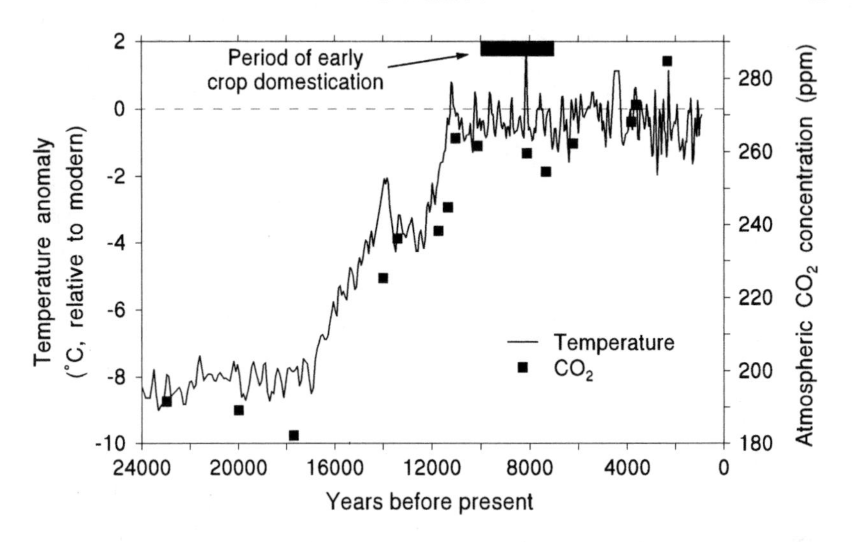

FIGURE 1.5. Atmospheric [CO_2] and temperature over Antarctica for the period 24,000 to 1000 years before present derived from measurements made on the ice core drilled at Vostok Station, East Antarctica. The horizontal bar along the top axis indicates the apparent period of domestication of many crops, i.e., 10,000 to 7,000 years before present. *Sources:* [CO_2]s from Figure 1.3; temperatures from Petit et al., 2000; Crop domestication from Evans, 1993; Smith, 2001.

In addition to playing a possible positive role in the onset and early evolution of agriculture, past climatic changes may have negatively impacted extensive areas of early agriculture. For example, increased aridity 4,000 to 4,200 years ago might have triggered crop failures in parts of the Middle East (Nissenbaum, 1994).

Temperature Changes During the Past 1,000 Years

The general pattern of global surface temperature during the past 1,000 years derived from "palaeoclimate proxy indicators" was one of modest cooling (perhaps 0.2°C in the northern hemisphere) from 1,000 to 100 years ago, followed by warming to the present (see, e.g., Figures 2.20 and 2.21 in Folland et al., 2001). The 1990s may have been the warmest decade of the past 1,000 years, and the 1900s was the warmest century (Palutikof, 2002). The so-called Medieval Warm Period during approximately 1,000 to 700 years ago now seems to have been geographically heterogeneous and of only moderate magnitude (Folland et al., 2001). Similarly, the Little Ice Age, from 400 to 200 years ago, now appears to have involved only modest cooling, and that cooling was not geographically uniform (Folland et al., 2001).

Measurements of air and sea-surface temperatures since the mid- to late 1800s provide records of recent decadal-scale warming patterns across Earth's surface (Easterling, Karl, et al., 2000; Jones et al., 2001). Globally averaged temperature changes were described by Easterling, Karl, and colleagues (2000, pp. 20, 101-120) as showing "a cooling of –0.38°C/century from 1880 to 1910, then warming of 1.2°C/century from 1911 to 1945, a period of no change to 1975, then strong warming of 1.96°C/century since the 1970s" (p. 102). Through 2002, the 1990s were globally the warmest decade of the modern instrumental period (since about 1856). The nine warmest years globally (annual mean temperatures), in descending order, were 1998, 2002, 2001, 1997, 1995, 1990 and 1999 (indistinguishable), and 1991 and 2000 (indistinguishable) (Jones et al., 1999, 2001; Palutikof, 2002; <lwf.ncdc.noaa.gov/oa/climate/research/2002/ann/ann02.html>, accessed February 5, 2003). The broad warming trend, on a globally averaged land-plus-ocean area basis, was about 0.4 to 0.8°C during the past 100 to 150 years (Folland et al., 2001).

The observed warming was not geographically or seasonally uniform. All seasons in the southern hemisphere experienced about 0.5°C warming during the record (Figure 1.6), and all seasons except September to November in the southern hemisphere experienced record warmth in 1998. But in the northern hemisphere, warming since the mid-1800s mainly occurred in autumn, winter, and spring (September to May); only in 1995, 1997, 1998, and 2000 (through 2000) were northern hemisphere summers (June to August) warmer than the warmest year of the mid-to-late 1800s (i.e., 1868) (Jones et al., 2001). When averaged annually, some regions experienced cooling during the period 1900-1998, such as parts of the southeast United States, equatorial West Africa, easternmost Brazil, central China, Venezuela, and southeastern Australia (Easterling, Karl, et al., 2000).

The warming over land may have exceeded the warming over ocean, and it is "virtually certain" that land air temperature increased 0.4 to 0.8°C since the late 1800s (Figure 1.7) (Easterling, Karl, et al., 2000, pp. 20, 108). The warming was manifested in observations

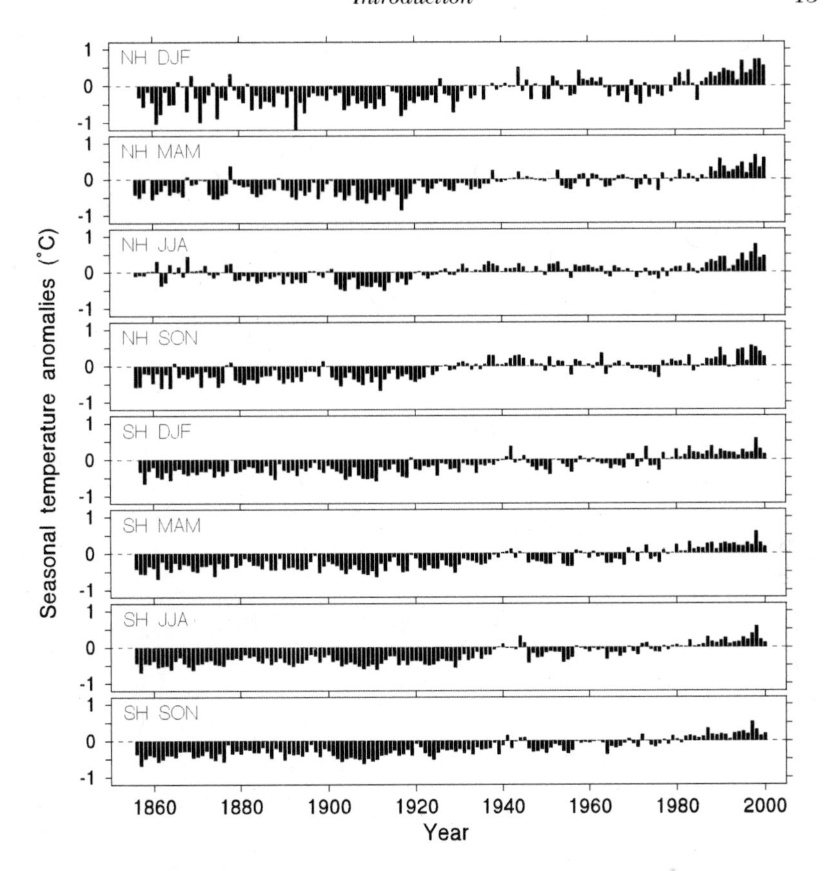

FIGURE 1.6. Seasonal hemispheric surface-temperature anomalies (relative to 1961-1990 temperature means) derived from measured land-surface air temperatures and sea-surface temperatures. Abbreviations: NH, northern hemisphere; SH, southern hemisphere; DJF, December (previous year)-January-February; MAM, March-April-May; JJA, June-July-August; SON, September-October-November. *Source:* Data from Jones et al. (2001).

such as later freeze dates and earlier thaw dates of northern hemisphere river and lake ice (Magnuson et al., 2000); warming is also manifested in increased heat content of the ocean (Barnett, Pierce, and Schnur, 2001). Whether warming during the past 50 to 100 years was caused (in part) by increasing levels of CO_2 and other greenhouse gases is actively debated, though most climate scientists believe the relationship is at least partly causal.

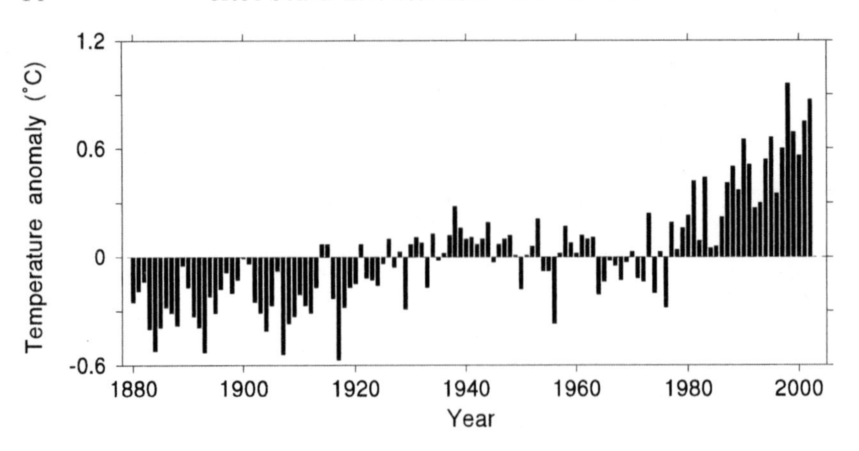

FIGURE 1.7. Global annual near-surface air temperature anomalies (relative to the mean temperature during the period) over land. *Source:* Data from <ftp. ncdc.noaa.gov/pub/data/anomalies/annual_land.ts>.

The diel temperature range (i.e., daily maximum minus daily minimum) over much of the earth's land has recently decreased because recent warming was greatest at night; exceptions are central Europe and parts of Venezuela, coastal United States, Korea, Japan, and northern parts of the North American Great Plains (Easterling, Karl, et al., 2000). Globally, the rate of increase of minimum (nighttime) temperature has been about double the rate of increase of maximum (daytime) temperature (Easterling et al., 1997; Easterling, Karl, et al., 2000). This may be related to increased cloud cover (Easterling, Karl, et al., 2000); daytime cloud cover reduces daytime surface warming and nighttime cloud cover reduces nighttime cooling.

The global hydrologic cycle is likely to be accelerated by global warming, which is consistent with increasing cloud cover. Although data on past precipitation are less authoritative than data on past temperature, annual precipitation over land apparently increased during the past 100 years in many regions, especially middle- to high-latitude areas (Easterling, Karl, et al., 2000). Decreases in subtropical precipitation have been observed, particularly in the Sahel region of Africa (Easterling, Karl, et al., 2000). Precipitation over parts of China may have declined since 1950 (Folland et al., 2001). Increased precipitation over middle- and high-latitude North America and Europe may have occurred mainly in autumn (Folland et al., 2001). In general, changes in precipitation have varied geographically and seasonally. Easterling, Karl, and colleagues (2000) noted evidence that extreme precipitation events are increasing in some areas, e.g., the conterminous United States, tropical Australia, Japan, and Mexico.

Future Changes

Physically based computer models of the earth's climate are the best tools available to predict or project regional and global climate during the coming decades and centuries. In particular, general circulation models (GCMs) of the coupled atmosphere and ocean that include sea ice and process-based submodels of land-surface energy, mass, and momentum exchanges are the most comprehensive of the tools for prediction. Although predictions by different GCMs vary in the details, the models are consistent in foretelling warming during the next 50 to 100 years because of increasing concentrations of CO_2 and other greenhouse gases. That warming, globally averaged and based on various estimates of future greenhouse gas emissions, could be on the order of 1.5 to 5.5°C (relative to 1990) by the year 2100 (Cubasch et al., 2001). The models indicate that warming will be greatest in winter, at night, and over mid- to upper-latitude land masses. GCMs also indicate an increase (perhaps 1 to 8 percent) in future global annual precipitation, but changes in precipitation may be geographically disparate. For example, although global precipitation is expected to increase, less precipitation may occur in southwestern and extreme eastern South America, Central America, Mexico and the southern United States, the Mediterranean region and the Middle East, and much of Australia (Cubasch et al., 2001). (Results of many GCM experiments, i.e., computer simulations or model runs, can be viewed at the IPCC Data Distribution Centre Web site at <ipcc-ddc.cru.uea.ac.uk>.) The frequency and "intensity" of extreme heat and precipitation events might increase in the future, though this is less certain than a general warming in the future. Finally, it is noted that many GCMs are now implemented with rather coarse spatial resolution (on the order of 300 to 500 km, or about the size of the state of Iowa) and many important aspects of weather and climate occur at smaller scales. The topic of finer-scale climate models is actively being addressed, and predictions of regional and local climatic changes are continually improving, though finer-scale models are usually run for shorter simulated time periods (typically 20 years or less), making century-scale projections difficult (Giorgi et al., 2001).

There are important implications of potential climatic changes for crops. Because temperature affects all biological processes, it is certain that any warming will affect crops. It is unclear at the outset, however, what *net* effect warming may have on overall crop physiology and, more importantly, yield. It will be the balance of effects on rates of transpiration, respiration, photosynthesis, translocation, growth, and perhaps most critically, development that will influence yield. Responses of farmers, including timing of planting and selection of species and cultivars or hybrids planted, will probably influence the effects of global warming on yield. Warming in spring and autumn, particularly at night, should reduce frost and other cold-temperature stresses in crops now subject to such stresses, but warming during normally hot periods could damage crops and limit yield. The potential trade-offs are complex.

The significance of potentially altered precipitation for yield is not obvious. Reduced precipitation, especially if coupled with warming, might reduce yield and increase demands on irrigation, especially in areas where yield is already limited by lack of water. More rain, at least if occurring at the right time, would presumably enhance yield in areas now lacking sufficient water for high yield. More precipitation in general might reduce drought frequency or extent, but more rain might also cause crop and soil losses from flooding and erosion. More frequent or intense rain could also interfere with crop management practices, such as planting and harvesting.

Tropospheric Ozone Concentration

Ozone is a natural component of the atmosphere. It is not emitted to the atmosphere, as CO_2 is, but rather is formed within the atmosphere from chemical precursors in the presence of solar radiation. Most ground-level O_3 is produced in the troposphere, but a small fraction of ground-level O_3 originates in the stratosphere (i.e., the "layer" of atmosphere above the troposphere with a height range of about 10 to 50 km), where it is produced by stratospheric photochemistry. The chemistry of O_3 formation, which involves addition of an oxygen atom to an O_2 molecule, is complicated; not one single set of reactions describes O_3 formation at different locations and times. The most important precursors of O_3 are nitrogen oxides (NO and NO_2), carbon monoxide (CO), methane (CH_4), and a number of hydrocarbons. These precursors are formed naturally, but importantly, their atmospheric concentrations increased during the past century because of human activities, most notably the same activities leading to atmospheric [CO_2] increase.

Past Changes

Because O_3 is highly reactive (its half-life is hours to days), it is not preserved in air trapped in ice, so direct records of past tropospheric [O_3]s are unavailable. Reliable measurements of ground-level [O_3]s have existed for only a few decades, but extrapolations further back in time have been attempted many times. Based on old (beginning in 1874) atmospheric chemistry measurements at the Pic du Midi Observatory in southwestern France (3000 m above sea level), Marenco and colleagues (1994) claimed that northern hemisphere tropospheric [O_3] increased from about 10 to 50 ppb (parts per billion) between 1890 and 1990. Based on other old measurements (beginning in 1876), Hough and Derwent (1990) estimated that ground-level [O_3] at mid- and high-northern latitudes more than doubled between 1890 and 1990. Other estimates indicate that *global* tropospheric [O_3] increased 36 percent from preindustrial times to the present (Prather et al., 2001). So although considerable uncertainty surrounds estimates of past changes in ground-level [O_3], it is a robust assumption that human activities caused significant increases in tropospheric [O_3] during the past 100 to 150 years.

Present Ozone Concentration Dynamics

Ozone production requires solar radiation and is favored by high temperature, so [O_3] is typically highest during hot, summer days, though it is produced all year. Of significance is the large diel and day-to-day changes in surface [O_3] (Figure 1.8). Concentrations at a site can also differ significantly from year to year. It is difficult to characterize the [O_3] at a site, and various averages, averages during the day (or midday and afternoon), maximum hourly average during a day, number of hours with [O_3] greater than a certain value (e.g., 50

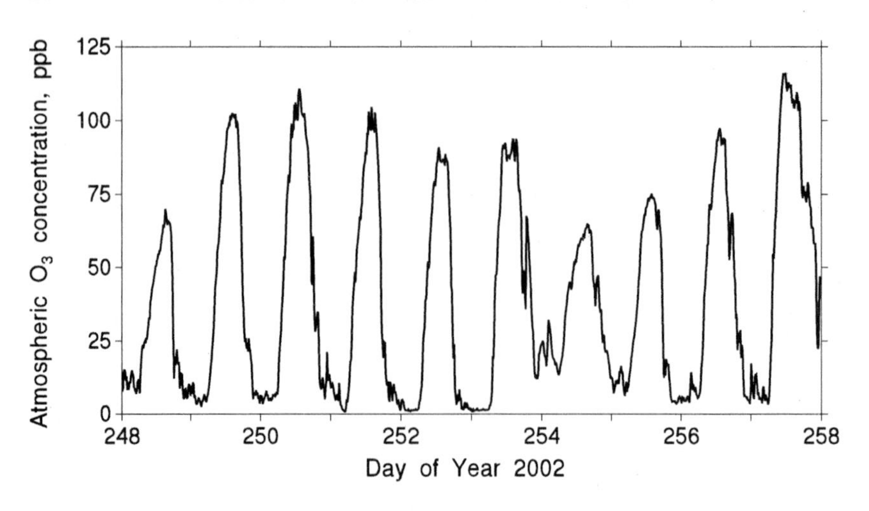

FIGURE 1.8. Near-surface $[O_3]$s near Urbana, Illinois (about 40°2'N, 88°17'W), during days 248 through 257 of the year 2002. The location is in the midst of a rural region with a high production of maize and soybean. Ten-minute mean values are shown. *Source:* Data courtesy of Timothy A. Mies and Stephen P. Long (University of Illinois).

ppb) during a season, and so on are used in experiments and atmospheric monitoring programs. In any case, daytime summer levels of 30 to 50 ppb O_3 are now common over large geographic areas, with episodic levels as high as 50 to 100 ppb existing in many developed and developing nations. These $[O_3]$s can negatively affect crop physiology and yield (Krupa et al., 2000).

An important feature of O_3 pollution is that it is not isolated to points of precursor emissions. In fact, $[O_3]$ is often higher downwind of metropolitan areas than it is in cities. As O_3 and its precursors spread spatially, more O_3 is produced through tropospheric photochemistry. The geographic extent of O_3 pollution is therefore regional (or larger; for example, fossil fuel combustion in eastern Asia could have a measurable effect on $[O_3]$ in the United States [Jacob, Logan, and Murti, 1999]), and large areas of cropland are, and will continue to be, exposed to damaging $[O_3]$s (e.g., Figure 1.8). Although other air pollutants (such as SO_2 [sulfur dioxide]) can damage crops, the extent of cropland affected by elevated $[O_3]$ makes it the key global air pollutant with respect to crops. (Other phytotoxic air pollutants are often point-source [i.e., local] pollutants emitted directly into the atmosphere rather than being formed from emitted precursors. As their transport occurs, atmospheric mixing dilutes the concentrations of point-source pollutants and, hence, their effects on crops.)

Future Ozone Concentrations

Because crop yields may be negatively affected by daytime $[O_3]$s of 40 to 70 ppb, tropospheric O_3 pollution already limits crop production in many areas (Heagle, 1989; Ashmore and Marshall, 1999). If projected increases of 10 to 50 percent in surface $[O_3]$s during the coming decades (Hough and Derwent, 1990; Chameides et al., 1994) are realized, or if high $[O_3]$s increase in areal extent (Chameides et al., 1994), the magnitude and areal extent of damage to crops may greatly increase in the future. Selection of O_3-tolerant genotypes is one avenue to reducing negative impacts of increasing $[O_3]$s, but whether that process can fully mitigate effects of further $[O_3]$ increases is unclear. In addition, efforts to develop and select crop varieties for O_3 tolerance may limit efforts to improve other aspects of crop physiology and yield by diverting limited scientific resources away from those other crop-improvement efforts.

Stratospheric Ozone Depletion and Ultraviolet Radiation

Another O_3 "problem" exists. Natural photochemistry in the stratosphere produces and maintains a significant amount of O_3 (about 90 percent of all atmospheric O_3 is in the stratosphere, with the remainder mostly in the troposphere). Stratospheric O_3 absorbs a significant fraction of solar ultraviolet radiation reaching Earth. In particular, O_3 strongly absorbs electromagnetic radiation in the 280 to 320 nm wave band, which is also called UV-B radiation. UV-B radiation can damage DNA and may disrupt several physiological processes in animals, plants, and microbes, so stratospheric O_3 acts as an important UV-B screen.

Depletion of stratospheric O_3 has been caused by emissions of several chemicals into the atmosphere, mostly during recent decades. Chlorine- and bromine-containing compounds such as CFCs were implicated as the chief culprits in stratospheric O_3 depletion (Molina and Rowland, 1974). The largest decreases in stratospheric O_3 have been observed over Antarctica, especially during September and October, when the amount of stratospheric O_3 is reduced as much as 60 percent relative to pre-1980 levels. Stratospheric O_3 has also been depleted over the Arctic, but because of differences in climate the reductions are not as large as they are over Antarctica. Stratospheric O_3 is not significantly depleted over the tropics (latitudes of 20°S to 20°N), but modest (about 5 percent) reductions over northern and southern midlatitudes have occurred since 1979 (World Meteorological Organization, 1998). Recent depletion of midlatitude stratospheric O_3 corresponds to increases in UV-B radiation reaching the earth's surface (World Meteorological Organization, 1998).

Increased UV-B radiation can affect crop photosynthesis, other aspects of crop physiology, and yield. On balance, data from field experiments indicate a small negative effect on yield of increased UV-B radiation. Field experiments relevant to actual surface UV-B radiation increases resulting from stratospheric O_3 depletion are technically challenging, however, and many experiments have been compromised by various factors. Moreover, yield of relatively few crops has been studied in elevated–UV-B field experiments (e.g., soybean, rice, pea, and mustard) though the physiology, rather than yield, of other crops also has been studied (Caldwell et al., 1998). In some experiments, increased UV-B radiation reduced crop disease, probably by damaging pathogens, while in other experiments disease was stimulated, perhaps due to changes in the crop that enhance infection or reduce tolerance to disease (Paul, 2000). As with some other environmental changes, increased UV-B radiation might at times be beneficial and at other times be detrimental to crops.

Although increased surface UV-B radiation is a potentially serious environmental change with respect to crop yield, little more will be written about it in this book. This is because many compounds that destroy stratospheric O_3 were banned by the international Montreal Protocol in 1987 (and its later amendments and adjustments in 1990, 1992, 1995, and 1997) and the releases of those compounds into the atmosphere have been markedly reduced. If full compliance with the international agreements is achieved, stratospheric O_3 depletion might be eliminated (i.e., stratospheric O_3 might be returned to pre-1980, or unperturbed, amounts) by the year 2050 (World Meteorological Organization, 1998). For more extensive coverage of research on effects of UV-B radiation on crops, the reader is directed to Caldwell and Flint (1994), Caldwell and colleagues (1998), Mazza and colleagues (1999), Papadopoulos and colleagues (1999), Stephen and colleagues (1999), Correia and colleagues (2000), Paul (2000), Yuan and colleagues (2000), Li and colleagues (2002), and Caldwell and colleagues (2003).

Soil Salinization

Lack of soil water is a main constraint on crop growth in arid and semiarid areas. Irrigation may remove that constraint, and FAO (1997, p. viii) estimates that, since 1960, global area of agricultural land "equipped to provide water to the crops . . . [including] . . . full and partial control irrigation, spate irrigation areas, and equipped wetland or inland valley bottoms" nearly doubled, to a present area of 272 Mha (Figure 1.9). Irrigated area will probably continue to increase as food demands increase.

An often serious drawback of irrigation is salinization of (concentrating of salts in) surface soil. All irrigation water contains dissolved salts, and if the irrigation water cannot (or does not) flow downward

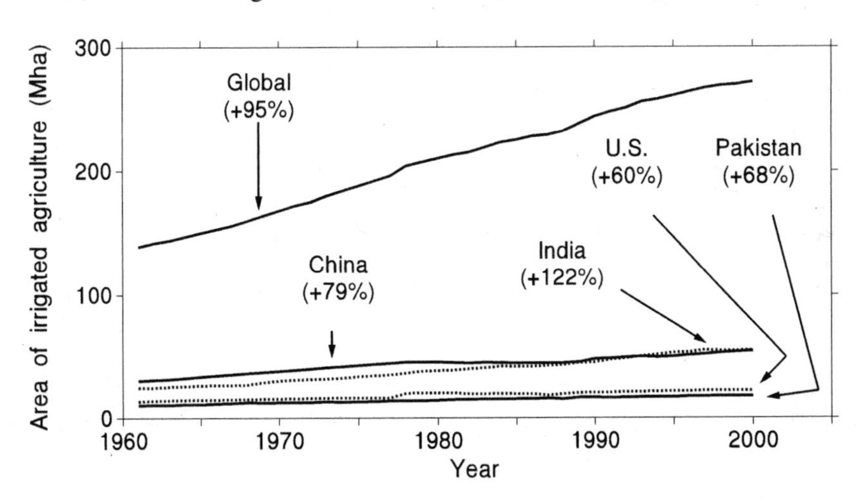

FIGURE 1.9. Global and selected national areas of irrigated agriculture. Values in parentheses are increases from 1961 to 2000. These areas include irrigated agricultural land outside cropland, such as orchards. *Source:* From FAO (http://apps.fao.org).

through surface soil, water evaporating from the surface leaves salt behind. In time, enough salt can accumulate to damage crops. Proper drainage and application of enough water to move salts through the soil can prevent, or even reverse, salinization and crop damage in some cases. Unfortunately, proper drainage is ignored in much of irrigated agriculture. In addition, in areas with salty groundwater and a water table near the crop root zone, excessive irrigation (even irrigation intended to leach salts from the surface soil) can raise the water table and expose crops to salty water. (Salinization is a natural process in many [semi-]arid soils, but as a natural process it involves long time periods. With irrigation, secondary [or anthropogenic] salinization [simply *salinization* hereafter] can occur rapidly and is an important environmental change.)

Salinization of cropland is apparently an old phenomenon, possibly affecting yield for millennia. For example, it was estimated that Mesopotamian cereal yields declined from about 2000 $kg \cdot ha^{-1}$ 4,400 years ago to about 1200 $kg \cdot ha^{-1}$ 4,100 years ago due to salinization over that period (Jacobsen and Adams, 1958). That salinization may have been related to climatic changes such as warming and/or drying (Nissenbaum, 1994). It is nearly dogma that the gradual, ancient replacement of wheat (relatively salt sensitive) with barley (relatively salt tolerant) in the Middle East was a required response to salinization from irrigated agriculture throughout the region several thousand years ago.

Largely because of population growth, the present extent of salinization is increasing rapidly, though estimates of the areal extent of salinization (Table 1.1) are imprecise. Although the absolute area of salinized land is small by some measures, irrigated agriculture is often highly productive, so "loss" of that area may result in a proportionally higher loss of national or global production.

In the early 1990s, FAO estimated that 1.5 Mha of additional land was salinized annually. At that rate, in 30 years more than 15 percent of present irrigated area would be salinized. A difficulty for interpreting these (and other) values for salinized area is that once soil becomes too damaged for economically viable crop production, it may no longer be counted as arable, and hence may be excluded from estimates of salinized cropland. Accounting systems also vary among countries, and the extent of salinization is often difficult to judge, so irrigated land not yet counted as salinized may still have accumulated enough salt to limit yield.

TABLE 1.1. Estimated cropped, irrigated, and salinized areas in 1985 and 2000 (selected nations and world totals).

	Cropland area (Mha)		Irrigated area (Mha)		Salinized area (Mha)	
	1985	2000	1985	2000	1985	2000
Argentina	25.0	25.0	1.6	1.6	0.5-0.6	0.5-0.7
Australia	47.2	50.3	1.7	2.4	0.1-0.2	0.1-0.3
China	121.0	124.0	44.6	54.4	6.7-7.0	7.0-8.0
Egypt	2.3	2.8	2.5	3.3	0.8-1.0	0.9-1.1
India	163.2	161.8	41.8	54.8	7-20	10-30
Iran	14.9	14.3	6.8	7.5	1.5-2.0	1.5-2.5
Pakistan	20.2	21.3	15.8	18.1	3.2-4.3	3.5-5.5
South Africa	12.4	14.8	1.1	1.5	0.1	0.1-0.2
Thailand	17.7	14.7	3.8	5.0	0.4	0.4-0.5
United States	187.8	177.0	19.8	22.4	4.1-5.2	4.5-6.0
USSR (former)	227.0	204.1	19.5	19.9	2.5-3.7	2.5-4.5
World total	1372.8	1364.2	225.1	271.7	45-60	62-82

Sources: Cropland and irrigated areas are from FAO (http://apps.fao.org). Estimates of salinized area (to various degrees of damage to soil) are crude and based on several sources (e.g., Postel, 1990; Ghassemi, Jakemand, and Nix, 1995) and present author extrapolations to the year 2000.

Note: Irrigated area is not limited to cropland, but includes all irrigated agricultural area; e.g., some orchards and pastures are irrigated. Because of this, Egyptian irrigated agricultural area exceeds Egyptian cropland area. Also, some damaged area resides outside present agricultural land (i.e., it is no longer used).

At the outset, the use of salt-tolerant crops such as barley as a viable solution to salinization can be dismissed. Without appropriate drainage, salinization of land will occur with irrigation in most cases, and the degree of salinization will increase over time. A gain in yield resulting from a switch to a more tolerant species or cultivar only delays the inevitable loss of productivity with continuing salinization. The long-term, sustainable solution is proper water management, not cultivar selection (Loomis and Connor, 1992).

CROPS AND FOOD SUPPLY

The importance of crops to humanity can be gauged in many ways. For example, domestication of crops was pivotal to the development of society during the past 10,000 years (Smith, 1995). Today, crops provide the bulk of human food supply. A ranking of all components of per capita food supply based on energy content (i.e., digestible calories), protein content, or fat content is one measure of the present importance of crops. Globally, rice, wheat, oil crops, sugar crops, maize, and tuber crops are especially prominent, with a few animal products (milk and cheese, pork, and seafood) supplying large amounts of protein and/or fat to humanity (Table 1.2). The two cereals rice and wheat alone furnished 39 percent of energy and 35 percent of protein to the 2000 global food supply. A similar situation exists in the United States, though relative rankings differ and animal products such as milk and meat make larger contributions to food supply (Table 1.3). (Data from the United States are prominent in this book because of author bias and because U.S. data are extensive and their quality is high.) It is striking that plants, mainly field crops, directly contribute a large fraction of energy, protein, and fat to global and national food supplies.

Although crops are clearly critical to food supply, Tables 1.2 and 1.3 do not specify their total contribution. In particular, a significant fraction of animal products is derived from crops grown for animal feed. Although some livestock production occurs by grazing land that is unsuitable for meaningful crop production, and that land would produce little human food if not grazed by animals, global livestock production does in fact rely on crops. The Council for Agricultural Science and Technology (1999) estimated that about one-third of global cereal grain harvested, and two-thirds of U.S. grain used in the United States (about 20 percent of the grain harvested in the United States is exported), is fed to livestock. Evans (1993) stated that by 1980, 44 percent of global cereal production was used as animal feed. In developed nations, 40 to 45 percent of energy consumed by livestock around 1980 was grain; the fraction was about 15 percent in developing nations (globally averaged), and generally increased with increased per capita gross national product (Evans, 1993). These fractions may be higher today, but in any case a notable fraction of energy and protein in the human food supply that is derived from animal

TABLE 1.2. Global per capita food supply in 1980 and 2000 (including waste)

	Energy (Cal·d^{-1})		Protein (g·d^{-1})		Fat (g·d^{-1})	
Food	**1980**	**2000**	**1980**	**2000**	**1980**	**2000**
Rice	516	576	9.5	10.7	1.3	1.4
Wheat	493	535	14.9	15.8	2.0	2.2
Plant oils[a]	180	252	0.0	0.0	20.4	28.4
Sugar and sweeteners[b]	236	243	0.1	0.1	—	—
Maize	145	157	3.4	3.8	1.1	1.3
Tubers and roots	157	148	2.2	2.2	0.3	0.3
Milk (includes cheese)	114	121	6.9	7.2	6.2	6.7
Pork	76	113	3.2	4.3	6.9	10.5
Other cereals	126	88	3.6	2.4	1.0	0.7
Fruits	62	77	0.7	0.9	0.4	0.5
Vegetables	43	66	2.2	3.4	0.4	0.6
Oil crops[c]	40	62	2.2	3.4	2.6	4.2
Alcoholic beverages	71	62	0.3	0.3	—	—
Animal fats[d]	71	61	0.1	0.1	7.9	6.8
Pulses	59	55	3.7	3.5	0.3	0.3
Poultry meat	22	44	2.1	3.8	1.4	3.0
Bovine meat	45	40	4.0	3.7	3.1	2.7
Eggs	22	32	1.7	2.5	1.5	2.2
Fish and other seafood	23	28	3.4	4.3	0.8	1.0
Other meat	12	13	1.0	1.1	0.8	1.0
Nuts	6	9	0.2	0.2	0.5	0.7
Edible offals	5	7	0.9	1.1	0.2	0.2
Spices	5	7	0.2	0.2	0.2	0.2
Cocoa, coffee, tea	4	6	0.4	0.4	0.2	0.3
Total plant products	2145	2347	43.5	47.5	30.7	41.2
Total animal products	390	459	23.4	28.1	28.8	34.0
Total food supply	2535	2805	66.9	75.6	59.4	75.2

Source: Derived from data collated by FAO (http://apps.fao.org.)

Notes: Food groups arranged in order of 2000 per capita energy content (digestible calories; Cal = 4.184 kJ). Double counting was avoided (e.g., barley grown for beer is included only in "Alcoholic beverages").

[a] Oil extracted from palm, peanut, rape, sunflower, soybean, and other plants.

[b] Includes honey, an "Animal product."

[c] Whole seed (or other organ) used, not just extracted oil as in "Plant oils."

[d] Includes butter, ghee, cream, raw animal fat, and fish (body and liver) oils.

TABLE 1.3. U.S. per capita food supply in 1980 and 2000 (including waste)

	Energy ($Cal \cdot d^{-1}$)		**Protein ($g \cdot d^{-1}$)**		**Fat ($g \cdot d^{-1}$)**	
Food	**1980**	**2000**	**1980**	**2000**	**1980**	**2000**
Sugar and sweeteners[a]	555	678	0.1	0.2	—	—
Wheat	508	616	16.5	20.0	2.1	2.7
Plant oils[b]	478	597	0.2	0.2	54.0	67.4
Milk and cheese	358	389	21.4	22.8	19.4	22.8
Poultry meat	103	187	9.5	17.2	6.9	12.5
Alcoholic beverages	185	154	0.8	0.7	—	—
Fruits	113	131	1.3	1.6	0.8	0.8
Pork	141	131	8.6	8.0	11.6	10.7
Animal fats[c]	127	122	0.1	0.1	14.3	13.7
Bovine meat	136	121	16.8	14.9	7.2	6.4
Roots and tubers	86	110	2.3	2.9	0.2	0.2
Maize	58	102	1.0	1.8	0.2	0.3
Rice	42	98	0.8	1.8	0.0	0.1
Vegetables	61	78	3.0	3.6	0.6	0.7
Oil crops[d]	31	56	1.2	2.3	2.7	4.8
Eggs	59	56	4.6	4.3	4.2	3.9
Pulses	28	41	1.9	2.7	0.1	0.2
Other cereals	35	33	1.2	1.3	0.4	0.4
Fish and other seafood	23	30	3.4	4.9	0.8	1.0
Cocoa, coffee, tea	12	18	1.2	1.4	0.4	1.0
Nuts	11	11	0.3	0.3	1.0	1.0
Other meat	8	6	0.7	0.6	0.5	0.4
Spices	4	5	0.1	0.2	0.2	0.2
Edible offals	5	2	0.9	0.4	0.1	0.1

Total plant products	2208	2729	31.9	41.0	62.6	79.8
Total animal products	960	1043	66.0	73.0	64.9	71.5
Total food supply	3168	3772	97.9	114.0	127.5	151.3

Source: Derived from data collated by FAO (http://apps.fao.org).

Notes: Food groups arranged in order of 2000 per capita energy content (digestible calories; Cal = 4.184 kJ). Double counting was avoided (e.g., barley grown for beer is included only in 'Alcoholic beverages').

[a]Includes honey, an "Animal product."

[b]Oil extracted from cotton seed, maize, olive, rape, soybean, and other plants.

[c]Includes butter, cream, and raw animal fat.

[d]Whole seed (or other organ) used, not just extracted oil as in "Plant oils."

sources is originally derived from crop grain production. Forage crops provide another significant fraction of animal feed, at least at the global scale.

Evaluating the importance of specific crops to food supply in some nations is complicated by the fact that some crops enter the food supply through multiple "Food" categories listed in Tables 1.2 and 1.3. A key example is U.S. maize. Only a small fraction of U.S. maize grain is used directly as human food, i.e., as corn kernels (included in "Maize" in Table 1.3 and "Cereals and other products" in Table 1.4). A majority of U.S. maize is used for animal feed, but other significant uses are production of fuel alcohol (not part of food supply), high-fructose corn syrup (HFCS), food starch (polymers of glucose), glucose and dextrose (dextrose is D-glucose), and beverage alcohol (Table 1.4). Maize used to provide HFCS and glucose is especially important in the United States because it supplies more than half the U.S. caloric sweetener amount, so it accounts for a majority of the largest single source of energy in the U.S. food supply, i.e., "Sugar and sweeteners" (Table 1.3). Conversely, the Table 1.3 entry "Maize" accounts for a relatively small fraction of the energy in the U.S. food supply.

Three additional points are significant with respect to Tables 1.2 and 1.3:

1. Global and U.S. per capita supplies of food energy, protein, and fat increased since 1980, indicating increased standards of living.
2. Values are for gross food supply and some of that supply is lost, spoiled, or discarded before or after it is served in meals (i.e., the values include waste). The fraction of food lost (wasted) is unknown, but may be in the range 5 to 50 percent of gross production, depending on food item and nation (Evans, 1993; Cohen, 1995).
3. Other uncertainties associated with food supply estimates exist. For example, data on global production and consumption of fish and of produce from home gardens are crude. Moreover, accuracy of crop-production data (a basis of food supply estimates) varies among nations and can be influenced by political objectives (Evans, 1993).

In addition to food, crops supply fiber (e.g., cotton), energy (e.g., ethanol in Table 1.4), and other valued products. For example, about 33 Mha of cotton are harvested globally each year, which produce about 18 Tg of fiber, or about 3 kg per capita (USDA NASS). Thus, total crop production exceeds that associated with food supply.

POPULATION, CROP YIELD, AND CROP PRODUCTION

Human population, crop yield, and crop production (i.e., yield × area harvested) are interrelated through demand, supply, and land-use factors. Population growth increases the number of people requiring food, and increased standards of living increase the amount of food needed per capita and/or the quality of that food. Past demands for increased food amount and quality were generally met with even larger increases in crop production and use of crops as animal feed, though many people remain undernourished, even in developed nations. Food supply could increase independently of crop production if fish, home gardens, and animals grazing on noncrop plants increase significantly as a fraction of food supply, though increased crop production was the main driver of past food supply increases.

Past crop production increases resulted mainly from four changes in cropping systems (Evans, 1993): (1) increased cropland area (see

TABLE 1.4. U.S. maize area harvested, yield, production, uses, and changes in stocks for marketing years ending August 31 of the stated year

	1980		2000	
Parameter	**Amount**	**Percent of grain**	**Amount**	**Percent of grain**
Area harvested (Mha)				
For grain	29.5	100.0	29.3	100.0
For silage	3.8	–	2.5	–
Yield and production				
Average grain yield ($kg{\cdot}ha^{-1}$)	5707	–	8590	–
Grain production (Tg)	168.6	100.0	251.8	100.0
Grain uses (Tg)				
Animal feed and residual[a]	107.5	63.8	148.2	58.9
Export	60.7	36.0	49.2	19.5
Alcohol (fuel, ethanol)	0.9	0.5	16.0	6.3
High-fructose corn syrup	4.2	2.5	13.6	5.4
Starch	3.8	2.3	6.3	2.5
Glucose and dextrose	4.0	2.4	5.6	2.2
Cereals and other products	1.4	0.8	4.7	1.9
Alcohol (beverage and manufacturing)	2.0	1.2	3.3	1.3
Seed (for future crops)	0.5	0.3	0.5	0.2
Change in U.S. grain stocks (Tg)	–16.3	–9.7	+4.6	+1.8

Source: Data from USDA NASS and Corn Refiners Association, Inc. (http://www.corn.org).

Notes: Yield and production are mass of harvested/stored grain (about 15% moisture, 25.4 kg per bushel). U.S. maize imports are minor (0.2–0.5 Tg $year^{-1}$). Exports are used mainly for animal feed, but also beverage alcohol, sweeteners, and foods.

[a] "Residual" means the small amount of grain not otherwise accounted for.

Box 1.2 definitions), (2) higher yield from a given crop species, (3) greater frequency of cropping on a given field (including growing more than one crop per year), and/or (4) replacement of lower-yielding crops with higher-yielding ones. Increased cropland area, and increased yield of specific crops, was especially important to increased crop production in recent centuries and recent decades, respectively.

BOX 1.2. Definitions of Land Use

Arable (adj): Fit for or used for the growing of crops.
Arable (n): Land fit or used for the growing of crops.
Arable land: Land used for the growing of crops. FAO (1997, p. viii) defines it as "land under temporary crops (double-cropped areas are counted only once), temporary meadows for mowing or pasture [i.e., *permanent* meadows and pastures are excluded], land under market and kitchen gardens and land temporarily fallow [less than five years]. The abandoned land resulting from shifting cultivation is not included in this category. '[A]rable land' [is] not meant to indicate the amount of land that is potentially cultivable." This book uses this subset of the dictionary definition.
Cropland: Same as arable land.
Cultivable land: Capable of being cultivated. Potentially arable land.
Cultivated: Prepared and used for the raising of crops. Preparation *may* include plowing, but "no-till" cropping systems are included.

Cropland Area

Data of historical cropland area are rare. Some relatively long-term statistics are available at local and national scales in developed nations, but estimates of pre-1950s regional and global cropland areas are crude (Robertson, 1956). Many values of pre-1950 cropland area are probably underestimates because (1) farmers and landowners underreported area to avoid taxes, (2) only area (and production) of major crops were reported or recorded, and/or (3) privileged (i.e., crown or religious) lands were excluded from reporting. Even today cropland area estimates in some countries are ambiguous. In particular, land cultivated in small blocks with crops that do not reach national or international markets may be untallied (Ajtay, Ketner, and Duvigneaud, 1979).

Based on coarse estimates, crops may have covered only 0.1 to 0.2 Gha globally in the year 1000, perhaps 0.3 Gha in 1700, and maybe 0.6 Gha by 1850 (Buringh and Dudal, 1987). Robertson (1956) suggested that cropland area increased about 90 percent from 1870 to 1950, while the analysis of Richards, Olson, and Rotty (1983) indicated that cropland area increased 0.43 Gha between 1860 and 1920 ("an informed and cautious estimate"), and then 0.42 Gha in the period 1920 to 1978. Cropland area increases prior to about 1950 roughly paralleled population increases, but not since then. FAO statistics (apps.fao.org) indicated that global cropland increased from 1.28 Gha in 1961 to about 1.40 Gha (or 9.6 percent) by 2001 (Figure 1.10). Meanwhile, global population increased more than 99 percent during the same period (according to FAO).

Relative changes in cropland area in the United States during the past 50 to 100 years differed from that of the world as a whole. The area of "principal crops" (see Appendix B) harvested in the United States peaked in 1932, and was 16 percent smaller by the year 2001 (Figure 1.10). (Area of principal crops accounted for about 84 percent of total U.S. cropland area [Appendix B].) Some of the largest year-to-year swings in area of principal crops harvested in the United

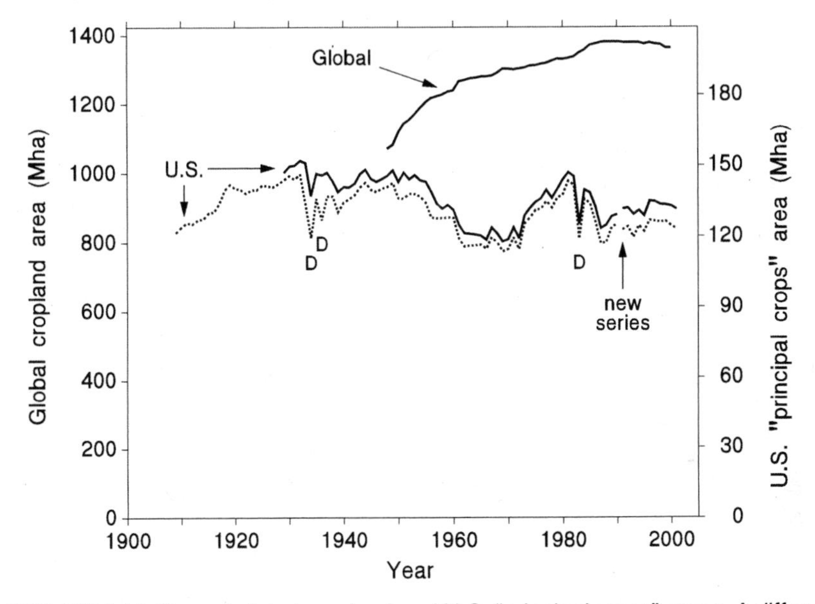

FIGURE 1.10. Recent global cropland and U.S. "principal crops" areas. A different set (i.e., "new series") of principal crops is shown beginning in 1991 (see Appendix B). The D's correspond to large-scale U.S. droughts in 1934, 1936, and 1983. Other significant droughts occurred since 1909, but are not shown. *Sources:* Global cropland area (1948-2000) is from FAO. Pre-1961 FAO values (from 1949 to 1965 annual editions of FAO *Production Yearbook*) combined cropland (arable) area with area of land under permanent crops (e.g., orchards); those values were modified here in an attempt to remove the permanent-crop area. Post-1960 FAO values are from <http://apps.fao.org>. Area of U.S. "principal crops" (see Appendix B) planted (—; 1929-2000) and harvested (- - -; 1909-2000) is from USDA NASS (http://usda.mannlib.cornell.edu/data-sets/crops/96120/trackrec2002.txt).

States during the 1900s were associated with droughts (Figure 1.10). It might therefore be expected that changes in frequency or intensity of extreme weather would result in changes in the interannual variability of area of U.S. crops harvested.

The fraction of land area that is cropped varies geographically. For example, only 5.5 percent of South America's 1.75 Gha land and only 6.1 percent of Africa's 2.96 Gha land were cropland in 2000, compared to 13 percent of China's land and more than 54 percent of India's land (Figure 1.11). About 20 percent of U.S., and 10 percent of global, land area is now cropped.

Recent changes in cropland area also varied greatly between regions and nations. In Africa, cropland area increased monotonically for a 28 percent gain from 1961 to 2000 (Figure 1.11). Chinese cropland area declined more than 6 percent from 1961 to 1980, but in-

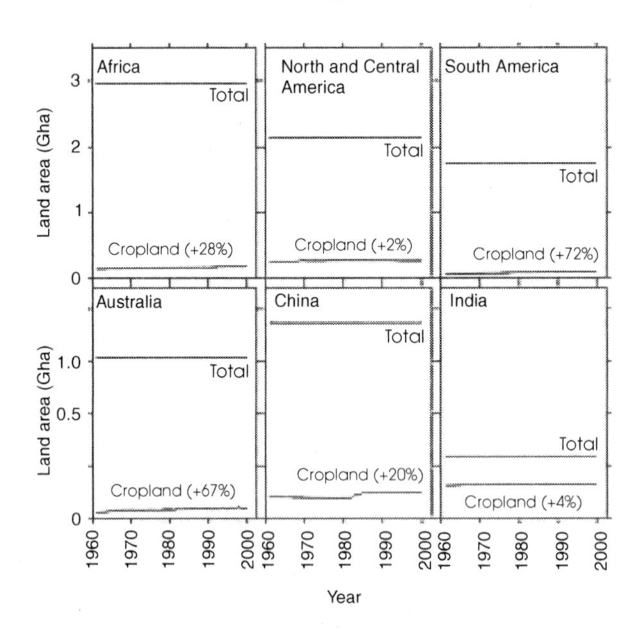

FIGURE 1.11. Total land and cropland areas in selected continents and nations. The top line in each panel shows total land area, whereas the bottom line in each panel shows cropland area. Values in parentheses on each panel are cropland area increases from 1961 to 2000. Cropland area values are based on individual nation reports, which vary in accuracy. Note that the *y*-axis scale on the three top panels (0-3.5 Gha) differs from the scale on the bottom panels (0-1.1 Gha). *Source:* From FAO (http://apps.fao.org).

creased 28 percent since then. For all six regions (or nations) shown in Figure 1.11, cropland area was larger in 2000 than it was in 1961, though cropland area apparently reached maxima in 1986 for North and Central America and in 1987 for India.

In many cases where cropland was a small fraction of land area in 1961, relative cropland area increases since then were large (e.g., Australia and South America; Figure 1.11).

It is cultivable rather than cultivated area (Box 1.2) that places a limit on future crop production. Best estimates of cultivable land are about 3.0 Gha (Buringh and Dudal, 1987), about double present global cropland area. Most of the difference between the cultivable 3.0 Gha and the presently cultivated 1.5 Gha is in developing nations (Buringh and Dudal, 1987). Even with a large cultivable area, future food demands might not be easily met by simple increases in cropland area. For example, with over 60 percent of global population in Asia (1999 FAO estimate) but with a potential increase in Asian cropland of less than 10 percent (Evans, 1993), increased Asian cropland area probably cannot meet future Asian food demands. Instead, increased yield (or increased frequency of cropping) will be required, or food imports will need to increase dramatically. It is also important to know that some high-quality cultivable land, in all regions, is being lost to other human land uses, including urbanization, and that much of the "best" cultivable land is already being cropped. Future cropland area increases, though potentially large in some regions, will compete with other land uses driven by population growth and increased standards of living.

Yield and Production

Crop yields increased greatly in the 1900s. Those advances may be attributed to increased availability of inexpensive fertilizers; crop breeding that improved harvest index (i.e., the fraction of total or aboveground biomass that is accounted for by the part of the crop that is harvested) and pest and stress resistance; breeding that allowed greater plant density (in response to greater nitrogen fertilization); use of herbicides and pesticides; advanced mechanization; and enhanced farmer knowledge (e.g., Loomis and Connor, 1992; Evans, 1993, 1997; Duvick and Cassman, 1999; Tollenaar, 1999; Miflin, 2000).

Rice grown in China, India, and Indonesia is an example of recent *yield* gains (Figure 1.12a). The large *production* increases in the same crops since 1961 (Figure 1.12b) were due to a combination of increased yield and increased area harvested (i.e., production increased more than yield), though yield gains were larger than area gains in all three countries. (According to data shown, Chinese rice yield and production increased during the early 1950s, but then declined an equal amount in the late 1950s.)

Wheat yields also increased in recent decades. In the United Kingdom and the United States, wheat yield increased only gradually from the mid-1800s to about 1940-1950, but after that period yield gains in both countries were rapid (Figure 1.13). In Australia, yield declined

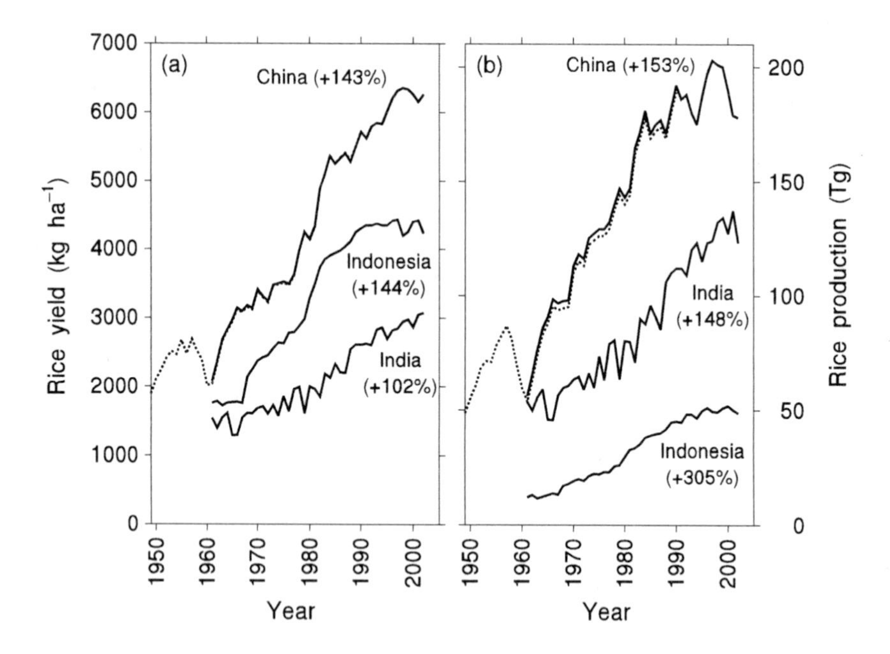

FIGURE 1.12. (a) Yield and (b) production of rice in China, India, and Indonesia, the three nations with largest production (combined, they accounted for more than 63 percent of global rice production in 2000). Values in parentheses following nation labels are percentage changes between the periods 1961-1965 and 1998-2002. *Sources:* Data are from FAO (http://apps.fao.org), except dashed line (- - -) for China (1949-1990), which is from Colby, Crook, and Webb (1992). FAO's data for China include Taiwan (Province of China), but Colby, Crook, and Webb's are for the People's Republic of China only.

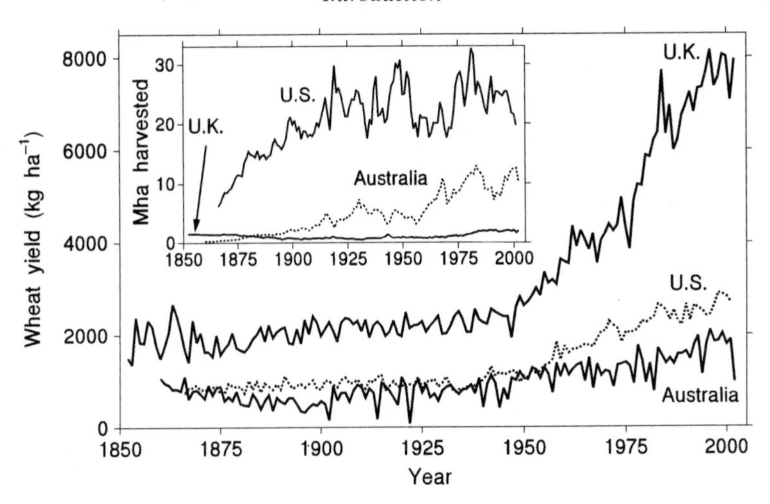

FIGURE 1.13. Australian, U.K., and U.S. wheat yields. Inset shows area harvested in each nation. *Sources:* Australian data for 1860 to 1960 are from Brennan and Quade (2000) and from 1961 to 2002 are from FAO (http://apps.fao.org). U.K. data are from Percival (1934) (for 1852-1865), Great Britain Ministry of Agriculture, Fisheries and Food (1968) (for 1866-1960), and FAO (for 1961-2002). U.S. data (1866-2001) are from the USDA NASS (http://usda.mannlib.cornell.edu/data-sets/crops/96120/).

between 1860 and 1900, perhaps due to depletion of available soil nitrogen and phosphorus, but since then large-percentage gains in yield occurred (Brennan and Quade, 2000). Recent Australian, U.K., and U.S. wheat-yield gains were not unique; Calderini and Slafer (1998) illustrated large yield gains for wheat during the 1900s in 21 nations (including all five continents with crops).

As for other crops, national average wheat yields vary among countries. For example, U.K. wheat *yield* consistently exceeded U.S. yield due to a favorable U.K. climate (for wheat) in combination with more intensive management. Nonetheless, U.S. wheat *production* is now about four times greater than that in the United Kingdom, owing to the larger area of U.S. crop harvested (Figure 1.13 inset). Indeed, for most years beginning in 1974, area of wheat harvested annually in the United States exceeded total land area of the United Kingdom. (Increases in area of U.S. wheat crops were made mainly prior to 1940 to 1950, whereas U.K. wheat area declined from 1850 to 1870 to the early 1900s, but then more than doubled after about 1970

[Figure 1.13 inset].) U.S. wheat yield is not especially low, however. Since 1961, U.S. wheat yield consistently exceeded Australian yield (Figure 1.13), because Australian wheat is grown mainly in dryer, less fertile soils.

These few examples for rice and wheat show recent large gains in yield and production. Many more examples are available (e.g., Evans [1993] and the FAO and USDA NASS databases).

Interannual variation in yield and production is often large; that is, yield (or production) versus time is not a straight line, but shows peaks and dips (Figures 1.12 and 1.13). Yield depends on many factors, including weather and the quality of land cropped. When economic conditions favor an increase in area devoted to a crop, that increase often includes marginal or less productive land. Similarly, when economic conditions favor a planned reduction in area of a crop, the land removed from cropping is often less productive than the mean. Thus, with *planned* changes in crop area, mean yield often changes, with the yield-versus-area-harvested relationship often negative. This differs from *forced* losses of harvested land due to extreme environmental events such as drought, flood, or extended high temperatures. In those cases, both area harvested and yield typically decline.

The challenge of increasing crop yield (and production) to match future food-demand increases is related to the question: When and at what levels will yield ceilings be reached for major crops? The question is vigorously debated (see Evans, 1993, 1998; Waggoner, 1997), with little resolution in sight, and is left unanswered in this book. It is important to note that even if yield limits have been reached in experimental crops (see Sinclair, 1994), differences between yield limits and average (i.e., obtained or actual) yields are still large (Waggoner, 1997). And although national average yields may never reach values obtained in good experimental crops, yield will surely continue to increase in many crops for some time. The rate and sustainability of those increases will depend on material, energy, economic, political, and scientific inputs. They will also, in many cases, be affected by ongoing and future environmental changes, which are the subjects of this book.

SUMMARY

A cursory study of inputs to the human food supply illustrates the preeminent importance of crops, including their uses as livestock feed. In addition to food and feed, crops are important sources of fiber, contribute to energy supply, and make other contributions to human endeavors. Increasing population places new demands on crop productivity (i.e., yield) and production (i.e., average yield × cropland area), as do desires for improved standards of living. Past demand increases were met with increased production. Indeed, global per capita crop production may have increased significantly during recent centuries. Prior to the mid-1900s, cropland area increases drove a large part of production increases. Since the mid-1900s, yield increase was dramatic, being responsible for much of the production increases. Relationships between yield and production have varied, and probably will continue to vary, among nations and regions.

Future population growth will place additional demands on crop production. There is scope for both increased yield and expanded cropland area, though that scope varies among regions. In addition to increasing food demand, both population and standard of living increases are bringing with them environmental changes at local, regional, and global spatial scales, and across temporal scales ranging from years to centuries. Many of those environmental changes—in particular increasing atmospheric [CO_2], the accompanying and projected climatic changes, increasing tropospheric [O_3], and soil salinization—have important implications, some positive and others negative, for future crop yield and production.

Crop breeding and management have the potential to mitigate the negative and reinforce the positive effects of environmental changes on crop physiology and yield. Adaptation by farmers and researchers may be the most important response to these changes. The success of such adaptation might be fostered by (or perhaps depend on) basic understanding of effects of environmental changes on crop physiology and growth. This is, at least, one view. In some cases, the degree to which adaptation will be successful, or even possible, may depend on the rate of environmental change (e.g., the rate of warming during coming decades). Unfortunately, certain knowledge of those rates will be obtained only with the passage of time, and by then it may be too late for effective action.

Chapter 2

Methods of Studying Effects of Environmental Change on Crops

Too many methods, and variations on those methods, are used to study effects of environmental conditions and environmental changes on crops to describe (or even mention) them all here. Therefore, only the most common approaches will be discussed. Two classes of study are distinguished: (1) statistical analyses of relationships between past yield and past environmental variation or change and (2) experimental manipulation of the environment coupled with measurements of crop physiology and/or yield. Each approach within these two classes has limitations (some more than others), but each can improve knowledge. Our view is that combining information gained from *multiple* statistical and/or experimental approaches is more effective in ascertaining likely effects of environmental change on the physiology and yield of crops than using information obtained with any single approach alone.

CORRELATIONS BETWEEN OBSERVED YIELD AND HISTORICAL PATTERNS OF ENVIRONMENTAL VARIATION AND CHANGE

Season-to-Season (Interannual) Environmental Variation

A powerful, longstanding method of studying effects of season-to-season (interannual) environmental *variation* on yield is to use correlation and/or regression analysis to relate observed yield to observed environmental variation.

> What is required is to measure the annual yields of a cultivar growing in the same soil and situation, with the same cultural

> treatments, over a period long enough to sample adequately the variation in weather [or other environmental variables] between years, and to keep meteorological [or other environmental] records over the same period. (Watson, 1963, p. 339; and see Wallace, 1920)

For example, yield at a location during an n-year period might be statistically analyzed in terms of annual, growing-season, or monthly air temperature and/or precipitation during those same n years (Chmielewski and Potts, 1995).

Limitations of this approach based on simple statistical or empirical (i.e., "black-box") models of effects of the environment on crops have been discussed many times (e.g., Katz, 1977). Watson (1963) claimed that effects of the environment on crops are too complex to be adequately described by simple regression equations involving only a few environmental variables and that important environmental factors may be left out of the analyses. Results can be difficult to interpret in terms of cause-and-effect relationships. Important environmental factors often covary (e.g., solar radiation received by a crop and air temperature are often positively related), and this can obscure relationships between yield and single environmental factors (e.g., Was it solar radiation or temperature or both that affected the crop?). Also, yield–environment combinations are limited to what the ambient environment provides.

Because of the long periods required to obtain long-term yield–environment records, space is "substituted" for time in some studies. In such cases, natural climatic gradients (e.g., over a range of latitudes) are used to "vary" environmental conditions within a relatively small number of years. Although results of such studies may be equivocal measures of the effects of weather and climate on crops, because other factors such as soil properties, management practices, pest and disease populations, and air pollution levels also vary between locations (Watson, 1963), use of climatic gradients can improve knowledge of the effects of environmental variation on crops. In addition, a range of planting dates can be used at a single location to obtain different early-season temperatures at that site, though the effects of daylength on some crop species can then become a complication.

Watson (1963) noted that a single cultivar (or hybrid) should be used to avoid difficulties of possible genotype × environment interactions. This approach, however, means that information is obtained for only a single cultivar, and a single cultivar may be unrepresentative of general effects of the environment on a crop species. (This problem applies equally to other approaches to studying the effects of the environment and environmental change on crops.) In addition, using a single cultivar over a long period may introduce yield trends to the extent that pest-cultivar relationships change over time. The approach described by Watson also holds crop-management practices constant, even though changes in management over time (e.g., changes in nitrogen fertilization) made significant contributions to past yield increases and might continue to do so in the future. When statistical analyses of relationships between past yield and past environmental conditions include data from sites in which cultivars (hybrids), crop management, and other technologies are allowed to change, it is necessary to account for the effects of those changes on yield variations over time (e.g., Thompson, 1970).

An example of a statistical analysis of yield–environment relationships conducted by Thompson (1970, 1986) indicated a strong effect of interannual variation in precipitation on yield of maize and soybean in the U.S. Corn Belt (Figure 2.1). For the preseason period (September-July), variation in precipitation was modestly related to yield, with maximum yield resulting from "normal" precipitation amount. Precipitation amount in July, however, apparently had strong, positive effects on yield of both crops over a wide range of monthly precipitation deviation. Yield of both crops was also positively related to August precipitation. These results indicate that U.S. Corn Belt maize and soybean yield was and/or is limited by water supply in both July and August.

A similar approach is to investigate whether significant deviations in yield or production from "normal" values occurred when weather was extreme (e.g., wetter/drier or warmer/colder than "normal") in a location or region. For example, the largest interannual "reductions" in U.S. maize production occurred in years of extreme weather (Figure 2.2). This result may portend reduced stability (sensu Marten, 1988) of U.S. maize yield and production if the frequency of extreme events increases in the future. Similarly, the lowest wheat yields in the Broadbalk field experiments at Rothamsted in southeastern England (e.g., Russell and Watson, 1940) from 1854 to 1967 "were always observed in wet/cold and wet/warm years" (Chmielewski and Potts, 1995, p. 60), indicating negative effects of extreme weather.

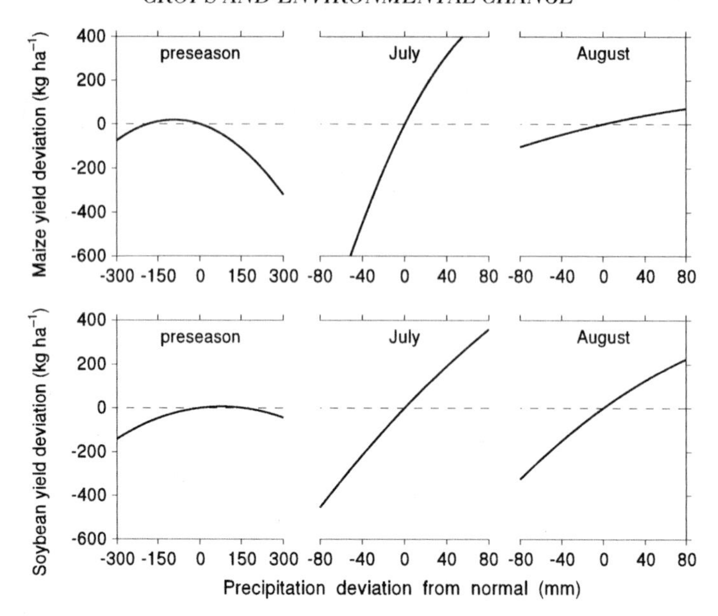

FIGURE 2.1. Quadratic regression relationships between precipitation deviation from "normal" and yield deviation from "normal" in maize (for the period 1930-1983) and soybean (for the period 1930-1968) in five of the U.S. Corn Belt states (Illinois, Indiana, Iowa, Missouri, and Ohio). Normal precipitation was the mean during 1891-1983 (maize) and 1930-1968 (soybean). Effects of technology on yield increases during the study periods were accounted for. Separate analyses were conducted for the preseason period (September-July), for July, and for August. *Sources:* Equations are from Thompson (1986) (maize) and Thompson (1970) (soybean).

The approach of correlating past yield with historical environmental conditions is limited to past conditions. Quantitative extrapolation to future conditions that may differ from the past (i.e., are outside the experience of previously studied crops) is problematic, especially when yield–environment relationships are nonlinear, and they often are. Extrapolation may also fail if normal covariation between different environmental factors changes in the future. For example, the relationship between solar radiation and air temperature may change with global warming (i.e., there will be a higher temperature for a given amount of solar radiation).

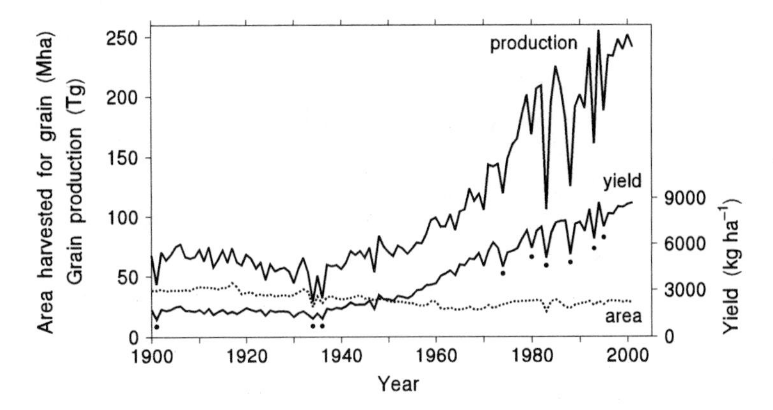

FIGURE 2.2. Area harvested, yield, and production of U.S. maize, 1900-2001 (http://usda.mannlib.cornell.edu/data-sets/crops/96120). Dots are shown below yield values in years with unusually low yield and/or production. The three years with lowest yield since 1866 (beginning of the data record) were 1901, 1934, and 1936. The midwestern United States was hot and dry during summer 1901 and, in addition to poor crop production, there were over 9500 heat-related deaths (http://www.weather.com/encyclopedia/heat/history.html). Especially low production was obtained in 1934 and 1936 because of "dust bowl" droughts. Six noticeable dips in yield after 1970 were attributed to climatic extremes (wet spring and early frost in 1974; droughts in 1980, 1983, and 1988; flood in 1993; and cool, wet spring with hot, dry summer in 1995) by the National Assessment Synthesis Team (2001). Note that interannual variation in production is often larger than variation in yield. For example, percentage drops in production in 1934 and 1936 (compared to 1933, 1935, and 1937) were much larger than percentage drops in yield because of considerable loses of area harvested in 1934 and 1936 due to droughts. Similarly, yield in 1983 was 28 percent below yield in 1982, whereas production in 1983 was reduced more than 49 percent compared to 1982 because of a concomitant 29 percent reduction in area harvested due to drought.

Long-Term Trends (Environmental Change)

Most analyses of historical environmental conditions and yield involve variation in environmental conditions rather than systematic *trends* in those conditions. This is because the primary environmental variables of interest are usually temperature and precipitation, and these factors vary "up" and "down" significantly from year to year (i.e., the weather is variable), with only small long-term trends (i.e., the climate is relatively stable).

Contrary to the small past increases in observed temperature, global atmospheric [CO_2] increased more significantly during the twentieth century (Chapter 1). Thus, it might be feasible to search for long-term trends in yield related to the observed [CO_2] trend. (A similar investigation might be carried out for increasing [O_3], though hourly, daily, seasonal, and interannual variations in [O_3] in a region can be large, and precise long-term [O_3] data in cropped areas are lacking.) Nonetheless, relating long-term trends in environmental conditions such as rising [CO_2] (i.e., environmental change) to yield is difficult because significant long-term trends in yield resulted from crop breeding and implementation of ever-advancing technology and farmer knowledge, and those effects might be mistaken for the effects of environmental change.

An exception to the complication of yield trends caused by advances in crop breeding, management, technology, and farmer knowledge over time occurs when a long-term agricultural experiment holds cultivars, management, and technology constant. For example, plots "3d" and "12d" of the Park Grass Experiment of permanent pasture started in 1856 by J. B. Lawes and J. H. Gilbert at Rothamsted received no lime, fertilizers, or intentional introduction of new species or cultivars during the experiment (which continues today). By using a regression model to "remove" effects of interannual variation in sunshine and rainfall on herbage yield in that experiment, Barnett (1994) searched for an effect on yield of the 20 percent increase in global atmospheric [CO_2] from the 1890s to the 1990s. The result: "no clear effect is evident. There was no statistically significant evidence of an increase in herbage yields with atmospheric CO_2" (Barnett, 1994, p. 176). This result is counter to expectations of positive effects of rising [CO_2] on plant growth, but it is nonetheless one of the few long-term experiments available that address this issue with management and other inputs held constant over time.

Use of statistical relationships (if any) between past environmental changes and yield changes derived from long-term experiments may be unapplicable to the future. For example, predicting responses to future elevated [CO_2] and/or [O_3] will be limited with this approach because effects of both gases on yield may be nonlinear (Chapter 9) and future concentrations of both gases will likely exceed past and present concentrations (Chapter 1). Also, use of new cultivars (or hybrids) in the future may affect yield through genotype × environment interactions. For these reasons and others, extrapolation from past statistical relationships into the future involves considerable uncertainty.

Regardless of any drawbacks of statistical relationships between crop yield and past environmental variation and/or change, such relationships can be enlightening. Indeed, a great deal has been, and still more can be, learned about effects of temperature, precipitation, solar radiation, and their combinations on yield. At the least, strong correlations between observed yield and environmental factors that may change in the future indicate environmental factors deserving experimental study. One clear advantage of this approach is that the "apparatus effects" or "experimental artifacts" that can occur with all experimental manipulations of the environment are circumvented. Moreover, whole fields or regions can be studied and normal crop management practices can be used, both of which are impossible for most experimental approaches.

EXPERIMENTAL CONTROL OF ENVIRONMENTAL CONDITIONS

Because it is difficult (or impossible) to disentangle effects on crops of individual environmental variables such as air temperature, soil moisture, solar radiation, [O_3], and [CO_2] by statistical analysis of historical yields and environmental conditions, it is necessary to experimentally manipulate (control) the environment to establish cause-and-effect relationships between crops and environmental conditions. In most cases, control of multiple factors (e.g., temperature *and* [CO_2] *and* [O_3]) is needed so that combinations of environmental conditions, and their possible interactions, can be studied. For present purposes, the control of environmental conditions should cover the full range of values possible as a result of continued (future) human alterations of the environment in any particular geographic location.

Facilities or instruments for experimental control of environmental conditions fall under the general headings of (1) leaf cuvettes, (2) laboratory (or growth) chambers (or cabinets), (3) glasshouses, (4) field chambers, and (5) open-air field fumigation systems.

Leaf Cuvettes

Leaf cuvettes are small chambers used to measure physiology (i.e., CO_2 and/or water-vapor exchanges) of individual leaves or parts of leaves. Many cuvette designs are used (Field, Ball, and Berry, 1989; Pearcy, Schulze, and Zimmermann, 1989; Allen et al., 1992; Jahnke, 2001). Leaf cuvettes can be used in laboratories, glasshouses, or the field. The principles of operation of leaf cuvettes are applicable to other chambers used to measure gas exchange of stem segments, (parts of) root systems, reproductive organs (e.g., pods or inflorescences), and sections of crop canopies (e.g., the crop covering a square meter of ground) (e.g., Musgrave and Moss, 1961).

Leaf cuvettes are useful in the study of short-term gas-exchange kinetics of individual leaves when light, [CO_2], temperature, and humidity inside the cuvette are controlled. Leaf cuvettes are not used to study plant growth directly, but they can be used in combination with approaches outlined in the following subsections for long-term control of the environment of whole plants in crop growth studies. Much of what is known about effects of environmental conditions on leaf physiology was derived from leaf cuvette measurements.

Laboratory Chambers, Glasshouses, and Phytotrons

A building containing a large number of controlled environments designed for the study of effects of environmental conditions on plants is called a phytotron (Morse and Evans, 1962). The controlled environments in a phytotron may consist of artificially lit growth "chambers" or "cabinets" (we call these laboratory chambers) and/or sunlit enclosures (i.e., some type of glasshouse) (Figure 2.3). An advantage of phytotrons, compared to smaller collections of laboratory chambers and/or glasshouses, is that a large number of chambers allows comprehensive factorial experiments with multiple treatment levels of multiple factors such as temperature, [CO_2], and [O_3]. But even though much research on effects of environmental change on crops is conducted in phytotrons, the full power of phytotrons has yet to be applied to this important topic. Instead, a small subset of the chambers or glasshouses in a phytotron is typically used, along with a limited number of environmental variables (treatments), treatment combinations, treatment levels, and chamber replicates (e.g., Gifford, 1977; Sionit, Hellmers, and Strain, 1980; Thomas et al., 1993). In-

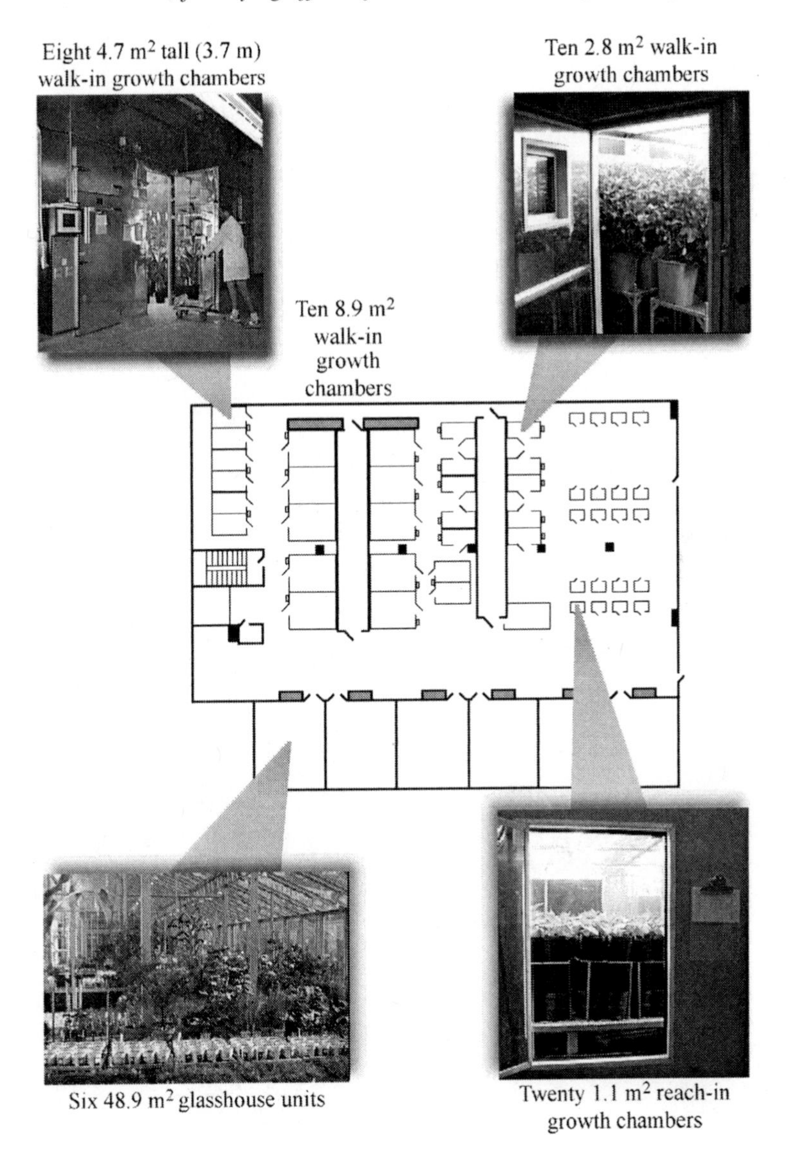

FIGURE 2.3. Floor plan of the National Phytotron at Duke University (Durham, North Carolina). This phytotron includes glasshouses and four types of laboratory chambers (see also Kramer, Hellmers, and Downs, 1970). (Courtesy of James F. Reynolds and David C. Tremmel, Duke University.)

deed, with all experimental approaches used, unreplicated and poorly replicated experiments are common in the study of effects of environmental change on crops.

Environmental conditions inside typical laboratory chambers used for environmental-change research differ significantly from conditions in a typical crop. For example, light levels are often low inside chambers, and the spectral quality of that light usually differs from that of the sun; air temperatures are often relatively high in chambers; air movement in chambers is slow; and the generally smooth diel cycles of environmental variables characteristic of our world are usually replaced with square waves in chambers (e.g., Lawlor and Mitchell, 1991). Moreover, the field often includes continuous but momentary fluctuations in environmental conditions superimposed on the more or less smooth diurnal cycles, and these short-term fluctuations may be important to crops, but they are ignored (lacking) in laboratory chambers and glasshouses (Evans, 1963). These departures from field conditions are not requirements of laboratory chambers (or glasshouses), but a result of how chambers are typically constructed and operated. The often significant differences between conditions inside chambers and environmental conditions in the field is a basis for concern or caution, although laboratory-chamber experiments can be invaluable for determining cause-and-effect relationships between crop physiology and the environment because only with full control of environmental conditions can unequivocal statements be made about the effects of particular environmental factors on crops.

Advantages of glasshouses compared to laboratory chambers include the availability of solar radiation (although glasshouse physical structures typically intercept a significant fraction of incoming solar radiation) and the larger available floor spaces to produce artificial plant communities. A disadvantage is that the diurnal course of solar radiation available at the time and location of an experiment may not suit the goals of the experiment (e.g., results of winter experiments may be difficult to relate to summer-grown crops). Thus, many glasshouses include artificial lights to extend daylength when needed, and to replace (some of) the solar radiation intercepted by the aerial structures of the glasshouse. Some glasshouses contain miniature glasshouses or sunlit-chambers to provide extra control on environmental conditions while using solar radiation as a light source. Several systems designed to study effects of warming in combination with elevated [CO_2] on crop growth and yield have been described (e.g., Lawlor et al., 1993; Gordon et al., 1995; Gorissen et al., 1996). A characteristic of some advanced laboratory chambers and glasshouses is the use of deep soil profiles.

When laboratory chambers or glasshouses are physically well sealed with respect to CO_2 and water vapor, they can be used to measure physiological rates (i.e., exchanges of CO_2 and/or water vapor) of whole plants or plant communities (Allen et al., 1992). Soil-plant-atmosphere-research (SPAR) systems (Phene et al., 1978) are special sunlit chambers, or well-sealed small glasshouses, used for short- and long-term studies of effects of temperature, humidity, soil moisture, and [CO_2] on crops and their physiology (Allen et al., 1992).

The ultimate approach to future studies of effects of environmental conditions on crops would involve facilities similar to the "Ecotron" designed, but never built, by Stanhill (1977). In addition to the properties specified by Stanhill, the ultimate facility would include control of [CO_2] and [O_3], and attention would be paid to the spectrum as well as light flux provided by the "solar radiation simulators." The facility would be sealed so that whole-crop physiological processes could be measured.

Field Chambers

Myriad chambers, and chamber designs, have been used to control the composition of the atmosphere in small plots of actual crops (e.g., Thomas and Hill, 1949; Moss, Musgrave, and Lemon, 1961; Heagle, Body, and Heck, 1973; Rogers, Heck, and Heagle, 1983; Hogsett et al., 1987; Fuhrer, 1994). Most early studies focused on air pollutants such as SO_2 and O_3, but more recent research includes many elevated [CO_2] experiments (e.g., Hertstein et al., 1999).

Field chambers are often flow-through devices in which air is forced into the chamber, passes through or over a crop, and then exits the chamber through various outlets. Chambers that are open at the top (i.e., open-top chambers or OTCs) generally rely on ambient precipitation, whereas covered (or closed-top) chambers (CTCs) require irrigation to make up for precipitation intercepted by the chamber. Chambers only *partially* open at the top (also called OTCs; e.g., Figure 2.4) require supplemental precipitation or irrigation treatments.

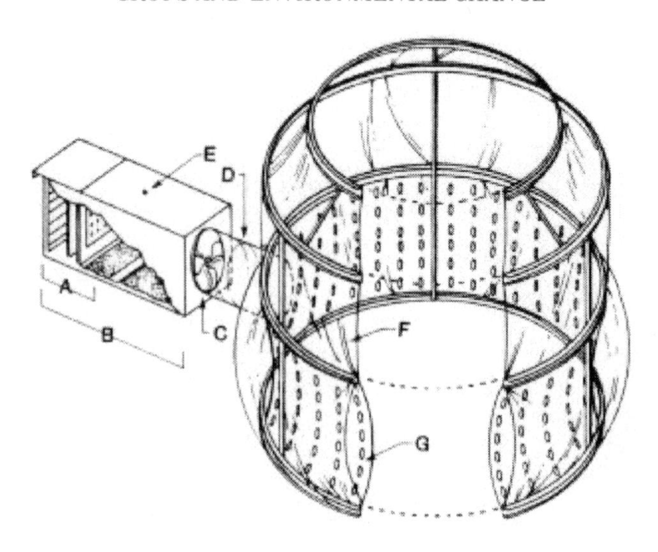

FIGURE 2.4. An open-top field chamber (OTC) and associated "fan box" suitable for controlling [CO_2] and/or [O_3] around plants (a radial segment of the chamber is removed in this drawing). The chamber can be placed over part of a crop growing in the field. The base of the chamber rests directly on the soil surface. During operation, air is pulled through particulate filters in section A (which could include charcoal filters to remove ambient O_3 if desired) and the rest of the plenum (B) by a fan (C) that forces air through D and into the chamber through holes in the inside layer of the transparent (80-94 percent transmission of solar radiation) double-layer wall making up the bottom of the chamber (G). Air forced into the chamber through section G exits the chamber through the large opening in the chamber top. The top two sections of the chamber are covered with a single layer of transparent material (F). This particular chamber has a 3.0 m frame diameter at the ground. About three changes of air inside the chamber occur each minute during the operation. Pure CO_2 and/or air containing "extra" O_3 can be metered through an external tube (not shown) inserted into the port E in the plenum to raise the [CO_2] and/or [O_3], respectively, inside the chamber. *Source:* Courtesy of Hugo H. Rogers, USDA National Soil Dynamics Laboratory, Auburn, Alabama.

Field chambers used to control [O_3] and/or [CO_2] alter environmental conditions other than the gas concentrations. In particular, air temperature, wind speed and turbulent structure, longwave and shortwave radiation exchanges, and humidity are often affected by field chambers. Normal movement of pollens, insects, spores, and animals is also affected. The degree of environmental alteration caused by field chambers varies between chamber designs and between the ways a single design is implemented (e.g., Fuhrer, 1994; Van Oijen et al., 1999). Because of this, when interpreting experimental results it is important to consider potential interactions between experimental [O_3]s and/or [CO_2]s and other environmental factors that may be altered by the chambers. An often significant problem is uncontrolled warming within field chambers. To control field chamber temperature, chambers can be fitted with coolers (e.g., Van Oijen et al., 1999), which has the added benefit of allowing studies of interactions of elevated [CO_2] and warming.

Environmental gradient tunnels are special chambers used either in the field (Hadley et al., 1995; Horie et al., 1995; Rawson, Gifford, and Condon, 1995) or within a glasshouse (Mayeux et al., 1993). Superambient temperature gradients can be established in tunnels when air moving from the tunnel entrance to the tunnel exit is heated by solar radiation during daytime and by heaters during nighttime (Horie et al., 1995; Rawson, Gifford, and Condon, 1995). Alternatively, a subambient-to-superambient temperature gradient can be established by a combination of coolers and heaters operating along the length of the tunnel during the entire diel cycle (Hadley et al., 1995). Elevated [CO_2] can be maintained in temperature-gradient tunnels by injecting controlled amounts of CO_2 into the tunnel (Hadley et al., 1995; Horie et al., 1995; Rawson, Gifford, and Condon, 1995). Alternatively, daytime subambient [CO_2] gradients can be created within tunnels by allowing photosynthesis to remove CO_2 from the airstream as it passes from the tunnel entrance to the tunnel exit (Mayeux et al., 1993). Subambient [CO_2] experiments can also be carried out in laboratory chambers and glasshouses (e.g., Gifford, 1977). Subambient [CO_2] experiments are valuable ways of studying crop responses to past changes in [CO_2].

Most field chambers are relatively inexpensive to build and operate, so they can be replicated well at individual sites. Moreover, field chambers can be used at many sites across climatic gradients (e.g., Jäger, Hertstein, and Fangmeier, 1999).

Open-Air Field Fumigation Systems

Advanced open-air (i.e., chamberless) systems were designed about two decades ago to fumigate crops with controlled levels of air pollutants in the field (McLeod, Fackrell, and Alexander, 1985; McLeod, 1993). The method was later adapted for elevated-CO_2 research. When fumigation is with CO_2, the approach is called FACE for free-air CO_2 enrichment (Allen, 1992; Hendrey and Kimball, 1994).

In simple terms, open-air systems release a gas (or gases) of interest from some type of tubing into the air upwind of treatment plots. The gas is then transported to the treatment plot through mainly ambient air movement. To accomplish this, the release points for the gas are changed dynamically as wind direction changes to keep the releases upwind of treatment plots. Considerable distance must be maintained between plots to avoid contamination from gas released at other plots. The present state of engineering for FACE is to release pure CO_2 through a large number of small holes under considerable pressure into the airstream upwind of the experimental plot (Okada et al., 2001).

Main advantages of open-air systems are that solar radiation received by the crops, air temperature, wind speed and turbulence characteristics, and humidity are mostly unaffected (McLeod and Long, 1999). At least one variant of the FACE approach, however, can significantly, in terms of crop development and yield, alter microclimate (Pinter et al., 2000). A disadvantage of open-air systems is that air temperature remains near ambient values, so the important combination of warming with elevated CO_2 is not studied. Irrigation water can be added to open-air systems, but precipitation reduction is not possible without imposing structures above the crop.

To date, FACE experiments have been, or are being, conducted with cotton in Arizona (Mauney et al., 1994), spring wheat in Arizona (Kimball et al., 1995; Pinter et al., 2000), potato in Italy (Miglietta et al., 1998), rice in Japan (Kim et al., 2001; Sakai et al., 2001), grain sorghum in Arizona (Ottman et al., 2001), a soybean-maize rotation in Illinois (Morgan et al., 2004), a rice-wheat double-crop system in China (K. Kobayashi, personal communication, 2002), grape in Italy (Bindi et al., 2001), and a ryegrass-clover pasture in Switzerland (Daepp et al., 2000). The soybean-maize study includes elevated [O_3] treatments to determine the effects of both gases on the crops.

Because CO_2 released by FACE systems is rapidly lost from experimental plots (especially with windy conditions), monetary costs of CO_2 are large. Thus, only modest CO_2 enrichment is usually used;

FACE treatments are generally about 200 ppm CO_2 above ambient, except in one potato plot and small grape plots, which experienced about 300 ppm CO_2 enrichment (Miglietta et al., 1998; Bindi et al., 2001).

Comparison of Methods

Each method of experimentally controlling environmental conditions has advantages and disadvantages. In general, laboratory chambers and glasshouses can exert superior control on temperature, humidity, gas concentrations, and other variables. Complex factorial experiments can be carried out in laboratory chambers to establish cause-and-effect relationships that can be used to generalize and predict effects of environmental change on crop physiology and yield. Unfortunately, many chambers and glasshouses used to study crop physiology and growth involve environmental conditions that differ significantly from those found in typical field settings, and results may therefore be unapplicable to crops in the field (Lawlor and Mitchell, 1991).

Field chambers have the advantage that temperature and light levels approximate those in the field, and good control of temperature and gas concentrations is possible. Another advantage of field chambers, and open-air fumigation systems, is that a normal soil profile is used, though some field-chamber experiments involve potted plants (e.g., Rogers, Thomas, and Bingham, 1983; Amthor, 1988).

It is more expensive to fumigate an experimental plot with an open-air system than it is with chambers because the added gas is more quickly "lost" in an open-air system. Because of this, more experimental replication is possible with chamber approaches for a given use of CO_2 and/or O_3. Also, the limited $[CO_2]$ range used, and lack of temperature control, in typical FACE experiments can be easily overcome with chambers. Conversely, open-air fumigation plots are usually larger than chamber plots so more research can be done with, and more researchers can work in, each plot in an open-air fumigation system.

An important consideration is whether different methods of controlling environmental conditions affect crops in similar ways. Unfortunately, few *direct* comparisons exist of the effects of environmental changes on crops imposed by the different approaches, and those comparisons usually ignore the most important crop variable, yield (e.g., Kimball et al., 1997; Dijkstra et al., 1999). An extensive *indirect* comparison of the effects of elevated [CO_2] on wheat yield using different methods to experimentally control [CO_2] derived from 80 published experiments indicated that the positive effect of elevated [CO_2] on yield was larger in laboratory chambers and glasshouses than it was in field experiments, though the 95 percent confidence intervals for all five methods studied overlapped (Table 2.1). Moreover, all three field methods—OTCs, CTCs, and FACE—gave nearly identical results, though the small number of FACE experiments and their limited [CO_2] range make it impossible to generalize about that method.

METHODS FOR SALINITY STUDIES

Saline soils are those containing high amounts of soluble salts. Soils are considered saline if the electrical conductivity of a saturated soil extract exceeds 4 $dS \cdot m^{-1}$ (www.soils.org/sssagloss/search.html), although sensitive crop species are often affected at electrical conductivity values of 2 $dS \cdot m^{-1}$ or less. Conductivity values greater

TABLE 2.1. Comparison of different experimental methods of controlling [CO_2] evaluated in terms of the effect of elevated [CO_2] on wheat yield.

Experimental approach	Effect of [CO_2] on yield (% increase per ppm CO_2 increase)	95% confidence interval	Number of experiments
Laboratory (growth) chambers	0.120	0.076-0.156	11
Glasshouses	0.140	0.071-0.210	9
Closed-top field chambers	0.080	0.048-0.112	20
Open-top field chambers	0.086	0.065-0.107	38
Open-air field fumigation	0.080	0.066-0.093	2

Source: Amthor (2001).

Note: The analysis applied to [CO_2] treatments in the range 350 to 750 ppm CO_2. In all the experiments considered, the [CO_2] treatments were applied for most or all of the crop life cycle. All experiments were conducted at near ambient [O_3], favorable temperature, full irrigation, and ample nitrogen fertilization (i.e., the experimental conditions were beneficial for wheat growth and yield).

than 8 ~$dS \cdot m^{-1}$ are indicative of strongly saline soils and can be tolerated by only the hardiest of crops.

The most common salts contributing to saline conditions are chlorides and sulphates of sodium, calcium, and magnesium. In most salinity experiments, however, crops are exposed to only a single salt, usually NaCl (Maas and Grieve, 1987). This approach might induce specific symptoms of NaCl toxicity that are uncharacteristic of crop responses to salinity caused by combinations of salts, as would normally be found in nature (even at equivalent electrical conductivity values). Some crops will tolerate high electrical conductivities when they are caused by a balanced composition of ions but are quite sensitive to similar electrical conductivity values owing to NaCl alone. Nearly all of the results reported in this book about crop responses to saline soils involve crop responses to NaCl alone. There are two reasons for this: (1) NaCl is the most prevalent salt in saline soils, and (2) very few experiments have been conducted using the balanced combinations of salts more reflective of the field soil environment.

Another problem faced when trying to interpret results of salinity experiments stems from the range of methods employed to expose crops to salts. Approaches range from field experiments, where only crude levels of control of soil salinity levels are possible, to laboratory hydroponics experiments. Hydroponically grown plants may behave differently compared with those grown in soil for at least three reasons: (1) nutrient relations often differ significantly for plants grown in sand or solution cultures compared to plants grown in soil in the field (Grattan and Grieve, 1994), and (2) roots develop and function differently in hydroponically grown crops compared to those grown in soil. Furthermore, in many laboratory studies on young crop plants, salinity levels are abruptly increased, which may cause osmotic shock and/or plasmolysis of root cells, effects that are not typically observed when salt treatments are introduced gradually (Munns, 2002).

The most appropriate approach to salinity experiments might involve (1) exposure to several salts in combination, (2) growing plants in soil instead of sand or hydroponics, and (3) minimizing confounding effects of osmotic shock and plasmolysis by gradually increasing the salt concentration of growth media.

SUMMARY

Many methods are used to study effects of environmental variation and change on crops. The methods can be grouped into (1) statistical relationships between historic yield and historic environmental variation or trends and (2) experimental control of one or more environmental conditions.

Statistically derived relationships between historic yield and historic environmental variation and change are useful for summarizing (or describing) past effects of environmental conditions on crops. They may lack explanatory power, however. That is, the statistical approach describes rather than explains the effects of the environment on crops. Nonetheless, the statistical approach has advantages, mainly related to its "real-world" nature and its potential applicability to large geographic areas.

Many methods of experimentally controlling environmental conditions are used with crops, and some of them combine properties of two or more of the general approaches outlined in this chapter. Experimental control of the environment provides opportunities to determine cause-and-effect relationships. Laboratory chambers can provide the greatest control on the environment and therefore have the potential to produce the most complete explanations and models of the effects of environmental change on crops. Laboratory chamber implementations are often weak, however, with respect to important environmental variables. The ultimate chambers for research have not yet been constructed, though good designs are available.

Field chambers bring modest control to field settings, but may include some artifacts and unwanted environmental changes. Open-air field fumigation systems control only gas concentrations. They are therefore constrained by ambient environmental conditions. Open-air fumigation systems can provide larger plots than typically found inside field chambers, but these are still only a tiny fraction of an individual field. In fact, all field experiments manipulating temperature, [CO_2], and/or [O_3] involve only small "islands" of treatments. Because of this, important regional-scale effects of environmental change on dispersal of weeds, insects, bacteria, fungi, and so on, as well as regional-scale meteorological feedbacks between vegetation and the atmosphere (see McLeod and Long, 1999) may not be faithfully represented in plot-scale manipulative experiments using any methodology.

In the end, a combination of statistical (i.e., observational) and several experimental methods, rather than a single approach, produces the most knowledge of the effects of environmental variability and change on crops.

Chapter 3

Cellular Responses to the Environment

The process of living is the process of reacting to stress.

Dr. Stanley J. Sarnoff,
Time, November 29, 1963

For a field crop, stress abounds. Crop stress may be defined as any factor that decreases growth and/or yield below the limits imposed by plant genotype and available solar radiation. Environmental changes such as rising atmospheric [O_3], warming, and soil salinization all represent stress because they reduce crop yields in many production areas. Ozone and high salinity are directly toxic to cellular metabolism. Atmospheric warming can reduce yield in two ways. First, warming may increase the percentage of time that crops are exposed to very high temperatures (acute cellular stress), and second, warming could reduce yields by speeding up phenology (chronic developmental stress). Of all the stresses a crop encounters in the field, catastrophic crop losses are most often caused by high temperatures, drought, and excess rainfall (Table 3.1). Effects of these and other environmental stresses on crops could intensify as the environment continues to change and as global populations rise, forcing agriculture to expand into areas currently considered unsuitable for crop production.

Effects of environmental changes on crop growth and yield are manifested mainly through effects of changes in physiology at the cellular level. Though connections between cellular and field-scale processes are typically difficult to establish, these links are nonetheless important. In fact, the ability to adapt to environmental changes could depend (in part) on an understanding of mechanisms that operate at the level of cells. In particular, cell membranes and proteins are especially sensitive to environmental conditions, and understanding

TABLE 3.1. Percentage of crop insurance indemnities attributed to specific abiotic and biotic environmental factors in the United States.

Crop	Start of data series[a]	Drought or excess heat (%)	Excess rain (poor drainage) or flood (%)	Frost, freeze, or other cold damage (%)	Hail, cyclone, tornado, wind, or hot wind (%)	Insects and disease (%)	All other (%)
Barley	1956	37	30	4	21	6	3
Bean	1948	15	22	25	34	3	1
Canola (rape)	1995	14	64	3	15	4	0
Cotton	1948	22	19	13	33	7	6
Forage	1979	38	27	28	4	1	2
Maize	1948	24	23	11	26	7	10
Maize seed[b]	1983	62	24	3	5	4	0
Millet	1996	67	12	0	21	0	0
Oat	1956	34	43	4	14	3	1
Pea	1963	46	16	14	20	4	0

Peanut	1962	45	20	7	3	22	3
Popcorn	1984	57	21	5	12	5	0
Potato	1962	23	22	25	8	21	1
Rice	1960	17	50	11	2	5	14
Rye	1980	41	36	10	12	0	0
Safflower	1964	27	31	30	10	2	0
Sorghum	1959	35	28	15	18	3	1
Soybean	1955	26	34	15	18	3	5
Sugarbeet	1965	10	35	15	21	19	1
Sugarcane	1967	16	7	19	4	21	33
Sunflower	1976	23	28	12	20	14	3
Tomato	1963	19	70	6	2	2	1
Wheat	1948	21	14	25	20	13	6

Source: Adapted from U.S. Department of Agriculture, National Agricultural Statistics Service (2002, Table 10-1).

[a]All data series end with the year 2001.

[b]Hybrid maize seed production.

how they are affected by environmental changes could be a key for developing suitable genotypes for the future.

MEMBRANE STRUCTURE AND FUNCTION

Biological membranes, only ~7 nm thick, have been called one of the most important structures on earth because they control the passage of substances into and out of cells and organelles (Reid, 1999), therefore making life possible. Membranes are comprised of a sheet-like bilayer of lipids in which many proteins, glycoproteins, and lipoproteins float about. The proteins embedded within membranes serve as carriers, channels, and pumps and also transduce environmental (and biochemical) signals from outside the cell to the cytoplasm. Membranes contain sterols as well, which evidently serve both functional and structural roles (Figure 3.1).

Phosphoglycerides, the major lipid component of most cellular membranes (except those of chloroplasts), consist of a glycerol backbone to which fatty acyl groups are linked at carbons 1 and 2. A phosphate group, sometimes called the phosphate head, is attached at carbon 3. A number of small, molecular-weight moieties may be found attached to phosphate head groups such as amines (including polyamines), amino acids, and carbohydrates.

In chloroplast membranes, the predominant lipids are glycosylglycerides or glycolipids, which do not possess phosphate head groups. They include sphingosides, galactolipids (e.g., MGDG, monogalactosyl diacylglycerol, and DGDG, digalactosyl diacylglycerol), and sulfolipids (e.g., SQDG, sulphoquinovosyl-diacylglycerol). Glycolipids contain a glucose or galactose at carbon 1 and acyl groups at carbons 2 and 3 of the glycerol backbone.

The membrane lipids previously described are amphipathic, which means they possess both hydrophobic (i.e., the acyl groups) and hydrophillic (i.e., the phosphate, amine, amino acid, or carbohydrate groups) characteristics. The types of molecules associated with hydrophilic phosphate moieties are especially important for cell function (Nilsen and Orcutt, 1996). For example, membrane charge, an important functional characteristic, can vary significantly depending on the molecules associated with the hydrophilic head groups (Nilsen and Orcutt, 1996).

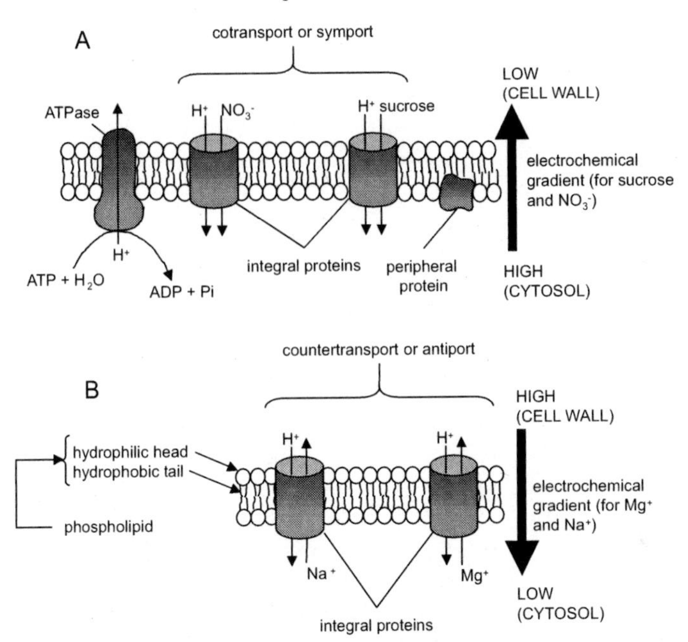

FIGURE 3.1. Diagram of membrane structure and mechanisms of active transport by plant root cells. Sterols and carbohydrates, both important components of biological membranes, have been omitted. (A) Energy for active transport is generated by pumping of H^+ ions from the cytoplasm into cell-wall regions by membrane-bound ATPases (primary active transport, left of panel). Pumping of H^+ ions creates a chemical (i.e., H^+ ion) as well as a charge (electrical) gradient. The electrochemical gradient powers simultaneous cotransport of anions (such as NO_3^- and sugars, as pictured here) and H^+ ions into cytoplasm. (B) H^+ ions may return to cytoplasm, but their movement must be coupled with simultaneous countertransport or antiport of a cation (such as Na^+ or Mg^+, pictured here). In general, cation absorption into root cells is favored by the more negative electrochemical potential typical of cytoplasm. Cations, therefore, are generally taken up passively through integral carrier or channel proteins (facilitated diffusion; not pictured here).

The characteristics of lipid acyl groups are also important. Fatty acids differ widely in length of their carbon chains as well as position and number of double bonds (i.e., their degree of unsaturation can vary). The structure and composition of these acyl groups influences membrane viscosity and permeability. The greater the viscosity of membranes, the lower their permeability (Nilsen and Orcutt, 1996).

Sterols are another important membrane component. They are structurally and metabolically distinct from glycolipids and phospholipids because they are derived from the mevalonate pathway (i.e., the pathway that synthesizes isoprenoids, including terpenoids and some hormones). Sterols can be found inserted between hydrocarbon tails of acyl groups. They function in stabilizing membranes by increasing vicosity and preventing thermal phase transitions.

Membrane Transport

Membranes are semipermeable; water and small gas molecules such as CO_2 and O_2 can easily pass through them, but electrically charged and polar uncharged solutes cannot. Some molecules pass through membranes passively (passive transport), whereas transport of others requires use of energy (active transport) (Figure 3.1).

Mode of transport across membranes is dependent on the thermodynamic gradient ($\Delta\mu$) present from outside to inside (or vice versa) the cell. Both chemical and electrical (i.e., membrane potential) components contribute to $\Delta\mu$, and are together referred to as the electrochemical gradient. The famous Nernst equation can be used to define $\Delta\mu$ (e.g., Nobel, 1999). Flux of nutrients "down" $\Delta\mu$ is passive while flux "up" $\Delta\mu$ is active. In (nearly) all cases, transport proteins associated with a membrane are involved in both passive and active uptake (Wignarajah, 1995; Cakmak and Engels, 1999). In total, there may be as many as 1,000 different transport proteins that are either peripherally (i.e., on the surface) or integrally (i.e., embedded in and spanning the lipid bilayer) associated with plant membranes (Reid, 1999).

Passive Transport

Passive transport (typical for K^+, NH_4^+, Mg^{2+}, and Ca^{2+}) involves transport across a membrane from higher to lower electrochemical potential. Simple diffusion down an electrochemical gradient through a membrane is too slow for effective cell physiology, however, and does not account for observed kinetics of cation uptake into plant cells. Probably, an integral protein (uniporter) forms a dynamic, perhaps selective, pore or channel through the membrane that facilitates cation movement down its electrochemical gradient (Clarkson, 1977; Marschner, 1995). The best explanation of passive cation uptake in plants is the process called uniport (or facilitated diffusion). In any case, no energy input (aside from that needed to construct and maintain the uniporter) is required for passive uptake.

Active Transport

While most cations can be passively taken up through uniporters, anions are actively taken up against electrochemical gradients by protein carriers. In addition, passive uptake of some cations from the soil solution into roots is probably followed by their active transport against a concentration gradient back out of roots into the soil solution (or active transport into vacuoles for storage within the root).

Secondary active uptake is a two-step process. The first step is the pumping of hydrogen ions across a membrane. This is an energy-requiring process powered by hydrolysis of ATP (forming ADP and Pi) or PPi (forming 2 Pi) by a H^+–pumping ATPase or pyrophosphatase, respectively, embedded through a membrane (Marschner, 1995). For uptake of nutrients from apoplast to symplast, the H^+ pump is embedded in the plasma membrane (also called the plasmalemma) and pumps H^+s from the cytosol (symplast) to the apoplast. The H^+ pumping produces a H^+ gradient across the membrane. That gradient then drives H^+ transport back into the symplast, although H^+ reentry is slow unless it is facilitated by a carrier or channel. Carriers and channels also couple H^+ reentry to simultaneous transport of mineral nutrients or other solutes (the second, non-energy-requiring step of active transport). For NO_3^- uptake (as an important example of anion uptake), a H^+ and a NO_3^- are simultaneously moved into the cell by a transporter called a symporter (the H^+ moves down its electrochemical gradient and the NO_3^- is "carried" along). The process is called cotransport (both ions moving in the same direction, into the symplast). Different symporters probably exist for different anions.

Carriers can also be used to simultaneously move H^+ into the cell and another cation (e.g., Na^+ or Ca^{2+}) out of the cell. Carriers that facilitate such countertransport are called antiporters. Antiporters are probably not channels. Countertransport of Na^+ out of root cells is an important response to soil salinity. Also, cytosolic Ca^{2+} levels are generally kept low in plant cells, and countertransport is the mechanism used to pump Ca^{2+} out of the cytosol. Along these lines, the vacuole can be used to store Na^+ and/or Ca^{2+} to keep cytosolic levels of those cations low. To do this, tonoplast (the membrane enclosing a vacuole) ATPases or pyrophosphatases pump H^+s into the vacuole, from which they return to the cytosol by countertransport with Na^+ or Ca^{2+}.

Abiotic Stress and Membranes

Abiotic stresses, including drought, high temperature, O_3 pollution, and salinity, can compromise the structure and function of membranes. Extensive oxidative damage to membrane components, alterations in membrane fluidity and transporter function, and depolarization of membranes are all common consequences of these environmental stresses (Reynolds et al., 1994; Sakaki, Tanaka, and Yamada, 1994; Guidi et al., 1999; Plazek, Rapacz, and Skoczowski, 2000). These effects decrease selectivity and increase permeability or "leakiness" of membranes, conditions that can lead to cell death. In fact, Fosket (1994) refers to membranes as "the eyes and ears" of the cell because they detect changes in environmental conditions. This explains why membranes are often the focus of studies aimed at elucidating the cellular mechanisms of stress resistance (Chang et al., 1996).

Homeoviscous Acclimation

Many plants adjust to local environmental conditions by altering the chemical properties of cellular membranes. Some of these changes include modifying acyl-group chain lengths, altering the extent of membrane-lipid unsaturation (number and location of double bonds), changing the proportion of lipid classes, or shifting lipid-to-protein ratios (Harwood, 1997). Furthermore, interconversions between free and conjugated sterols, in addition to shifts in sterol structural characteristics, also may play a role in acclimation to environmental changes. The process whereby plants shift membrane composition in order to maintain fluidity in fluctuating environments has been termed homeoviscous acclimation (Nilsen and Orcutt, 1996).

Homeoviscous acclimation is well characterized for plants exposed to sub- and supraoptimal temperatures. Lowering growth temperatures, for example, often results in an increase in the proportion of unsaturated lipids, while increasing temperature causes the opposite (Klueva et al., 2001). Shifts in the degree of unsaturation ensure that membranes function properly under different thermal conditions (Harwood, 1997).

One example of homeoviscous acclimation involves chloroplastic trienoic fatty acids (18:3 and 16:3). Trienoic fatty acids have three double bonds, a trait which allows them to remain fluid at low temperatures. As we would expect, it is more prevalent in chloroplasts of plants adapted to cool climates than in those adapted to warm climates. Furthermore, in plant species adapted to warm climates, temperature increases are often accompanied by decreases in trienoic fatty acids. Recently, four lines of transgenic *Arabidopsis* plants were created in which a gene required for the synthesis of trienoic fatty acids was "silenced" (Murakami et al., 2000). Those transgenic plants were able to withstand much higher temperatures than were the wild-type controls, which is a clear example of the importance of membrane integrity for plant survival in high or low temperatures. It also underscores the potential to genetically alter crop genomes in order to produce varieties better adapted to altered environmental conditions. It is already well established that screening crop genotypes for membrane thermostability is an effective way to select for heat-tolerant genotypes in breeding programs (Saadalla, Quick, and Shanahan, 1990; Reynolds et al., 1994; Ismail and Hall, 1999; Blum, Klueva, and Nguyen, 2001).

Significant changes in lipid composition have also been reported for crops exposed to O_3 pollution. These shifts, however, probably reflect damage instead of acclimation. For example, changes in various lipid and fatty acid contents, including decreased phospholipids and sulfolipids, increased free fatty acids, and increased sterols, have all been observed (Runeckles and Chevone, 1992). In general, oxidative stress also leads to a decrease in the level of unsaturation of fatty acids (Navari-Izzo and Rascio, 1999). Many of the detrimental effects that high $[O_3]$ has on lipids are thought to result from degradation of the ester bonds that link fatty acids to their glycerol backbone. Similar damage to membranes can also occur in water-stressed crops.

Salinity also affects membranes (Chang et al., 1996; Kerkeb et al., 2001). For example, several studies have shown that salinization changes the weight ratio of lipids:proteins. This shift is probably caused by upregulation of membrane proteins that adjust the flow of ions between different cellular compartments in order to counteract the negative effects of salinity (Singh et al., 1985). Furthermore, salt stress leads to upregulation of heat-shock proteins, which may provide some degree of protection to membrane-bound proteins. Changes in membrane proteins are clearly important for rendering plants salt-adapted.

Evidence also indicates that salinity induces changes in root lipid composition and that these changes are necessary for salt tolerance (Kafkafi and Bernstein, 1996; Kerkeb et al., 2001). Membrane characteristics such as the size and charge of polar head groups, length of fatty acid chains, and sterol composition are all important for normal uptake of nutrients. The ability of a crop to make the appropriate adjustments in these membrane parameters when confronted with saline conditions may be correlated with salt tolerance (Kafkafi and Bernstein, 1996). Specifically, increases in the ratio of sterols to phospholipids in cell membranes as well as decreased fatty acid unsaturation may both lead to membrane stabilization and greater salinity tolerance (Kerkeb et al., 2001).

ENZYMES

Enzymes reduce the energy of activation required for chemical reactions to proceed in any cells, and outside them too. In doing so, enzymes greatly accelerate biochemical reactions and allow physiology to proceed. Functioning of enzymes is highly dependent on their three-dimensional conformation, which is controlled partly by hydrogen bonding between functional groups present on different amino acids and partly by sulfhydryl bridges.

Environmental stress can change enzyme conformation by disrupting the chemical bonds that maintain them in their functional conformation. Ozone and other oxidants, for instance, can oxidize components of cellular proteins, thereby reducing their functionality. Soil salinity interferes with proteins in two ways: (1) cytosolic Na^+ may directly inactivate enzymes (by binding to inhibitory sites) or may cause enzyme dysfunction by displacing K^+ from activation sites; and (2) the charge to mass ratio of Na^+ is much greater than K^+, which is thought to change the way in which proteins interact with water molecules, ultimately leading to enzyme destabilization (Amtmann and Sanders, 1999).

Temperature also has a profound influence on enzymes. Ordinarily, enzyme activity increases with temperature up to some optimum (which varies among enzymes) and then declines. As temperature increases, the kinetic energy of reacting molecules increases, causing enzymes and substrates to couple with greater frequency. But supraoptimal temperatures disrupt chemical bonding in proteins, leading to denaturation and loss of enzyme activity.

Q_{10} values are often applied to enzymes to generalize their responses to temperature. The proportional change in enzyme activity for a 10°C increase in temperature is the Q_{10}. As an example, if an enzyme has a Q_{10} of 2, it means that the activity of this enzyme would double for a ten degree increase in temperature. Table 6.4 illustrates Q_{10} values for several enzymes important in carbon metabolism and partitioning. Warming will have an effect on enzyme kinetics in plants. The extent to which this could influence agriculture is unknown, however, because different enzymes have different temperature optima.

CELLULAR MECHANISMS OF STRESS RESISTANCE

Several mechanisms perceived by many cell biologists to be particularly important in guarding against stress-induced damage to membranes, proteins, and other cellular machinery involve antioxidants, polyamines, and heat-shock proteins. Other mechanisms of avoiding or reducing stresses are discussed later in this book, such as osmotic adjustment (Chapter 4), production of pathogenesis-related proteins and secondary compounds (Chapter 10), and hormones (Chapter 6).

Oxidative Stress and Antioxidants

Reactive Oxygen Species

Production of active oxygen species (AOS), including singlet oxygen ($1O_2$), superoxide ($O_2{\cdot}^-$), hydrogen peroxide (H_2O_2), and the hydroxyl radicle ($\cdot OH$), is an inevitable consequence of life in an oxygen-rich environment (Polle et al., 1990). These molecules are toxic because they oxidize lipids, nucleic acids, and proteins. Oxidation of lipids, for example, results in membrane dysfunction within chloroplasts, mitochondria, and cytosol. Effects on proteins can be equally debilitating to cell function. (Animals, humans included, also produce powerful oxidants within their cells as a result of normal metabolism, and their cellular effects are similar to those observed in plants. In humans, oxidants have been linked to aging, cancer, and many other diseases.) In plants, AOS have been linked to programmed cell death (e.g., the hypersensitive response; see Chapter 10) as well as organ senescence (Pell, Schlagnhaufer, and Arteca, 1997).

Although AOS are generally associated with stress, they are produced even under optimal conditions. This raises some important questions. First, if it is accepted that these molecules are indeed incompatible with cellular metabolism, why and how are they produced, and how are their toxic effects counteracted in plants? How do current and, equally important, how might future levels of environmental stress (be they greater than or less than levels of today) affect the balance between production and detoxification of these molecules? How important might a robust antioxidant system prove to be for crop production in the future? These questions are the subject of this section.

In plants, AOS are formed by the excitation of O_2 forming singlet oxygen or by the reduction of atmospheric oxygen ($3O_2$). Singlet oxygen ($1O_2$) is produced mainly by the absorption of electromagnetic energy from photoexcited compounds such as pigments (e.g., chlorophyll) and other components of electron transport systems (e.g., quinones) (Navari-Izzo and Rascio, 1999). Singlet oxygen leads to the peroxidation of membrane lipids and oxidizes a number of amino acid residues of proteins (Bowyer and Leegood, 1997).

Another important AOS is formed by the stepwise (one electron at a time) reduction of O_2. The complete reduction of O_2 requires four electrons, the end product being water (see Equations 3.1-3.4) (Navari-Izzo and Rascio, 1999; Minkov et al., 1999). The first step in the reduction of O_2 results in the production of the superoxide anion radicle ($O_2{\cdot}^-$) via the one electron reduction of ground-state oxygen (Equation 3.1). The reduced form of ferredoxin (Fd_{red}) is probably the major electron donor in this reaction (Figure 3.2).

$$O_2 + 1\ e^- \longrightarrow O_2^- \qquad (3.1)$$

Superoxide radicles, aided by the catalytic properties of superoxide dismutases (see below), are further reduced (and protonated) to form

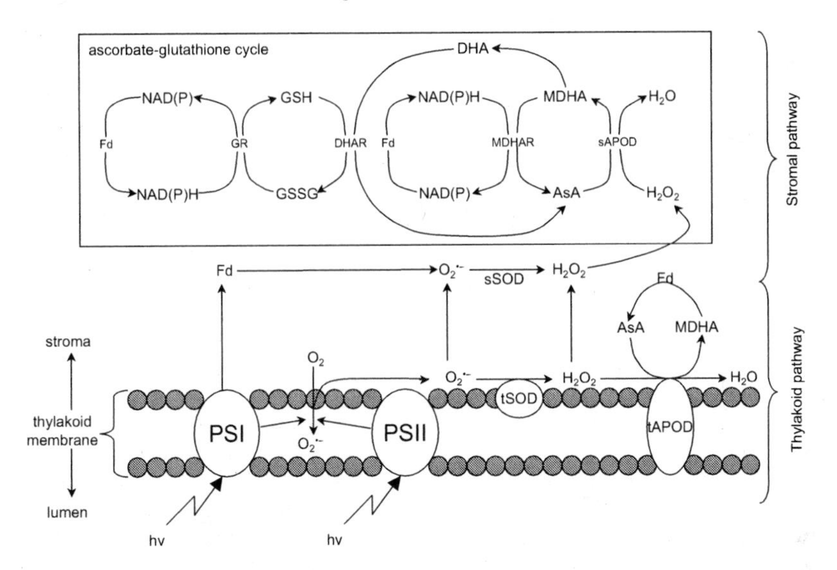

FIGURE 3.2. Within chloroplasts, there appear to be both thylakoid and stromal pathways for detoxifying hydrogen peroxide (H_2O_2) and the superoxide radical (O_2^-) generated by PSI and PSII. The superoxide radical generated by the photosynthetic apparatus is quickly dismutated by either thylakoid bound (tSOD) or stromal (sSOD) superoxide dismutases. Hydrogen peroxide produced by this reaction is then reduced (to water) by ascorbate (AsA) in a reaction catalyzed by either thylakoid bound (tAPOD) or stromal (sAPOD) ascorbate peroxidase. In the thylakoid pathway, the monodehydroascorbate radical (MDHA; oxidized ascorbate) is then reduced back to ascorbate by (photoreduced) ferredoxin (Fd). H_2O_2 that escapes the thylakoid pathway and diffuses into the stroma is reduced to water by sAPOD. The reduced form of ascorbate is then regenerated by the ascorbate-glutathione pathway as illustrated in the box. Briefly, MDHA is reduced to AsA in a reaction catalyzed by monodehydroascorbate reductase (MDHAR). Some MDHA spontaneously converts into DHA (dehydroascorbate reductase) which is reduced to AsA by glutathione (GSH) catalyzed by DHAR (dehydroascorbate reductase), forming the oxidized form of glutathione (GSSG). The regeneration of reduced glutathione (GSH) then requires NAD(P)H and the enzyme glutathione reductase (GR). Although glutathione is especially important for its role in supplying/regenerating reduced forms of ascorbate, it may also directly scavenge AOS. *Source:* Adapted from Navari-Izzo and Rascio (1999).

hydrogen peroxide (H_2O_2 [Equation 3.2]). Hydrogen peroxide represents the first stable form of reduced oxygen with radical properties (Navari-Izzo and Rascio, 1999).

$$O_2^{\cdot-} + O_2^{\cdot-} + 2H^+ \longrightarrow H_2O_2 + O_2 \quad (3.2)$$

Hydrogen peroxide can convert to the hydroxyl radicle (OH·) through the transfer of one more electron (Equation 3.3). Fe^{2+} serves as the electron donor in this reaction:

$$H_2O_2 + 1e^- \longrightarrow OH\cdot + OH \tag{3.3}$$

The hydroxyl radicle is certainly the most reactive of the AOS (it reacts within microseconds). It is therefore only capable of diffusing a few molecular diameters before oxidizing cellular constituents (Bowyer and Leegood, 1997). Reduction of the hydroxyl radicle ultimately forms H_2O (Equation 3.4).

$$OH\cdot + 1e^- + H^+ \longrightarrow H_2O \tag{3.4}$$

Although some are formed in cytoplasm, mitochondria, and microsomes, the majority of AOS are produced within chloroplasts. Chloroplasts are particularly vulnerable to AOS for several reasons. First, chloroplasts contain high levels of oxygen, which they both produce (water-splitting reaction of photosynthesis; see Chapter 5) and consume. Second, the absorption of light energy and passage of electrons down an electron transport chain in the presence of high O_2 makes electron leakage to oxygen inevitable. In other words, absorption of light energy not only drives CO_2 assimilation, it also provides the energy for the excitation of O_2, which leads to production of AOS. Electron leakage to oxygen (Equation 3.1) is particularly high when the rate of reduction of ferredoxin by photosystem I exceeds use of NADPH in the photosynthetic reductive pentose phosphate pathway. These conditions are met, for example, when plants are in high-light environments or when chloroplastic CO_2 concentrations are low (as can occur during water stress, for example). Enhanced production of AOS accompanies many forms of stress, including drought, salinity, air pollutants such as O_3 and SO_2, and heavy metals (Hausladen and Alscher, 1993).

Infection by pathogenic microbes also elicits a surge in AOS production. This surge is often referred to as the "oxidative burst" and is typified by the rapid induction of relatively large quantities of hydrogen peroxide and superoxide. This burst of AOS production can often be measured less than three minutes after exposure to an elicitor (Low and Merida, 1996) and is thought to stimulate a number of changes within plants (Levine et al., 1994; Tenhaken et al., 1995): (1) cross-linking of cell-wall proteins, (2) stimulation of phytoalexins, (3) up-regulation of various defense genes, and (4) initiation of the hypersensitive response. These changes are thought to limit further spreading of disease. Exposure to O_3 pollution often causes a response similar to that induced by pathogen infection (Chapter 10).

Antioxidant Systems

To reduce cellular oxidative damage, plants produce both enzymatic and nonenzymatic antioxidative compounds. The enzymes involved include superoxide dismutase (SOD), catalases (CAT), peroxidases (POD), ascorbate peroxidase (APOD), dehydroascorbate reductase (DHAR), monodehydroascorbate reductase (MDHAR), and glutathione reductase (GR). The primary nonenzymatic components include ascorbate (vitamin C) and glutathione (a tripeptide). Some osmolytes such as mannitol, fructans, trehalose, ononitol, proline, and glycinebetaine also scavenge AOS (Zhu, 2001). In addition, α-tocopherol (vitamin E) and ß-carotene, both present at high concentrations in chloroplast thylakoid membranes, are also important antioxidants. Mechanisms and molecules involved in scavenging the superoxide radical and hydrogen peroxide produced in chloroplasts as a byproduct of photosynthesis are outlined in Figure 3.2. This figure illustrates how superoxide and hydrogen peroxide produced in chloroplasts are scavenged either by the thylakoid or stromal pathways. Other enzymes such as catalase and nonspecific peroxidases may also play a significant role in scavenging AOS in the cytosol and mitochondria but are not important in plastids.

Effects of Environment on Antioxidant Systems

Drought. Water stress leads to stomatal closure and reduced chloroplastic [CO_2]s, a condition that is conducive to electron leakage to O_2 and, consequently, AOS production. Up-regulation of antioxidant systems in water-stressed crops appears to be an important mechanism of drought tolerance, but depends on the rate at which water stress is imposed as well as the magnitude of the stress (Navari-Izzo and Rascio, 1999). When water stress is brought about slowly (as occurs in the field), plants seem to acclimate and produce fewer AOS compared to situations in which the imposition of water stress is more abrupt (as in many laboratory experiments). Also, the activity of AOS-scavenging enzymes often increases at mild to moderate water stress, but decreases at severe water stress (presumably as a result of the impairment of protein synthesis by acute water stress).

Elevated atmospheric ozone concentration. Ozone, a powerful oxidant, represents a significant oxidative stress to crops. Ozone enters the leaf through open stomates and then diffuses/dissolves into cell walls. Antioxidants localized in cell walls immediately intercept a portion of the O_3 before it can react with the plasmalemma or enter the symplast (Moldau, Bichele, and Hüve, 1998; Long and Naidu, 2002) (Figure 3.3). In broad bean, for example, up to ~30 to 40 percent of ozone immediately reacts with cell-wall localized ascorbate (Turcsányi et al., 2000). The remaining O_3 quickly reacts with other cell-wall components, potentially forming other AOS, including the hydroxyl radical, superoxide, singlet oxygen, and hydrogen peroxide. If either O_3 or its reactive products are not immediately neutralized (chemically reduced) in the apoplast, they may lead to cell membrane peroxidation and enter the symplast, where they can react with cellular membranes and other components of the cytosol.

Ozone resistance (tolerance) is often attributed to high antioxidative capacity in plant cells (Azevedo et al., 1998; Robinson and Britz, 2000). For example, mature leaves of an O_3-tolerant variety of soybean had higher levels of GR (+30 percent), APOD (+13 percent), and SOD (+45 percent) compared with an O_3 -sensitive variety (Chernikova et al., 2000). But, since most antioxidants are located in chloroplasts, and O_3 is thought to react principally within cell walls and at the plasmalemma (i.e., little O_3 or its oxyradical derivatives actually make it all the way into chloroplasts), there are probably other factors involved (Kangasjärvi et al., 1994). In fact, responses elicited by O_3 closely parallel plant response to other forms of stress, including both wounding and pathogen attack (see Chapter 10). So, O_3 resistance or susceptibility is more likely influenced by a suite of factors, not just antioxidative capacity.

Leaf biochemical responses to O_3 depend on both the severity and duration of O_3 exposure. In most cases, O_3 stimulates production of signal molecules such as ethylene, jasmonate, and salicylic acid, which in turn trigger a suite of stress-response mechanisms, including up-regulation of antioxidative systems (Figure 3.3). The hormone ethylene also likely down-regulates expression of photosynthetic

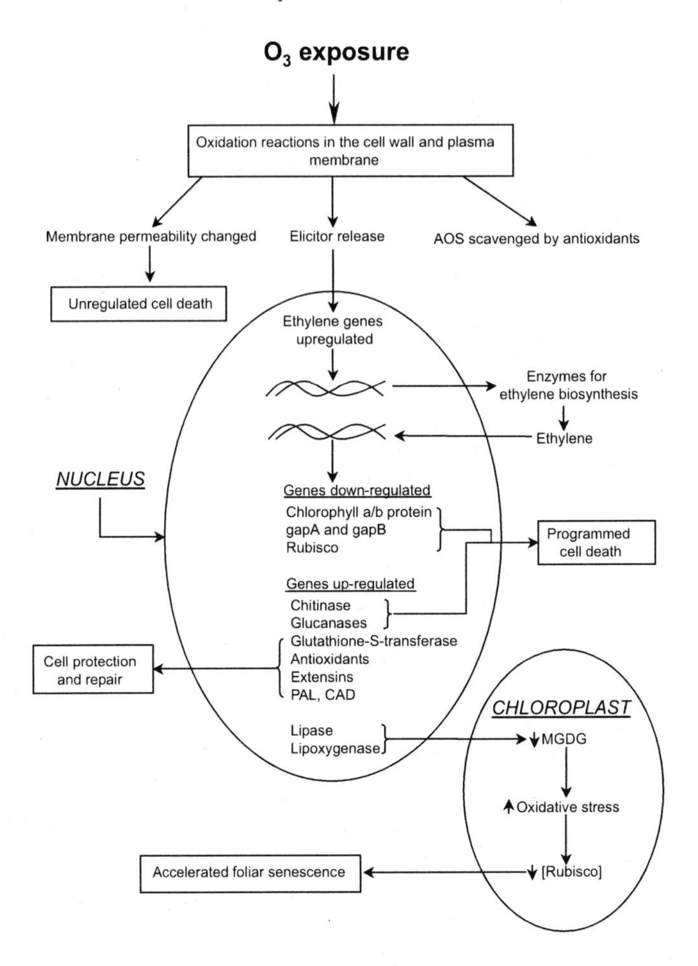

FIGURE 3.3. Model illustrating proposed effects of ozone on cells and entire leaves. Both resistance and toxicity mechanisms are shown. Acute exposure is more likely to result in unregulated or programmed cell death and/or damage followed by cell protection and repair. Lesions are a likely result of acute ozone damage. Chronic exposure is more likely to result in the scavenging of AOS and possibly induction of lipase and lipoxygenase genes. These enzymes are thought to react with MGDG (monogalactosyl diglycerides) in chloroplasts forming free fatty acids, which in turn act as a substrate for lipid peroxidation. These events may result in increased oxidative stress, decreased rubisco concentrations, and eventual foliar senescence. Abbreviations: AOS, active oxygen species; gapA and gapB, genes encoding glyceraldehyde-3-phosphate; PAL, phenylalanine lyase; CAD, cinnamyl alcohol dehydrogenase; MGDG, monogalactosyl diglycerides. *Source:* Modified from Pell, Schlagnhaufer, and Arteca, 1997.

genes, and this is one known trigger of premature leaf senescence. Ethylene formation is much greater in plants subjected to acute doses of O_3 than those exposed to chronic, but lower, doses (Mehlhorn, O'Shea, and Wellburn, 1991). This difference may partly account for variable responses of plants subjected to chronic, compared to acute, O_3. Exposure to acute O_3 leads to leaf necrosis (i.e., the typical bronzing or flecking of leaves), whereas chronic exposure leads to reductions in photosynthesis and accelerated foliar senescence (Peñarrubia and Moreno, 1999). In one line of thought, acute O_3 exposure is more likely to increase membrane permeability, followed by localized cell death (i.e., leaf necrosis), whereas chronic exposure is more likely to result in a decrease in chloroplast photosynthetic enzyme content, followed by hastened leaf senescence (Pell, Schlagnhaufer, and Arteca, 1997).

Elevated atmospheric carbon dioxide concentration. As atmospheric [CO_2] increases, generation of AOS by plant cells may decrease, and this should lower basal oxidative stress and reduce the necessity of antioxidants in crops. Theoretically, there are several reasons why higher CO_2 levels should decrease production of AOS. Each of these mechanisms is described in greater detail in Chapter 5.

1. Increased [CO_2]/[O_2] ratios within chloroplasts should lower superoxide production by decreasing electron leakage from PSI to O_2 (see also Chapter 5).
2. Increased [CO_2]/[O_2] ratios also decrease the oxygenase activity of rubisco in favor of carboxylation (carbon fixation). This will lower photorespiration and resultant H_2O_2 production (in C_3 plants). But photorespiration can also function as an alternative pathway for excess light energy and reduction potential.
3. Rubisco activity often decreases when C_3 plants are grown with high [CO_2], and this should lower photorespiration even more.

Recent data have shown that antioxidative systems are indeed down-regulated in plants grown in high [CO_2] (Pritchard et al., 2000). Presumably, these reductions in antioxidants reflect fewer oxyradicles in need of scavenging (Polle, 1996).

Interestingly, however, several studies have reported increased DHAR and MDHAR activities in response to growth in high [CO_2] environments. Since these two enzymes are involved in regenerating reduced forms of ascorbate and glutathione (via the ascorbate-glutathione pathway), higher ascorbate turnover times in CO_2-enriched plants has been suggested (Badiani et al., 1993; Schwanz and Polle, 1998; Pritchard et al., 2000). The reasons for these changes and the implications for crop stress resistance, growth, and yield are unknown.

Salinity. High salt concentrations often impair electron transport in chloroplasts and mitochondria resulting in generation of reactive oxygen species (Gueta-Dahan et al., 1997; Sreenivasula et al., 2000). Direct toxic effects of Na^+ and/or Cl^- ions on membrane structure and functioning (Seeman and Sharkey, 1986; Gossett, Millhollon, and Lucas, 1994; Meneguzzo, Navarri-Izzo, and Izzo, 1999), in addition to stomatal closure and the corresponding reductions in chloroplastic [CO_2] (Hernández et al., 2000; Hoshida et al., 2000), are both implicated in this response. Whatever the mechanisms, oxidative stress resulting from NaCl stress often stimulates antioxidative systems. A positive correlation between antioxidant activity and salt tolerance has in fact been established for a number of species, including rice (Dionisio-Sese and Tobita, 1998; Hoshida et al., 2000), pea (Olmos et al., 1994; Hernández et al., 2000), cotton (Gossett, Millhollon, and Lucas, 1994), potato (Benavídes et al., 2000), and foxtail millet (Sreenivasula et al., 2000).

In light of the susceptibility of cellular membranes to oxidative damage, and because the ability to tolerate high salt concentration depends largely on the efficiency of subcellular Na^+ compartmentation (which is dependent on membrane function), it is easy to understand why an effective antioxidative system is so crucial for plants growing in salt. To put it another way, if unchecked, AOS generated as the result of salt stress can lead to extensive damage to cellular membranes, increase their "leakiness," and interfere with Na^+ transport into, and retention within, vacuoles or cell walls. Leakage through membranes also leads to cell death and can contribute to premature leaf senescence, and both of these conditions are typical of salt-stressed plants. There is little doubt that, particularly in the absence of other resistance mechanisms (such as NaCl exclusion or compartmentation), an effective antioxidant system has the potential to enhance salinity tolerance.

Warming

Theoretical reasons, but few experimental data, indicate that high-temperature stress might affect production of AOS in plant cells (Edreva et al., 1998; Raychaudhuri, 2000). High-temperaure stress does affect membrane integrity, for example, and this may favor electron leakage to oxygen and, hence, production of AOS. Furthermore, rubisco is quite thermolabile, which means that heat stress may decrease CO_2 fixation, a situation that is also expected to result in electron leakage to oxygen. Nevertheless, there is still insufficient evidence in higher plants to conclude that antioxidants play a significant role in conferring thermal tolerance; at least not to the extent that yield could be affected. Besides, reports of a heat-stress effect on production of AOS or antioxidants can probably be linked to conditions accompanying the thermal stress, namely dehydration and stomatal closure, and not the thermal stress itself.

It is perhaps important that antioxidative enzymes in different species typically require different temperature ranges to function, and this dependence is correlated with the environment where the crop originated. For example, warm-climate crops have a higher optimal temperature range for glutathione reductase function than do cool-climate crops (Burke, 1990). So, in some species, high temperatures may compromise scavenging of free radicals.

Polyamines

Polyamines (PA) are polycationic nitrogen-containing molecules, the most common of which include putrescine (Put; a diamine), spermidine (Spd; a triamine), and spermine (Spm; a tetraamine) (Figure 3.4). Polyamines occur in every living plant cell, either in free form, conjugated with small molecules such as phenolic acids, or bound to macromolecules such as proteins (Bouchereau et al., 1999). Since they are positively charged, they readily react with anionic membrane components such as phospholipids, phosphate groups of DNA and RNA, and with cell wall components, including pectins (Kakkar, Bhaduri, et al., 2000; Kakkar, Nagar, et al., 2000). As a result, polyamines are thought to play important roles in synthesis and maintenance of membrane lipids, DNA, and proteins. Their role in DNA replication, transcription, and translation as well as other facets

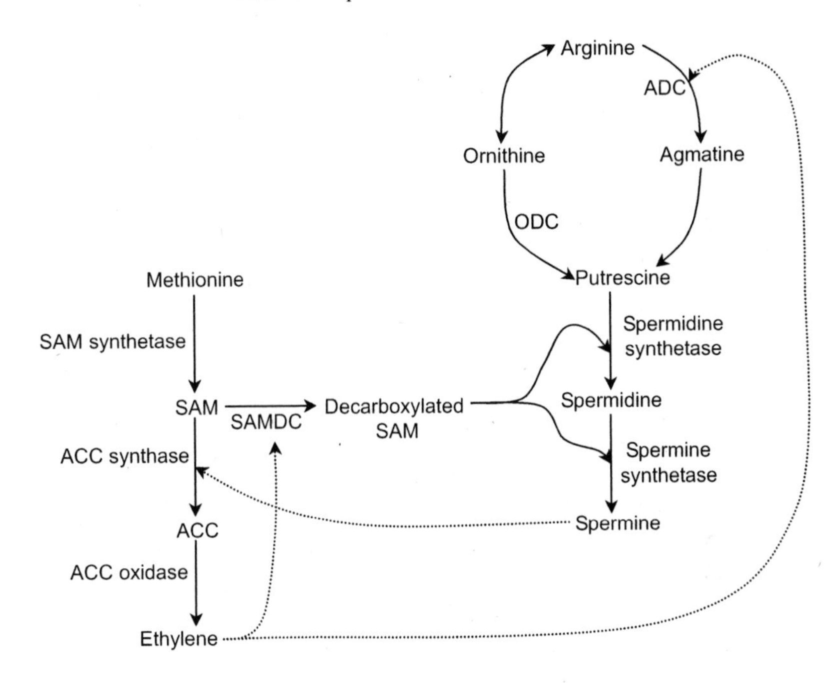

FIGURE 3.4. Pathway for polyamine and ethylene biosynthesis. Putrescine can be produced directly from ornithine in a reaction catalyzed by ODC, or it can be produced from arginine in a reaction catalyzed by ADC (agmatine is an intermediate in this reaction). Putrescine can then be converted into spermidine by the addition of an aminopropyl group derived ultimately from methionine. Spermidine can be converted to spermine by the addition of another aminopropyl group. SAM is an intermediate for both polyamine formation as well as formation of the important stress hormone ethylene. Dotted lines indicate known inhibitory interactions. Abbreviations: SAM synthetase, S-adenosylmethionine; ACC, 1-aminocyclopropane-1-carboxylic acid; SAMDC, S-adenosylmethionine decarboxylase; ADC, arginine decarboxylase; ODC, ornithine decarboxylase. *Sources:* Modified from Kangasjärvi et al. (1994) and Kumar et al. (1997).

of cell division, cell expansion, and organ morphogenesis are perceived as particularly important (Wink, 1997).

Polyamines have a putative role in plant stress resistance. Expression of enzymes involved in PA synthesis, as well as accumulation of PAs themselves, are quite sensitive to environmental changes. Salinity, high temperatures, and air pollutants such as O_3 and SO_2 alter PA metabolism. Variation in their cellular concentrations brought about either by exogenous application or endogenous synthesis has "profound" effects on growth, development, and stress resistance (Liu et al., 2000). Their ability to ameliorate stress is most often attributed to their stabilizing effects on membranes. Conjugated forms of PAs also have antioxidant properties comparable to ascorbate, which could also contribute to their importance in stress resistance. Recently, polyamines were shown to modulate stomatal movement by regulating voltage-dependent inward K^+ channels in plasma membranes of guard cells (Liu et al., 2000). Spm and Spd were also recently shown to regulate fast-activating vacuolar channels of mesophyll cells, a function which could have significant implications for ion compartmentalization under conditions of salt stress (Brüggemann, Pottosin, and Schönknecht, 1998).

All polyamines are not created equal; different polyamines apparently fulfill different roles within plants. The diamine Put, for example, responds to stress differently, and also elicits very different plant responses, than do Spd and Spm. Putrescine is often positively, whereas Spd and Spm are negatively, correlated with senescence. In fact, sensitivity to stress has been positively correlated with Put accumulation in a number of crops (Mansour and Al-Mutawa, 1999). Symptoms of osmotic stress, for example, are often accompanied by the accumulation of Put or the inability to maintain adequate Spd and Spm levels. This was the case in salt-tolerant varieties of rice, tomato, and sorghum whose resistance to high salt was attributed to shifts in PA contents from Put to Spm and Spd (Galiba et al., 1993; Santa-Cruz et al., 1997; Bouchereau et al., 1999; Mansour and Al-Mutawa, 1999). Similarly, application of endogenous Put to barley, cotton, and pea led to symptoms characteristic of extreme salinization, including loss of turgor and leaf necrosis. Conversion of Put to Spd and Spm also was associated with O_3 tolerance in wheat and rice (An and Wang, 1997; Hur et al., 2000). Therefore, the ratio of Put to (Spd + Spm) is probably a better metric of the functional significance of shifts in polyamine metabolism than is total polyamine content (Bouchereau et al., 1999).

The common polyamines Spd and Spm as well as uncommon long-chain PAs such as thermine and caldine have been linked to plant thermotolerance (Roy and Ghosh, 1996; Murkowski, 2001). In fact, in rice, unusual long-chain PAs were found in heat-tolerant but not heat-sensitive varieties (Roy and Ghosh, 1996; Bouchereau et al.,

1999). In another example, exogenous application of Spd improved heat tolerance in two tomato cultivars (Murkowski, 2001), probably by alleviating high-temperature inhibition of pollen germination and pollen tube growth (Song, Nada, and Tachibana, 1999), processes particularly sensitive to high temperature (Chapter 9). Although the presence of long-chain PAs (e.g., caldine and thermine) have been linked specifically to heat-stress resistance in a number of crops, their specific functions are yet to be elucidated.

In addition to their stabilizing effects on membranes, amelioration of stress symptoms by PAs may be linked to the antagonism between polyamine and ethylene biosynthesis (Gallardo et al., 1996; Kakkar and Rai, 1997; Tiburcio et al., 1997). Polyamines and ethylene (an important stress hormone) share a common metabolic intermediate, S-adenosylmethionine (SAM; Figure 3.4) and many data support the hypothesis that PA and ethylene biosynthetic pathways compete for this precursor. Moreover, Spm interferes with conversion of SAM to ACC (1-aminocyclopropane-1-carboxylic acid), and ethylene is thought to directly interfere with conversion of arginine to Put. Since a chief result of ethylene is to trigger senescence, decreases in ethylene production should slow stress-induced cell death. This appears to be the case in several studies (Bouchereau et al., 1999). For example, PA levels, especially Spd, tended to be highest in young tissues and then declined as organs aged and then senesced (Galston and Sawhney, 1990).

Unfortunately, the ubiquity of polyamines in plant cells, and their involvement in so many aspects of cell function, has complicated the analysis of their precise role (Kumar et al., 1997). Exact mechanisms whereby stress leads to up-regulation of PA metabolism are unknown. Different functions of different polyamines have also created some confusion. A better grasp on PA metabolism and function will likely be achieved in the near future because of recent progress made locating and characterizing genes involved in polyamine biosynthesis (Kumar et al., 1997).

Heat-Shock Proteins

Twenty years after the heat-shock response was observed in fruitflies in the 1960s, it was discovered that seedlings exposed to temperatures 5°C or more above optimal synthesized a suite of proteins that appeared to confer greater tolerance to subsequent high-temperature stress (Key, Lin, and Chen, 1981). In other words, pre-treatment with high, but sublethal, temperature triggered synthesis of proteins that apparently extended the temperature range in which cells could function and plants could survive. It also enabled plants to recover after periods of high-temperature stress (Schöffl, Prändl, and Reindl, 1998). This phenomenon is commonly referred to as acquired thermotolerance, and the proteins responsible are called heat-shock proteins (HSPs) (Vierling, 1991). Temperate crops including soybean, pea, maize, and wheat generally begin to synthesize HSPs when their tissue temperatures exceed 32 to 33°C (Ougham and Howarth, 1988).

Many of the world's present crops are exposed to high-temperature stress at some stage of development (Stone, 2001). Wheat, for instance, the most important field crop worldwide, must endure periods of heat stress in 40 percent of irrigated wheat-growing areas (Reynolds et al., 1994). If global temperatures rise as projected by climate models, the proportion of time spent under these conditions would likely increase and the heat-shock response would presumably assume greater importance (Figure 3.5).

Several classes of HSPs have been identified in plants (Table 3.2): HSP100, HSP90, HSP70, HSP60 (numbers after HSP indicate the approximate molecular weights of these proteins in kilodaltons), and low-molecular-weight (small) HSPs. Plants synthesize as many as 30 different small HSPs in response to heat stress, which evidently are a more important component of the heat-shock reponse than higher molecular weight HSPs (Yeh et al., 1994; Nakamoto and Hiyama, 1999).

Heat stress affects cells in many ways. Disruptions of the cytoskeleton (e.g., microtubules), fragmentation of golgi bodies, swelling and dysfunction of mitochondria, disruption of normal protein synthesis, disappearance of polysomes, altered mRNA and rRNA processing, changes in nucleic acid synthesis and assembly, and membrane alterations have all been observed in heat-stressed plants (Nakamoto and Hiyama, 1999). Most cellular damage is either directly or indirectly linked to effects on membranes or disruption of enzyme function as a result of changes in protein conformation (denaturation), and thus these are the key sites of heat damage. A few key thermolabile enzymes identified to date include rubisco (Chapter 5),

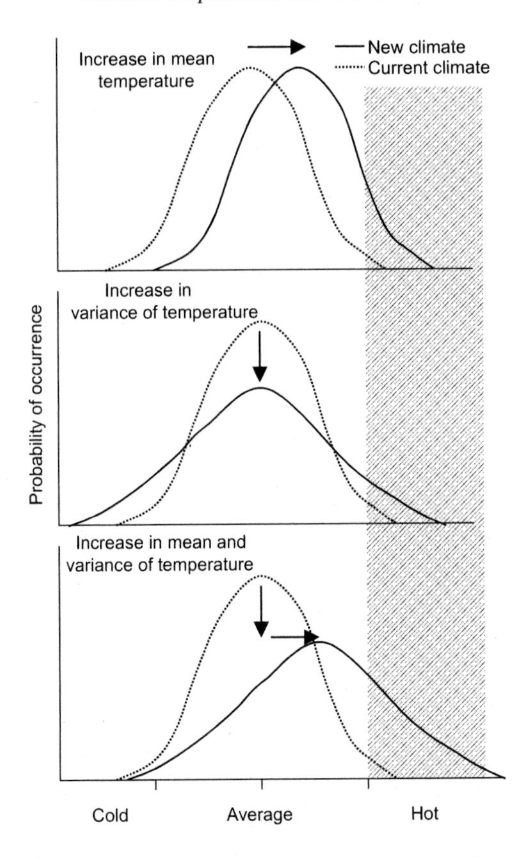

FIGURE 3.5. Global warming will involve either an increase in mean temperatures, an increase in variance of temperatures, or an increase in both mean and variance. All three of these potential scenarios will likely increase the amount of time that crops are subjected to hot temperatures (Figure redrawn from National Assessment Synthesis Team, 2001).

sucrose synthase, catalase, and SOD (Burke, 1990; Klueva et al., 2001).

Heat-shock proteins appear to function as molecular chaperones, proteins that bind and stabilize other proteins during folding, assembly, translocation across membranes, and degradation. Most HSPs either prevent the aggregation of proteins or facilitate the proper refolding of proteins that have already been denatured by high temperatures (Lee and Vierling, 2000).

TABLE 3.2. Summary of heat-shock proteins (HSPs), their functions, and the crops for which these genes have been cloned.

Class[a] assimilated	Subcellular localization	Putative functions[b]	Other inducers	Cloned in crops[c]
HSP100	cytoplasm nucleus	ATPase; prevents protein disaggregation and denaturation	development ABA (?)	soybean
HSP90	cytoplasm ER	signal transduction	constitutive cold stress pathogen attack	maize, rice, tomato, barley
HSP70	cytoplasm nucleus ER mitochondria chloroplasts	ATPase and peptide binding; facilitates trafficking and translocation of materials in cells; aids in protein folding	constitutive cold stress heavy metals pathogen attack cold stress	*Brassica,* carrot, maize, pea, rice, soybean, spinach, tobacco, tomato, barley, potato, pumpkin
HSP60 cpn60	mitochondria chloroplasts	ATPase translocation, folding, and assembly with cpn10	constitutive	*Brassica,* maize, potato, pumpkin, pea, wheat, rye, castor bean

smHSP	cytoplasm ER mitochondria chloroplasts	chaperone activity; prevents protein aggregation; facilitates protein reactivation; protection of PSII	embryogenesis water stress heavy metals ABA cold stress fruit maturation	alfalfa, barley, *Brassica,* carrot, maize, millet, pea, rice, soybean, tobacco, sunflower, tomato, wheat, parsley
HSP10 cpn10	mitochondria chloroplasts	protein translocation and folding	constitutive	pea, spinach

[a]HSP, heat-shock proteins; cpn, cognate proteins (proteins that are structural homologues of HSPs that are expressed constitutively at normal temperatures). The numbers after HSP and cpn represent their approximate molecular weight in Kilodaltons.

[b]The functions of these proteins were assembled from the literature by Klueva et al. (2001)

[c]Information on cloned crops was assembled by Klueva et al. (2001) based on the following databases: Genbank, European Molecular Biology Laboratory (EMBL), and DNA Bank of Japan (DDBJ).

The metabolic costs associated with mounting a heat-shock response are poorly understood. Several authors have estimated that HSPs alone require as much as 10 percent of a plant's total nitrogen (Heckathorn et al., 1996). This is a significant cost, one that requires diversion of much nitrogen away from production of enzymes and essential pigments. High costs, combined with breakdown of existing photosynthetic enzymes, may explain much of the photosynthetic decline observed at supraoptimal temperatures (Chapter 5).

Heat-shock proteins are not exclusively associated with high-temperature stress. Some are expressed constitutively. They are needed to establish and maintain normal protein configuration and functionality for many, if not most, cellular proteins. A number of HSPs are also developmentally regulated (Nieden et al., 1995), while others are induced by other stresses such as heavy metal exposure, drought, salinity, and O_3 (hence, the name heat-shock protein may be a misnomer). In fact, any condition leading to protein damage or inactivation is expected to elicit HSP induction. It follows that exposure to salinity, O_3, and a few other stresses also tends to increase plant thermotolerance. Likewise, exposure to high, but sublethal, temperature often brings about changes that increase plant tolerance to other abiotic stresses (termed cross-tolerance). Although this evidence may lead one to believe that HSPs are simply general stress proteins, the expression of the full suite of HSPs specifically requires high-temperature stress (Vierling, 1991).

THE ROLE OF CALCIUM IN STRESS RESPONSES

Environmental stress often triggers transient Ca^{2+} signals, which are temporary periods of high cytosolic [Ca^{2+}]. It appears that increases in membrane permeability and shifts in polarity caused by stress often result in Ca^{2+} flux from apoplast and vacuoles into cytosol. Membrane-bound ion channels and pumps probably play a significant and specific role in generating Ca^{2+} signals during transduction of stress (Rudd and Franklin-Tong, 2001; Xiong and Zhu, 2002). Many environmentally regulated hormones also trigger an increase in cytoplasmic Ca^{2+} (Fosket, 1994). In fact, so many environmental conditions lead to a surge in cytosolic [Ca^{2+}] that it is difficult to understand how plants keep these signals sorted out (Smith and Gallon, 2001).

Elevated cytosolic [Ca^{2+}] may combat stress in at least two ways. First, Ca^{2+} activates calcium sensors, including calmodulin (a small calcium-binding protein), calcium-dependent protein kinases (CDPKs), and calcium-regulated phosphatases. When activated by free cytosolic Ca^{2+}, these molecules trigger various intracellular changes that ultimately lead to adaptive physiological responses. Second, divalent cations, especially Ca^{2+}, bind and link adjacent phospholipid head groups, causing membranes to become more viscous (in this regard, Ca^{2+} probably functions similarly to polyamines). The role of Ca^{2+} as an activator of stress-resistance mechanisms, coupled with its stabilizing effect on membrane lipids, probably explains why high soil, tissue, or cell [Ca^{2+}] can sometimes alleviate the deleterious effects of salinity (see Chapter 7), O_3 (Runeckles and Chevone, 1992), and high temperature (Gong et al., 1997).

The Role of Ca^{2+} in Salt Tolerance

Apparently, uptake of some Na^+ by crops in salinized soils is unavoidable because this ion competes so efficiently with K^+ for uptake through common protein channels (Chapter 7). Therefore, salt tolerance must also involve the ability to efficiently compartmentalize Na^+ in cell walls or vacuoles in order to avoid its toxic effects in the cytoplasm. Na^+ begins to inhibit enzyme function at 100 mM (Munns, 2002). Generally, halophytes and relatively salt-tolerant glycophytes exclude a large portion of salt and the salt that is absorbed into the plant is compartmentalized (Munns, 2002). These species possess transporters in cell and vacuolar membranes that maintain cytoplasmic Na^+ concentrations at levels that do not interfere with metabolism.

High [Na^+] triggers the influx of Ca^{2+} into cytoplasm from cell walls, which in turn triggers a cascade of reactions involving the salt overly sensitive pathway (genes *SOS1, SOS2,* and *SOS3;* Figure 3.6). SOS3 is a Ca^{2+} binding protein with a myristoylation site in the N terminus that may function as the Ca^{2+} sensor. After activation by Ca^{2+}, SOS3 activates SOS2 kinase (Hasegawa, Bressan, and Pardo, 2000). Activated SOS3 and SOS2 in turn up-regulate SOS1 by some unknown transcription factor. SOS1 is a plasma membrane Na^+/H^+ antiporter that functions in transporting Na^+ from symplast to apoplast in both roots and shoots (leaves). Presumably, up-regulation of

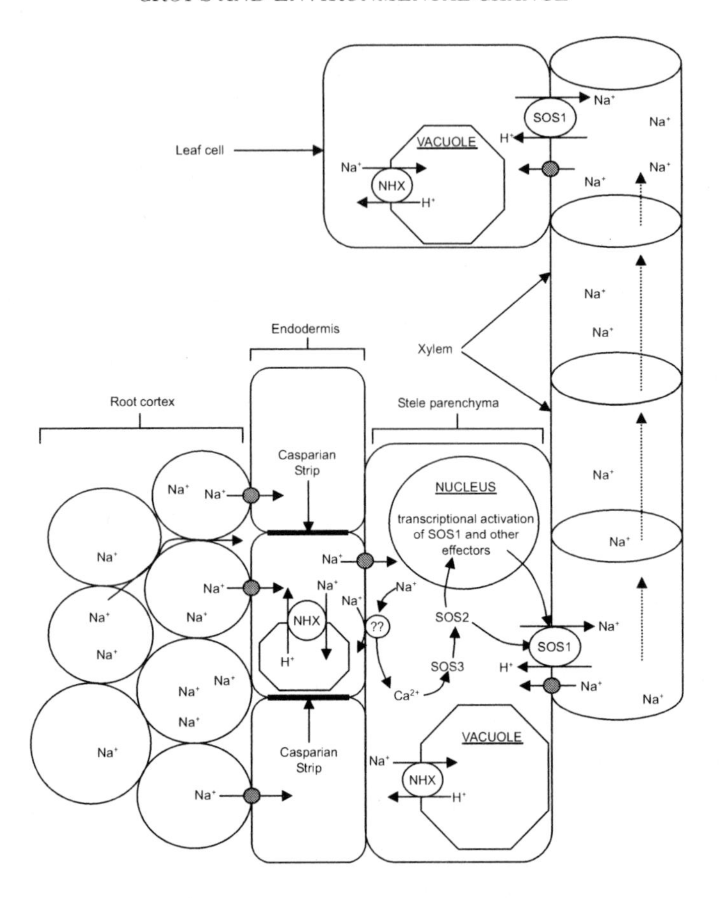

FIGURE 3.6. Representation of the SOS (salt overly sensitive) pathway. The SOS pathway mediates salt tolerance by controlling Na^+ flux by SOS1. Cells not drawn to scale. Na^+ diffuses from salinized soil across the root cortex either apoplastically or symplastically. Upon reaching the endodermis, solutes (including Na^+) can no longer move apoplastically because radial walls of the endodermis are suberized (forming casparian strips) prohibiting movement through cell walls. Na^+ competes with K^+ for uptake through common membrane carriers (shaded circles). Cells sense high Na^+ concentrations (by some unknown mechanism) and respond by elevating free cytoplasmic [Ca^{2+}]. SOS3 binds to Ca^{2+}, which in turn activates SOS2. The SOS3/SOS2 complex is then required for activation of SOS1 (this may also up-regulate expression of SOS1 in the nucleus). This cascade is thought to enable plants to better regulate ion flux (such as Na^+ and K^+), thereby increasing salt tolerance. NHX are tonoplast/vacuolar membrane proteins that transport Na^+ into vacuoles by an antiport process. *Sources:* Drawing adapted from Hasegawa, Bressan, and Pardo (2000) and Zhu (2000).

the Na^+/H^+ antiporter could enable plants to more efficiently compartmentalize Na^+ in the apoplast, alleviating many of the negative effects associated with salinity. And indeed, cultivars with higher expression of this antiporter have been shown to be more salt tolerant (Sreenivasula et al., 2000). Overexpression of SOS1 in particular lowers shoot Na^+ content and improves salt tolerance in transgenic *Arabidopsis* (Zhu, 2001).

Although the SOS pathway appears consistent with many data, there are probably other models, reflecting different mechanisms of salinity tolerance, equally capable of explaining salt tolerance. In fact, other protein kinases have been implicated as intermediates in the response of plants to salt stress such as CDPK, and mitogen-activated protein kinases (MAPK) (Sheen, 1996; Hasegawa, Bressan, and Pardo, 2000). These other signaling pathways may also lead to activation of Na^+ transporters. Regardless of the role played by these other molecules, recent data on the SOS pathway represent a positive step toward understanding mechanisms of salt tolerance and may prove useful for efforts to engineer salt-tolerant crop varieties.

SPECIFICITY, CROSS-TALK, AND CROSS-TOLERANCE

It is becoming clear that stress-induced signaling cascades are not always stress-specific and do not function in isolation. They are instead components of complex signaling networks; networks in which significant overlap exists among branches or metabolic pathways (Knight and Knight, 2001). Many common genes are up-regulated by different types of stress.

Convergence between signaling pathways associated with different types of inducers is often called "cross-talk." Molecular cross-talk can result in different pathways reaching a common outcome, or cause different pathways to change the course or outcome of other pathways (Knight and Knight, 2001). Cross-talk may also involve systemic signals (e.g., hormones, AOS) that carry messages from cells under stress to other nearby cells, or even to organs or tissues long distances away. Such responses are important for coordinated responses of tissues, organs, and whole plants (Karpinski et al., 1999).

It is quite likely that, in some instances, plants are unable to differentiate between different types of stress. For example, many types of biotic and abiotic stresses are accompanied by oxidative damage. Regardless of the cause, plants generally respond to an oxidative threat by up-regulating antioxidant systems. AOS, especially H_2O_2, are associated with many different types of stress and appear to be important signals for triggering adaptive plant responses. Because distinct types of stress often share common cellular "symptoms," plants often become resistant to one type of stress when exposed to another, different stress.

Cross-tolerance is resistance to one environmental stress afforded by exposure to a different stress. Cross-tolerance implies that different stresses undermine "normal" plant function in similar ways and hence require similar protective actions. This is reflected by the common observation that plants acclimated against one stress are often more resistant to others. Cross-tolerance is important for crops because it opens up the possibility of simultaneously breeding or engineering plants (albeit at a significant metabolic cost) that are resistant to multiple environmental stresses (Bowler and Fluhr, 2000).

In spite of all the cross-talk existing between signaling branches, and cross-tolerance afforded by one type of stress for another, specific responses to many abiotic stressors also exist. Induction of only those mechanisms of resistance that precisely target stress-specific damage may often be advantageous, compared to mounting a more general stress response. The main reason involves the metabolic costs of mounting a general stress response compared with induction of only the specific and appropriate responses necessary to alleviate the specific stress encountered (Knight and Knight, 2001). For example, although HSPs are induced by several types of environmental stress, why induce the full suite of these proteins if they divert ~10 percent of total plant nitrogen away from growth and reproduction and less metabolically expensive resistance mechanisms could be employed? Receptor molecules that specifically detect osmotic stress have been identified (Urao et al., 1999), and it is likely that other receptors dedicated to other specific stresses also exist.

Specificity, for example, may be built into transient Ca^{2+} signaling (Bowler and Fluhr, 2000; Rudd and Franklin-Tong, 2001). Abiotic stress often leads to increased cytosolic free Ca^{2+} levels, and those changes in $[Ca^{2+}]$ lead to a number of adaptive plant responses. It now appears that cytoplasmic free Ca^{2+} signals may possess subtle differences that encode information about the type of stress inducing the signal, which then cues the appropriate metabolic response (Rudd and Franklin-Tong, 2001). For example, the source (i.e., apoplast, chloroplast, mitochondrion, etc.), magnitude, and duration of the Ca^{2+} signal may depend on both the type and severity of the stress, and these differences may elicit different cellular responses. Furthermore, the organ, tissue (e.g., cortex, epidermis, phloem), or cell type (e.g., epidermal, endodermal, mesophyll) in which the Ca^{2+} signal occurs may also be important for stress recognition. Differences between Ca^{2+} signals are referred to as "calcium signatures."

Characterization of the Ca^{2+} signatures elicited by specific environmental changes is just beginning, and understanding of how these signals are decoded by the plant remains limited. Ozone exposure, for example, is accompanied by a brief free cytoplasmic [Ca^{2+}] peak followed by a more prolonged period of high concentrations (Clayton et al., 1999). These signals then trigger up-regulation of antioxidants. Calcium signatures have also been characterized for *Arabidopsis* exposed to both drought and salinity (Kiegle et al., 2000). Drought-induced ABA (abscisic acid) signals stomates to close by triggering a calcium signature consisting of a series of oscillating spikes of increased [Ca^{2+}] in guard cells (Rudd and Franklin-Tong, 2001). This type of information may prove to be important for adapting to many facets of environmental change.

SUMMARY

Because plants are sessile organisms, they must adapt and acclimate to the environment in which they exist or they might perish. Mechanisms of adaptation and acclimation may be architectural (e.g., modification of leaf number, surface characteristics, size, or orientation), allometric (e.g., preferential growth of some organs over others), or physiological/biochemical (e.g., change in complement of enzymes present). There is currently a great deal of interest in exploiting the function of specific stress-resistance mechanisms through breeding and genetic engineering, not only to close the gap between average and record yields but also to increase the potential environmental range in which crops can be grown. Similarly, understanding and exploiting stress-resistance mechanisms is relevant for developing cultivars suited for future environments.

The overall effects of environmental changes on the cell physiology of crops will depend on the balance of the amelioration of stress by rising [CO_2] and the imposition of stress by O_3, soil salinization, and global warming. The severity of environmental stress on cell metabolism is mediated by the activity of resistance mechanisms such as antioxidants, HSPs, polyamines, and membrane modification (Table 3.3). Antioxidants neutralize O_3 and are therefore an important aspect of O_3 tolerance. Robust antioxidative systems might also protect crops against drought and salt stress. Polyamines apparently increase tolerance to both O_3 and salt damage. Damaging effects of acute high temperatures are sometimes alleviated by HSPs as well as membrane adjustments. HSPs protect against protein denaturation at high tem-

TABLE 3.3. Summary of relative responses of stress-resistance mechanisms to environmental change.

	Environmental changes				
Resistance mechanisms	**Salinity**	**Ozone**	**Heat stress**	**Elevated CO_2**	**Drought**
Antioxidants	++	+++	?	–	++
Polyamines	++	+++	++	?	++
HSPs	+	+	+++	?	++
Leaf wax	+	0	+	?	+++
Membrane modifications	+	+	+++	?	++

Notes: Plus signs indicate a positive relationship between resistance to an environmental change and a specific stress mechanism (+++ > ++ > +), a negative sign indicates a negative relationship, and 0 indicates no relationship. A question mark indicates that not enough information is currently available to determine if there is, or is not, an effect. We should caution that in many cases, these generalized responses can be modified significantly by interactive effects of other environmental conditions. For example, elevated carbon dioxide may lead to stomatal closure, which sometimes alleviates drought stress; but stomatal closure may also increase leaf temperatures, leading to periods of high-temperature stress. Furthermore, long-term exposure to an environmental perturbation can change whole-plant carbon and nutrient budgeting, leading to alterations in growth and developmental patterns. These whole-plant changes can then change both direction and magnitudes of responses of resistance mechanisms to an environmental change. Short-term and long-term acclimatory responses to the environment may differ substantially.

peratures and membrane properties dictate how well membranes function at high temperatures.

Membranes are particularly sensitive to environmental conditions. In many cases, the ability to sustain cell function during periods of stress hinges on the resilience of membranes to maintain their fluidity, semipermeability, and selectivity. Antioxidants, heat-shock proteins, polyamines, and divalent cations such as Ca^{2+} all play roles in protecting membranes during stress. In addition, plants often acclimate to a fluctuating or changing environment by altering the chemical properties of membranes themselves in order to maintain fluidity.

Enhancing resistance to abiotic stress is clearly an important facet of crop production, one that promises to become even more important in the future as environmental change progresses. But mechanisms of resistance are not without costs. Indeed, in nature, many stress-resistant plants are small, slow growing, and partition carbon in ways that foster survival, not production of fruits or tubers (i.e., the extent of stress resistance is often inversely related to growth). Over thousands of years, crop breeders have methodically sacrificed some degree of stress resistance in favor of faster-growing, higher-yielding crops. If the future environment becomes more stressful, it may become time to trade back some of that growth and yield potential for greater tolerance to environmental stress. Alternatively, beneficial changes in the environment (e.g., rising atmospheric [CO_2]) might allow an even greater sacrifice of stress resistance (tolerance) in favor of greater growth in some crops.

Chapter 4

Water Relations

Water, water, everywhere, nor any drop to drink.

—Samuel Taylor Coleridge
The Rhyme of the Ancient Mariner

Availability of fresh water constrains crop production more than any other resource. Some plants are well adapted for dry (or saline) environments, but unfortunately field crops are generally not among these. Most crops use tremendous quantities of water; a single maize plant transpires more than 15 liters of water per week and a maize crop may transpire more than 1,325,000 liters per acre (0.4 ha) during the growing season. Worldwide, nearly two-thirds of all the water that cycles from land to atmosphere in a year passes through living plants (Holbrook and Zwieniecki, 2003).

Breeding efforts over the past 50 years have been successful at producing new, higher-yielding crop varieties, but along with yield increases have come equivalent increases in crop water-consumption rates. In fact, growth and yield are often linearly related to the amount of water transpired by a crop throughout the growing season (Lu et al., 1994; Bassil and Kaffka, 2002; Serraj and Sinclair, 2002).

There are three explanations for the linear relationship between yield and crop water consumption: (1) carbon and water vapor share a common diffusive pathway between the atmosphere and the interior of plant leaves; consequently, stomates must be open, and water transpired, in order for crops to assimilate carbon from air surrounding leaves; (2) both water loss and photosynthesis are driven by absorption of light (Amthor, 1999); and (3) evaporative water loss significantly cools leaves and canopies, which alleviates high-temperature stress (Lu et al., 1994). In other words, the very adaptations that favor fast growth in crop plants—such as leaves with large surface areas, short diffusive pathways from the leaf interior to the atmosphere, and high stomatal conductance values—also favor water loss.

The overall influence of environmental changes on water use by a crop will be manifested mainly as a change in total evapotranspiration (Box 4.1). Evapotranspiration is a combined measure of plant transpiration, evaporation from soil, and interception losses. Interception losses represent rainfall that evaporates directly from external surfaces of vegetation or from soil-surface residues without ever entering either soil or plant and can be significant in dense canopies (Helvey and Patric, 1965; Amthor, 1999). Dense crop canopies lose most water to transpiration, but sparse canopies, where significant ground area is exposed to light and wind, can lose much water through evaporation from the soil surface.

BOX 4.1. Definitions

Evapotranspiration: Total water lost from a crop. Equivalent to plant transpiration + evaporation (from leaves, stems, residue, and soil); abbreviated ET.

Interception losses: A component of total ET, interception losses represent rainfall that evaporates directly from leaves, stems, and soil-surface residue without entering the soil; can be significant in dense canopies.

Transpiration: Evaporative water loss from stomata of leaves and stems.

Stomata: Leaf structures consisting of a pore surrounded by two guard cells. Crops regulate their water use by adjusting the numbers or movements (opening or closing) of stomata.

Stomatal conductance: The proportionality constant that relates transpiration to water vapor concentration difference between substomatal cavities and the air outside the leaf/stem.

Boundary layer: A layer of still, relatively unmixed air surrounding a surface such as a leaf or a crop canopy; resistance to water vapor flux imposed by a boundary layer is proportional to the thickness of the layer of still air; boundary layer thickness is in turn affected by wind speed, as well as the roughness of the surface. Plants can modify the boundary layer of leaves by altering the structure of leaves such as size, orientation, and by constructing trichomes or wax structures.

Apoplast: Continuum of plant compartments external to plasma membranes, which includes cell walls, intercellular air spaces, and xylem.

Symplast: Continuum of plant compartments bounded by the plasma membranes of individual cells, connected together by plasmodesmata.

Environmental changes are also likely to affect leaf, plant, and crop water-use efficiency (WUE). An important measure, WUE tells us something about the stoichiometry of the CO_2/water exchange that occurs at the leaf-atmosphere interface. It can be measured and expressed at several different scales. For example, leaf level instantaneous WUE is equal to the ratio between photosynthesis and transpiration and is ordinarily measured with a cuvette or, less often, a plant chamber. These measurements, however, often do not accurately represent patterns of water use in the field because they neglect canopy-scale feedbacks. It is superior to express the relationship between water use and photosynthesis at the plot or canopy level. At these scales, efficiency of water use can be expressed in a more relevant way, such as total biomass or grain yield per unit of ET for a given area of ground surface.

THE FUTURE OF FRESH WATER RESOURCES

Of the earth's total water, only about 0.014 percent is found in lakes, rivers, streams, groundwater, and water vapor and is therefore available for agriculture (Miller, 2002). A significant fraction of this freshwater has already been or is currently being polluted; approximately one of every three medium to large lakes in the United States suffers from cultural eutrophication (excessive accumulation of plant nutrients from agricultural, industrial, and sewage runoff) and up to 25 percent of U.S. groundwater is polluted. Water pollution is often not directly critical to agriculture, but as water quality declines, competition among different economic sectors, and between different nations, for clean water will become more and more contentious. Agriculture could prove particularly sensitive to water shortages where they occur, because water to irrigate crops is normally assigned lower priority than drinking water and water used for industrial purposes.

Another significant consideration for planning long-term irrigation practices involves ongoing depletion of groundwater. A well-documented and often discussed example is the Ogallala aquifer. This aquifer is the largest in the world and provides irrigation water for a large agricultural region in the central United States that spans from South Dakota through Nebraska, western and central Kansas, eastern Colorado, western Oklahoma, eastern New Mexico, and into the Texas panhandle. This region accounts for 20 percent of all U.S. agricultural output.

In many areas, water is pumped from this aquifer eight to ten times faster than the natural recharge rate and it is projected that one-fourth of the Ogallala's water will be gone by the year 2020 (Miller, 2002). Groundwater depletion is also evident in Saudi Arabia, China, northwest and southern India, northern Africa, southern Europe, the Middle East, Mexico, Thailand, and Pakistan (Miller, 2002).

EFFECTS OF CLIMATIC CHANGE ON GLOBAL HYDROLOGY

It would be useful to predict how climatic change will influence the global hydrologic cycle and regional precipitation patterns in order to understand and plan for potential effects on agriculture. A change in global hydrology would influence availability of water resources and therefore food production. Aside from direct effects on regional precipitation patterns, which could be significant, climate change might also influence the ratio of precipitation to evaporation, thereby influencing water levels of lakes and the runoff and flow of rivers. Based on historical data, it was estimated that precipitation has already increased 5 to 10 percent during the past century due to more intense rainfall events and that this has contributed to even greater increases in stream-flow during this same period (www.usgcrp.gov/usgcrp/Library/nationalassessment/16WA.pdf). In fact, it is likely that climatic change has already contributed to observed changes in surface runoff for several major river basins recorded over the past 30 years (Evans, 1993).

Unfortunately, predicting future changes in water resources with global climatic change and global hydrological models is not yet possible. Uncertainties regarding the fate of the hydrological cycle are effectively illustrated by examining the differences in precipitation changes predicted by two climate models. The Canadian model predicted that precipitation would generally increase in California and the U.S. Southwest but would decrease in the southern half of the United States east of the Rocky Mountains (Climate Change 2001 Synthesis Report). In contrast, the Hadley model predicted more modest increases in rainfall for the Southwest and California, but indicated that precipitation was likely to increase throughout much of the United States. Although these model predictions differed, they generally agreed that some crop-production regions will likely get dryer and some will get wetter. They also agreed that, at least averaged over large regions, the incidence of heavy rainfall events could increase in the United States over the next 50 years.

More rain coupled with the effects on crop physiology of rising atmospheric CO_2 and temperature have led some researchers to predict a 5 to 10 percent decrease in demand for water by irrigated agriculture by 2030 and a 30 to 40 percent decrease by 2090 (www.usgcrp.gov). While wetter conditions could boost crop yields in many regions it will also increase flooding and soil waterlogging (Rosenzweig et al., 2002). Other researchers predict that demand for irrigation water will *increase* because of greater evaporative demand, reduced rainfall, and longer cropping seasons (Rosenzweig and Hillel, 1998). Obviously, in spite of the fact that over 900 articles dealing with the effects of climatic change on water resources have been published (www.pacinst.org), climate change scientists are less certain about potential changes in precipitation patterns than they are about temperature changes.

EXCESS PRECIPITATION

Although water shortages are currently the single most important cause of crop failure (Polley, 2002), losses from waterlogging are also high and may get worse. According to the USDA, 23 percent of all crop insurance indemnities are paid out for crop yields lost to waterlogging or flooding (USDA Agricultural Statistics, 2002; Figure 3.1). Models indicate that losses to the U.S. maize crop resulting from excess precipitation could double during the next 30 years as a result of climatic change (Rosenzweig et al., 2002).

Some percentage of financial losses from flooding result from the inability to harvest the crop, but flooding can also damage the plants themselves by creating anoxic soil conditions (Blackwell and Wells, 1983). Adaptations that enable plants to cope with waterlogged soils include the development of aerenchyma tissue in roots. Almost nothing is known about how environmental changes will affect crop resistance to excess precipitation. But, in light of climatic change predictions, crop resistance to temporary flooding may become a higher priority to crop breeders in the near future.

DROUGHT

If warming drives up evapotranspiration more than rates of precipitation, then a decrease in soil-water availability could result, leading to more frequent and severe droughts. Lack of water could force producers to switch to more drought-tolerant crop varieties or species. Among warm-season cereals, sorghum and pearl millet are generally more drought tolerant than maize and rice. For legumes, cowpea is very drought tolerant, peanut and soybean are of intermediate tolerance, and common bean is drought sensitive. The cool-season crop barley is more drought tolerant than either wheat or potato. Adaptations that allow for maximum growth and reproduction during dry periods include rapid development, C_4 photosynthesis, high cuticular and stomatal resistance to water loss, deep rooting systems, minimal leaf area, and leaf rolling, flagging, and self-shading (Machado and Paulsen, 2001).

WATER UPTAKE, TRANSPORT IN XYLEM, AND TRANSPIRATION

Water passage from the soil through roots, stems, and leaves is driven by a water-potential gradient along the soil-plant-atmosphere continuum (SPAC; Box 4.2). Typical water-potential values along the SPAC are –0.3, –0.6, –1.0, and –95 MPa for soil, root, leaf, and atmosphere. Soil water enters roots and passes through the root cortex along this gradient in the apoplast until reaching the endodermis. Here water is forced to pass through membrane channels (aquaporins) of endodermal cells because the apoplastic route is blocked by casparian strips (see Figure 3.6). Water enters the xylem for long-distance transport shortly after traversing the endodermis. Passage of water across the endodermis represents considerable (radial) resistance while (axial) resistance in the xylem proper is generally low.

Water contained in xylem conduits of the stem is pulled upward to leaves, where it eventually evaporates from the moist walls of mesophyll cells and then diffuses out from intercellular spaces into the atmosphere. Water loss from leaves is replenished by water diffusing in from stems and roots whose water potential values are less negative. Water use by crops is ultimately controlled by leaves, and more

BOX 4.2. Definitions

Hydraulic conductivity: A measure of the physical potential for water conductance through plant structures; conductance is determined by resistance to water passage imparted by protein water transporters within cell membranes (i.e., aquaporins) coupled with resistance to water flow from anatomical features of cells and tissues (such as xylem vessel numbers and diameters).

SPAC: Gradient of water-potential values along the pathway water takes as it moves from soil into roots, through stems, and from leaves into the atmosphere. The water potential along this pathway must become increasingly negative in order for plants to maintain water uptake (water moves from an area of high to an area of low water potential); typical values (in MPa) along the SPAC are –0.3 in bulk soil, –0.6 in xylem of roots, –1.0 in leaf mesophyll cells, –7.0 in air within leaf intercellular spaces, –70 in the atmosphere just outside stomata, and –95 beyond the leaf boundary layer.

Cohesion tension theory: Explains how water moves from the soil through the plant and into the atmosphere. Water molecules form a continuous column from soil through the plant body and into the atmosphere. This column is held together by hydrogen bonding (cohesion) between water molecules. Evaporation of water from leaves places tension on this water column, drawing water upward from roots by bulk flow.

Water potential: Free energy (ability to do work) of water per unit volume (expressed as ψw often in units of MPa); $\psi w = \psi s + \psi p + \psi g$ where ψs, ψw, and ψg represent the effects of dissolved solutes, pressure, and gravity on the free energy of water. Solutes reduce the free energy of water ($\psi s = 0$ in pure water but is negative whenever solutes are present); ψp is positive in most living plant cells (also called turgor pressure or pressure potential) but may be negative in the xylem; ψg is negligible and generally can be ignored (except in tall trees).

specifically by stomates because these are the pores through which water vapor is actually lost to the atmosphere. Figure 4.1 is a simple compartment model that shows the important pools and fluxes of water for a crop (Loomis and Connor, 1992).

THE PLANT SURFACE

The plant surface is the interface between the interior of the plant and the external environment. As such, its physical and chemical characteristics greatly influence water loss. The function and density

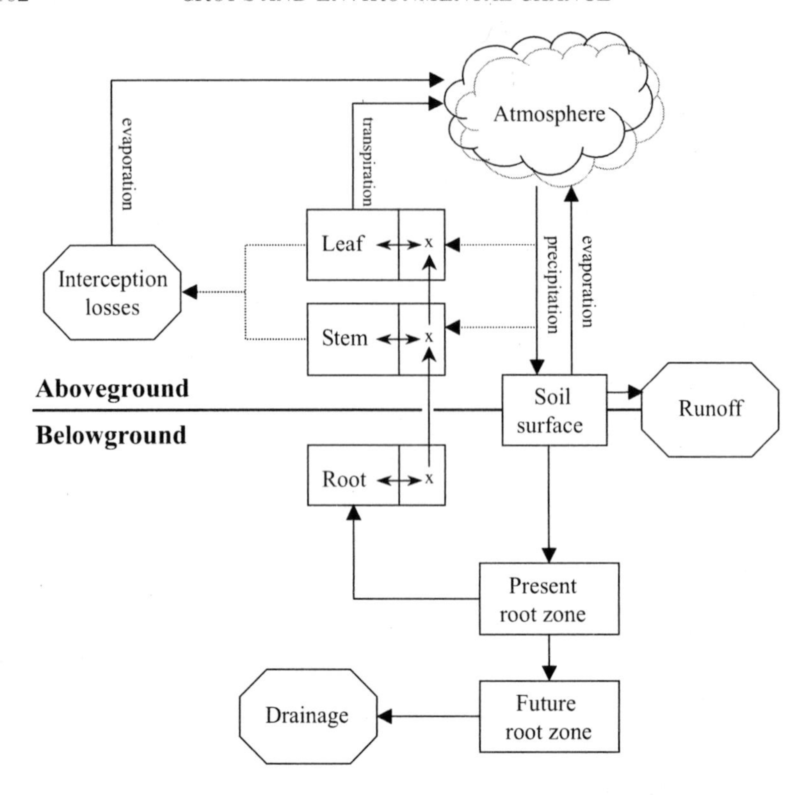

FIGURE 4.1. Model illustrating the distribution of water in the soil-crop-atmosphere system. Within the crop plants themselves, water flows from roots through stems and into leaves through the xylem (x). Water and solutes can pass back and forth between shoots and xylem and leaves and xylem. Interception losses represent precipitation that evaporates directly from surfaces of leaves and stems without entering either soil or plant. *Source:* Modified from Loomis and Connor, 1992.

of stomates on leaf surfaces, leaf surface wax density and structure, and the presence or absence of leaf hairs are all important (Figure 4.2).

Stomates

Stomates consist of a pore surrounded by two guard cells (Figures 4.2 and 4.3), and are quite numerous; abaxial (bottom) surfaces of oat and tomato leaves, for example, have as many as 2,500 and 13,000

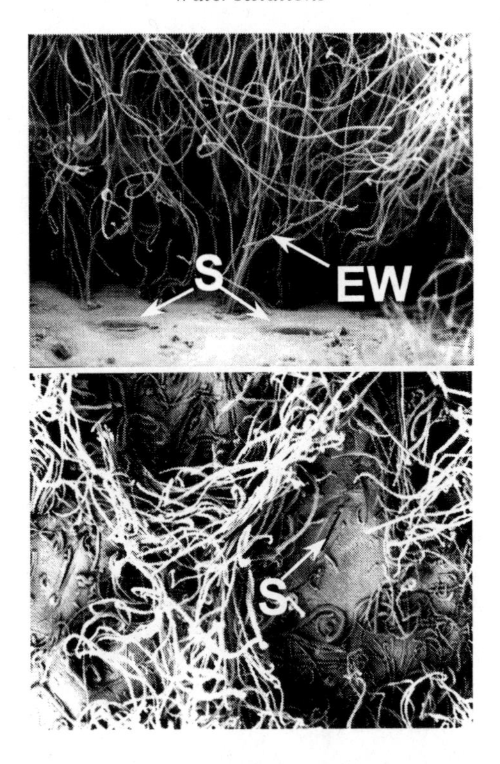

FIGURE 4.2. Scanning electron micrographs of the epicuticular wax of sorghum. The top photograph is a cross-sectional view showing two stomates (S) and epicuticular wax deposits (EW). The bottom photograph is a view from directly above the plant surface. As seen here, epicuticular wax can modify the physical characteristics of the leaf surface, impacting water relations and leaf energy balance. *Source:* Photographs kindly provided by Matt Jenks, Purdue University.

while common bean has 28,000 stomates cm^2. Monocotyledonous crops have stomates distributed more or less equally between top and bottom leaf surfaces, while dicotyledonous crops have most stomates on bottom leaf surfaces (Box 4.3). By adjusting the number (density) of stomatal pores, plants can exert long-term (weeks to months) control over water use. Plants growing under water-stress conditions, for example, sometimes develop leaves with fewer stomates as compared to plants grown with plenty of water (Pritchard et al., 1998). In the short term (minutes to days), plants respond to environmental fluctuations mainly by opening and closing stomatal pores.

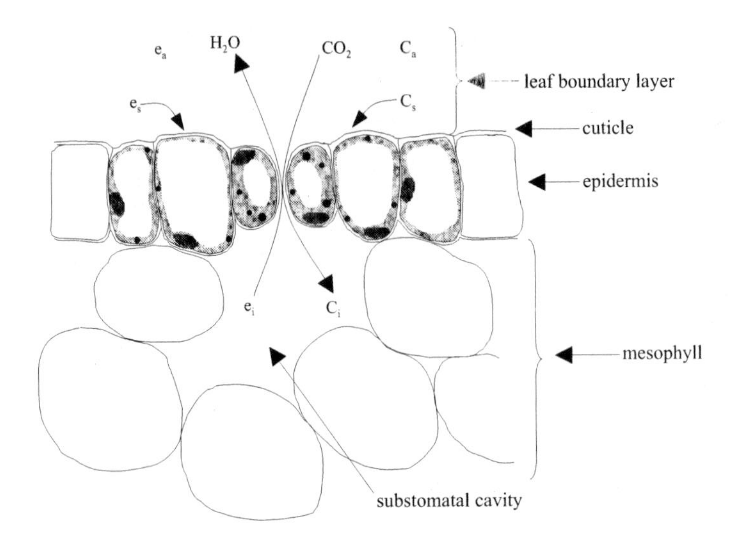

FIGURE 4.3. Illustration of a cross-sectional view of a stomate. Pathways of CO_2 and H_2O diffusion between the leaf and atmosphere are shown. Definitions: e_a, e_s, e_i: atmospheric, surface, and internal humidity; C_a, C_s, C_i: atmospheric, surface, and internal CO_2 concentrations. Diffusion of water vapor into the leaf interior is dependent upon the concentration gradient between the leaf interior and the atmosphere. Similarly, diffusion of CO_2 into leaves depends on the concentration gradient between the atmosphere and the interior of leaves. Gas exchange between the leaf and atmosphere is influenced by resistance to diffusion imparted by the boundary layer and stomatal conductance.

BOX 4.3. Monocots and Dicots

Cotyledon: The first leaf, or one of the first pair (or whorl) of leaves, developed by the embryo of a seed plant (i.e., spermatophyte) or of some lower plants, such as ferns.

Monocot: Short for monocotyledon. A seed plant that produces an embryo with a single cotyledon. Monocots usually have parallel-vein leaves. The group includes the grain crops barley, maize, millet, oat, rice, rye, sorghum, and wheat. It also includes sugarcane and yam.

Dicot: Short for dicotyledon. A flowering plant (i.e., an angiosperm; part of the group Spermatophyta) that produces an embryo with two cotyledons (i.e., the first two leaves develop as a pair). Dicots usually have leaves with reticulate venation. The group includes the seed crops bean, chickpea, cowpea, flax, lentil, pea, peanut, pigeon pea, rape, safflower, sesame, soybean, and sunflower. It also includes cassava (tapioca), cotton, potato, sugarbeet, and sweet potato.

Stomates open when guard cells become turgid (i.e., the size of the aperture increases), and close when guard cells become flaccid. Fluctuations in turgor are brought about by the active transport of K^+ and Cl^- ions back and forth between epidermal and guard cells (Kearns and Assmann, 1993). Ion accumulation lowers the solute potential of guard cells, which results in inward flow of water from surrounding cells and ultimately stomatal opening. Presumably, hydrolysis of guard cell starch granules into simple sugars and production of malate also play a role in stomatal opening. During stomatal closing, ions are released from guard cells, malate levels drop, and turgor is relaxed.

Many factors, including intercellular $[CO_2]$ (C_i), light levels, humidity, temperature, soil-moisture availability, nutrient availability, and environmental stress affect stomatal opening and closing and therefore water use by the plant. The precise mechanisms whereby stomata sense environmental conditions are not known but are likely to involve Ca^{2+} signaling (Chapter 3; Figure 4.4).

Light levels up to ~1000 μmol m^{-2} s^{-1} generally induce stomatal opening. This response may be linked directly to K^+ pumping in re-

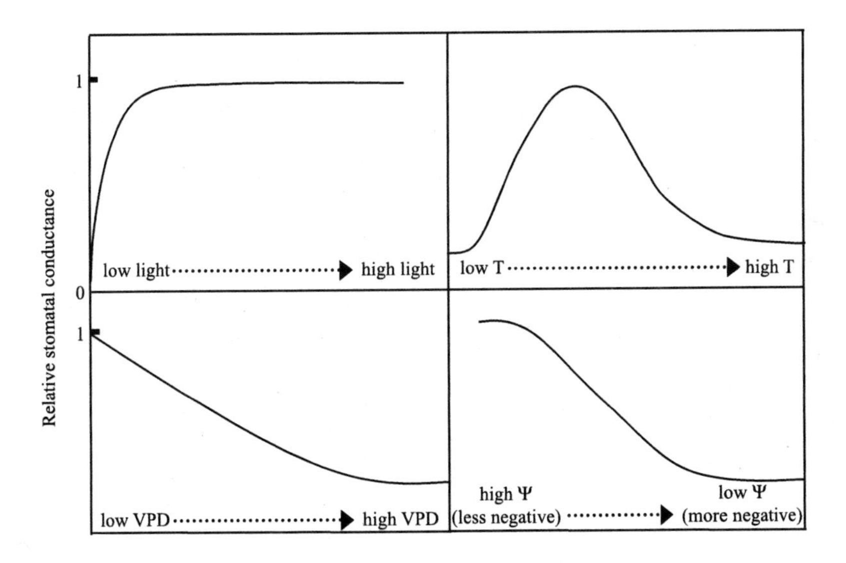

FIGURE 4.4. Generalized effects of light, temperature, humidity, and leaf water potential on stomatal conductance (T = temperature; VPD = vapor pressure deficit). *Source:* Adapted from Jones (1992).

sponse to the light itself and/or may be an indirect response mediated through the effects of light on photosynthesis (Hall, 2001).

Stomates are also sensitive to intercellular [CO_2] (C_i). When C_i is high, stomates close, and as C_i drifts lower, stomates open. This system effectively links photosynthetic demand for CO_2 with stomatal apertures, since C_i is ultimately controlled by atmospheric CO_2 concentrations coupled with the rate of CO_2 assimilation by mesophyll cells (Morison, 1998). As will be discussed, one of the primary ways rising atmospheric [CO_2] affects crop water relations is through its influence on stomatal movements. Again, the exact sensing mechanism to C_i is not known but may be mediated through effects on photosynthesis.

The hormone ABA links soil conditions with stomatal function (Tardieu and Davies, 1993). In response to drying soils, roots synthesize ABA, which is then transported through the xylem to leaves where it causes stomates to close. This mechanism enables plants to maximize stomatal conductance and photosynthesis by matching leaf water use to soil water availability (Hall, 2001).

Stomates close when subjected to dry air and open when humidity is increased. How plants detect and respond to such changes is not known.

The Leaf Boundary Layer

The effects of stomatal apertures and densities on gas exchange are mediated by the effects of leaf and canopy boundary layers (Box 4.1). A boundary layer is a layer of still, unmixed air that surrounds a surface such as a leaf or an entire crop canopy. Resistance to water diffusion across a boundary layer is proportional to its thickness. Boundary layer thickness is influenced by wind speeds and by the roughness of the boundary layer zone. For example, leaves with complex three-dimensional wax structures or with plentiful leaf hairs have thick boundary layers that increase resistance to water loss compared to smooth leaves.

Leaf Surface Wax

Surfaces of aboveground plant parts are coated by waxes, cutin, and suberin (Figure 4.2). Surface waxes are hydrocarbon chains 20 to 35 carbons long, made up of complex mixtures of ketones, alcohols, acids, esters, and paraffins. Surface wax components are extruded onto leaf surfaces, where they combine and crystalize to form rods, scales, threads, or crusts (Figure 4.2) (Gunning and Steer, 1996). These materials are commonly referred to as epicuticular waxes. The epicuticular waxes form a layer external to another water-impervious layer called the cuticle. The cuticle is made up of cutin and suberin, fatty acid molecules that are 16 to 18 carbons in length.

Cuticular and epicuticular wax layers serve multiple functions, not the least of which is to deter pathogens such as bacteria, viruses, and fungi from entering the plant interior. But of more relevance to our current discussion, these materials also reduce both plant water and thermal stress (Jefferson, Johnson, and Asay, 1989). The presence of a prominent epicuticular wax layer increases WUE by reducing cuticular transpiration and increasing leaf boundary layer effects, and it reduces leaf and canopy temperatures by reflecting solar radiation (Jefferson, Johnson, and Asay, 1989). In fact, this waterproof coating is so important, most plant biologists believe the evolution of shoot-surface waxes was a key event in the transition of plants from aquatic to terrestrial lifestyles (Gunning and Steer, 1996).

Surface wax production has been correlated with increased yields in drought- and heat-stressed crops (Jefferson, Johnson, and Asay, 1989). For example, positive relationships between leaf epicuticular wax production and WUE have been noted for cultivars of sorghum, wheat, and alfalfa. In fact, of several physiological traits measured in 34 different wheat cultivars, grain yield under drought conditions was most highly correlated with leaf waxiness (Fischer and Wood, 1979).

Epicuticular wax production is generally a constitutive characteristic, but high light conditions (Orcutt and Nilson, 2000), and water and temperature stress can all stimulate wax production (Jordan et al., 1984; Jefferson, Johnson, and Asay, 1989; Ashraf and Mehmood, 1990; Premachandra et al., 1991). Although increased wax is a useful drought and heat-stress adaptation, it can pose problems for crop management by inhibiting the passage of pesticides, herbicides, plant growth regulators, and chemical defoliants into leaves. In cotton, water deficits increased cuticle thickness by 33 percent, which led to 34 percent less defoliant entering leaves compared to a well-watered control (Oosterhuis, Hampton, and Wullschleger, 1991).

We know little about how [CO_2] or ozone will affect wax production or how wax production will affect plant response to these environmental changes. There is some evidence that thick cuticles resist ozone uptake compared to thinner cuticles, but in light of the fact that ozone flux through the cuticle is only about 0.00001 times the flux through open stomata, this effect is probably of little biological importance (Kerstiens and Lendzian, 1989). Elevated CO_2 will likely increase epicuticular wax production through its effects on the availability of carbon-based wax precursors (short carbon skeletons). So far, this idea has been supported by studies with soybean (Thomas and Harvey, 1983) and grass species (Tipping and Murray, 1999).

Trichomes

Leaf hairs can alleviate stress associated with high light, drought, and high temperatures by altering the exchange of gases and energy between atmosphere and plant canopy (Ehleringer and Björkman, 1978; Ehleringer, 1981). Densely pubescent (hairy) soybean leaves, for example, reflect more solar radiation, are characterized by greater boundary-layer resistance to leaf water loss, and therefore transpire less than smooth soybean leaves (Zhang et al., 1992). Selecting for pubescence represents one mechanism (albeit modest) for reducing heat and water stress in the future. In addition, because trichomes increase boundary-layer resistance to gas flux through stomata or the cuticle, they might decrease the flux of atmospheric pollutants such as ozone into the interior of a leaf (Elkiey and Ormrod, 1980).

It is unlikely that trichomes have much of an effect on salinity tolerance, or that salinity has much of an effect on trichome density or structure in field crops. It is worth mentioning, however, that a number of nondomesticated halophytes do possess specialized trichomes that function in salt storage and/or secretion. The presence of these specialized trichomes enables them to successfully grow and reproduce in high-salt environs. However, because of the apparent complexity of this adaptation, it is difficult to imagine the suites of genes encoding salt bladders or hairs from halophytes being transferred into crops. Nevertheless, it is noteworthy and tantalizing that some halophytes do cope with high-salt stress through compartmentalization and excretion of salt through leaf trichomes.

ROOTS AND HYDRAULIC CONDUCTANCE

Root properties such as total length, depth, and lateral spread are important components of water relations. Deep roots enable plants to acquire water from a large soil volume and high root-length densities (root length per cm^{-3} of soil) enable plants to efficiently mine water from a given soil volume. Root architecture (topology) is important too, as herringbone-type roots with few branches seem to function more efficiently when soil resources are low, while more highly branched root systems may be more efficient when soil resources are plentiful (Fitter, 1996).

Obviously, root systems of different species have different strategies for maximizing water uptake and different suites of strategies are generally best suited for a given set of soil conditions. For example, it is clear that access to deeper water sources can sometimes delay water stress and prolong active plant growth during dry periods. On the other hand, large root systems sometimes use up all the available soil water before critical reproductive growth stages! The complex problem of breeding crop varieties for maximal root characteristics has not been resolved but remains an important topic. Details of the important effects of environmental changes on growth and development of root systems are discussed elsewhere (Chapter 8).

It is not just the physical dimensions of roots that are relevant; hydraulic conductivity is also an important factor for water conduction and is often influenced by the environment (see Chapter 7). Hydraulic conductivity, a measure of the resistance imposed by plant tissues to water flow, ultimately controls the passage of water through the plant (Steudle and Peterson, 1998). In roots, hydraulic conductivity is influenced by xylem diameters as well as radial resistance to water flow across the cortex and endodermis. Mounting evidence indicates that water flow across roots and into xylem is controlled mainly by channels embedded in membranes of endodermal (and other) cells called aquaporins.

Aquaporins are present in membranes of cells where rapid flux of large volumes of water is required as is found in stomates and endodermal cells of roots (Schäffner, 1998; Tyerman et al., 1999; Johansson et al., 2000). Function of these membrane proteins is evidently crucial for water conduction; for instance, tomato roots treated with $HgCl_2^-$, a chemical shown to inhibit aquaporin activity, exhibited a 60 percent decrease in hydraulic conductivity (Maggio and Joly, 1995). Similar results were obtained with barley roots (Johansson et al., 2000). It is highly likely that the effects of the environment on root hydraulic conductivity, and perhaps also stomatal function, are mediated through changes in aquaporin number or activity (Steudle and Peterson, 1998; Johansson et al., 2000). The hormones gibberellic acid and abscisic acid as well as water and salinity stress, have all been shown to mediate expression of aquaporin genes, which in turn affects hydraulic conductivity (Joly, 1989; Johansson et al., 2000).

THE CROP CANOPY

In most environmental change experiments, in which crops are grown in well-mixed air in controlled environments, transpiration is proportional to stomatal conductance. In the field this relationship diminishes, and sometimes disappears altogether, because of canopy and regional-scale feedbacks. For example, stomatal closure often decreases evaporational cooling, leading to higher leaf temperatures. Higher leaf temperatures drive up water vapor density inside leaves and a steeper vapor density gradient between leaves and air is generated. Such an effect would tend to cancel out the initial drop in transpiration caused by stomatal closure. However, stomatal opening could result in the opposite effect.

Stomatal control over canopy water use is mediated by boundary-layer effects of leaves and canopies. Changes in stomatal apertures have little effect on canopy transpiration when resistance to water flux imposed by the canopy boundary layer is high. For example, we would not expect stomatal opening and closing to drastically change canopy transpiration if the air was calm, the canopy was humid, and the canopy boundary layer thick.

Water use by a crop canopy also changes during development. Early in the season, evaporation is controlled almost entirely by rainfall patterns and soil properties, and transpiration is controlled largely by stomatal conductance. But as a crop matures and the canopy closes, evaporation often becomes insignificant and ET becomes more and more dependent upon canopy conductance. These few examples illustrate the importance of air humidity and air, leaf, and canopy temperatures for determining water flux from a crop. Light intensity, day length, cloudiness, altitude, slope, aspect, wind, and the overall energy balance of the crop are also important.

ENVIRONMENTAL CHANGES

Elevated Atmospheric Carbon Dioxide Concentration

As discussed previously, when [CO_2] in the air surrounding a leaf increases, stomatal apertures decrease, lowering stomatal conductance and transpirational water loss (Figure 4.5). Although this issue needs to be resolved in crop species, many plants grown with CO_2-enrichment construct leaves with fewer stomates, a long-term response that contributes further to less transpiration (Pritchard et al., 1999). Consequently, for plants grown with CO_2-enrichment (ambient +360 ppm) in controlled environments with well-mixed air, stomatal conductance and transpiration typically decrease 20 to 90 percent (Drake, Gonzàlez-Meler, and Long, 1997; Morison, 1998; Kang et al., 2002). These effects, coupled with enhanced photosynthesis, enable most crop species to grow bigger and produce more yield per unit of water transpired by leaves. In other words, CO_2-enrichment drives up WUE (Figure 4.6, Table 4.1).

Although transpiration decreases with increasing [CO_2] at the leaf level, a proportional water savings is not necessarily realized by whole plants, canopies, or crop regions because of feedbacks operat-

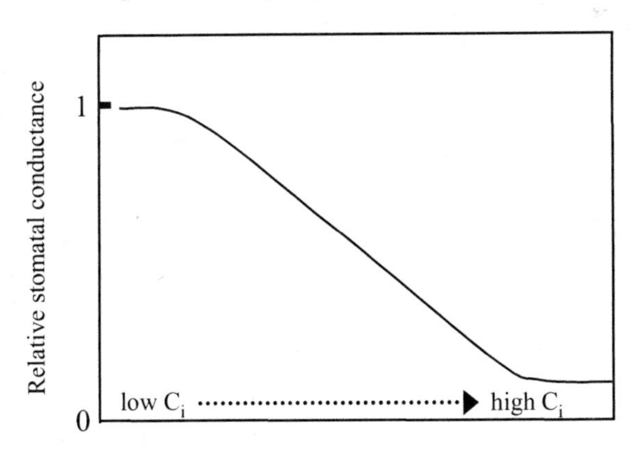

FIGURE 4.5. Generalized effect of [CO_2] on stomatal conductance.

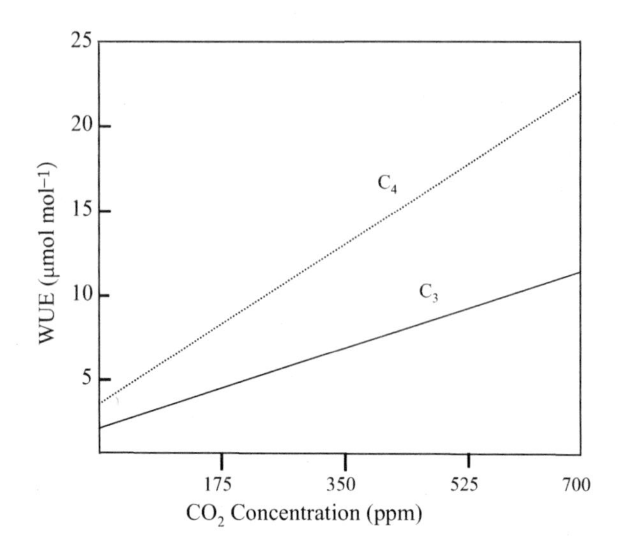

FIGURE 4.6. Potential responses of WUE of both C_3 and C_4 plants to $[CO_2]$. WUE was calculated assuming a ratio of intercellular to air CO_2 concentrations of 0.7 for C_3 species and 0.04 for C_4 plants. The mole fraction water vapor gradient between the leaf interior and the atmosphere of 12×10^{-3} mol mole^{-1} was assumed for all $[CO_2]$s. *Source:* Adapted from Polley (2002).

ing at these scales. Reduced transpiration on a leaf-area basis is often negated at the whole-plant level by higher leaf area index (LAI) (Pritchard et al., 1999; Kimball, Kobayashi, and Bindi, 2002). Furthermore, at the stand level, the relationship between stomatal conductance and transpiration tends to disappear (Polley, 2002). One reason for this is that as transpiration decreases (because of CO_2-induced stomatal closure), the air in and directly above the canopy becomes dryer, increasing the driving force for evapotranspiration from the canopy. Furthermore, high-CO_2-induced stomatal closure will increase leaf temperature, which will tend to increase canopy transpiration. The direct effects of CO_2 on leaf-level gas exchange may also be diminished by regional-scale feedbacks (i.e., between regional-scale vegetation and the mixed air immediately above it) (Amthor, 1999). Finally, because rising CO_2 will promote faster closing and denser crop canopies, interception losses could increase at the expense of soil water content.

TABLE 4.1. Percentage increases in WUE of crops grown with CO_2-enrichment.

Species	% increase in WUE	% change in ET	Experimental apparatus	Experimental $[CO_2]$	Source
Sorghum					
wet plots	9	−10	FACE	570 ppm	A
dry plots	19	−4	FACE	570 ppm	
Spring wheat					
high N	23	0	FACE	570 ppm	B
low N	10	0	FACE	570 ppm	
	21	0	GC	680 ppm	C
wet	75	—	GC	700 ppm	D
medium	66	—	GC	700 ppm	
dry	79	—	GC	700 ppm	
Sunflower	26	+11	Mesocosm	746 ppm	E
Maize					
wet	20	—	GC	700 ppm	D
medium	20	—	GC	700 ppm	
dry	72	—	GC	700 ppm	
Cotton					
wet	87	—	GC	700 ppm	D
dry	93	—	GC	700 ppm	
wet	29-39	—	GC	570 ppm	F
dry	19-37	—	GC	570 ppm	

Sources: (A) Conley et al. (2001); (B) Hunsaker et al. (2000); (C) Mitchell, Mitchell, and Lawlor (2001); (D) Kang et al. (2002); (E) Hui et al. (2001); (F) Mauney et al. (1994).

Notes: In all cases, WUE was calculated from either canopy gas-exchange measurements or represents the ratio of biomass (either grain or total) to water lost.

Abbreviations: WUE, water-use efficiency; ET, evapotranspiration; FACE, free-air CO_2 enrichment; GC, growth chamber.

Because of the many feedbacks discussed previously, rising CO_2 levels may have only small, if any, influence on total water consumption by a crop. Furthermore, increases in WUE will be smaller than predicted based on measurements of isolated plants. This is supported by several recent studies that utilized FACE technology to expose intact canopies to high CO_2 (ambient +200 ppm); WUE of field-grown sorghum was increased by 9 percent and 19 percent in wet and dry plots, but cumulative evapotranspiration was reduced only 10 percent and 4 percent (Conley et al., 2001). Similarly, WUE of field-grown wheat was enhanced 21 percent (high N treatment) and 10 percent (low N treatment) but total evapotranspiration was not changed by CO_2-enrichment (Hunsaker et al., 2000). Sunflowers grown in CO_2-enriched mesocosms (746 ppm CO_2) used 11 percent more water but exhibited a 26 percent increase in WUE compared to sunflowers grown in a mesocosm with ambient air (Hui et al., 2001). These data indicate that increases in canopy WUE associated with higher atmospheric CO_2 are mainly driven by enhanced rates of photosynthesis, not less water use (Kimball, Kobayashi, and Bindi, 2002).

The proportional (but not absolute) enhancement of growth and yield resulting from CO_2-enrichment is sometimes higher when crops are subjected to mild to moderate water stress. This effect is generally attributed to partial stomatal closure and higher WUE. Growth in elevated CO_2 influences a wide range of other plant attributes, such as carbon assimilation and allocation, growth, and nutrient relations, and these effects influence water use too (Morison, 1998). Root growth in particular is often stimulated by CO_2-enrichment and sometimes the root-to-shoot ratio is higher (Figure 6.2). Effects on roots include greater horizontal and vertical growth and higher root-length densities (Pritchard and Rogers, 2000). More roots per unit of soil volume and growth of roots deeper into soil could provide better access to soil water and might delay water-stress symptoms during droughts. Furthermore, exposure to supra-ambient [CO_2] decreased hydraulic conductivity along the pathway from roots through stems and into leaves in soybean, alfalfa, and sunflower (Bunce, 1996; Huxman, Smith, and Neuman, 1999). Higher resistance to water flow through the plant might limit water uptake and perhaps productivity in wet soils but might contribute to plant success in dry soils.

A doubling of atmospheric [CO_2] will have negligible effects on total water consumption by crops (total ET could change by – 0 percent to +10 percent). Crop productivity per unit of water consumed, however, is likely to increase 20 to 30 percent, suggesting that dryland agriculture may benefit and demands for irrigation water might drop. There are likely to be interactive effects on crop water use for plants subjected to multiple environmental changes, however. For instance, warming is likely to cancel out some proportion of the increase in WUE associated with rising [CO_2] (Figure 4.7). Stomatal closure induced by high [CO_2] might increase crop resistance to O_3 pollution and high salinity.

Warming

The water vapor content of saturated air increases exponentially with rising temperature (Young and Long, 2000). For example, air saturated with water vapor holds 6.8 $g \cdot m^{-3}$, 17.3 $g \cdot m^{-3}$, and 39.7 $g \cdot m^{-3}$ at 5, 20, and 35°C (Nobel, 2000; Figure 4.8). The water content of air in intercellular spaces of leaves is generally very close to saturation regardless of temperatures. Therefore, as canopies warm, the water density of the air within leaves will rise and so will the water den-

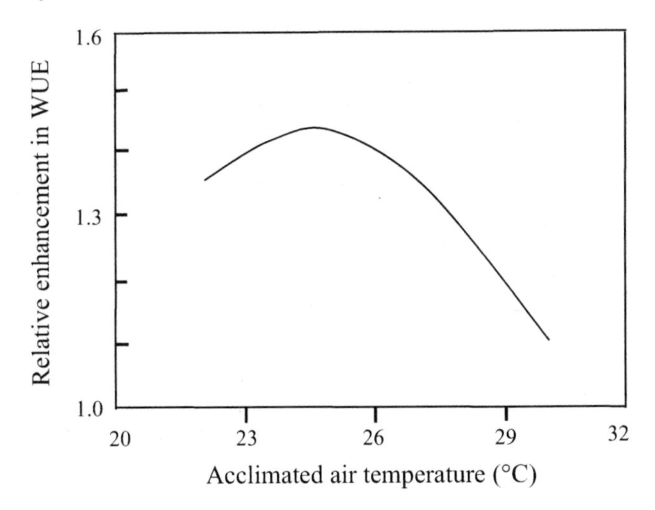

FIGURE 4.7. Interaction between daily mean temperatures and relative enhancement of WUE caused by exposure to elevated (700 ppm) [CO_2] (compared to rice grown at 350 ppm CO_2). Plants were grown in temperature gradient chambers. *Sources:* Figure adapted from Nakagawa et al. (1997) and Horie et al. (2000).

sity gradient between crop canopies and the atmosphere (which is rarely saturated). Furthermore, over a certain range of temperatures (which probably varies from crop to crop) temperature increases cause stomata to open, increasing conductance even when vapor density gradient from leaf to air is held constant (Hall, 2001; Figure 4.4). This suggests a direct effect of temperature on stomatal opening. Even assuming constant stomatal apertures and densities, transpiration is 4.8 times higher at 20°C compared to 5°C, and 2.7 times higher at 35°C compared to 20 (Nobel, 2000; Figure 4.8). These data clearly indicate that global warming will drive canopy and landscape scale ET upward (Young and Long, 2000). In an experiment on soybean over the temperature range of 28 to 35°C, for instance, transpiration increased by ~4 percent for each degree C.

Warming is also expected to decrease WUE for both C_3 and C_4 crops (Figure 4.7). In the case of rice grown at ambient and elevated CO_2 (700 ppm) in a temperature gradient tunnel, CO_2-enrichment enhanced WUE by 40 to 50 percent between 24 and 26°C, but that increase in WUE dropped to ~20 percent at 30°C (Horie et al., 2000). Unfortunately, very few data concerning temperature effects on field-scale transpiration, ET, or WUE exist.

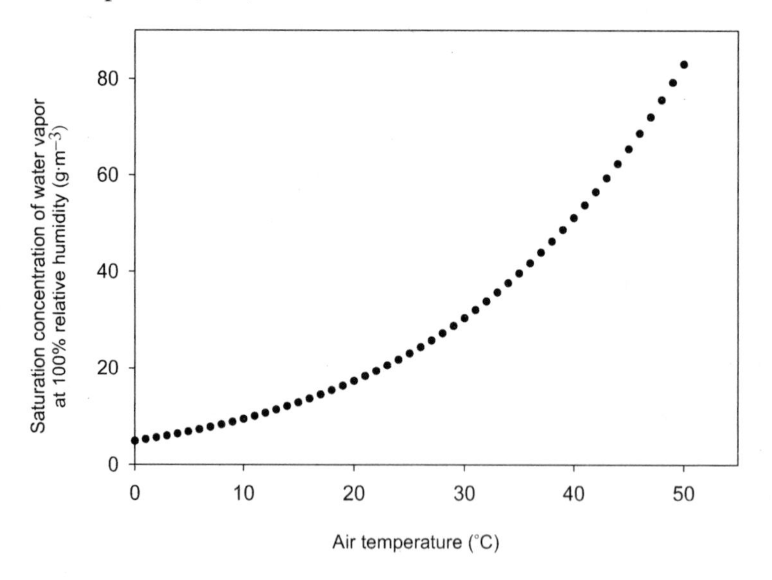

FIGURE 4.8. Effects of air temperature on the concentration of water vapor in saturated air.

The extent to which transpiration and WUE will be affected by an increase in air temperature will depend a great deal on stomatal properties, atmospheric humidity, leaf temperatures (which are a function of both air temperatures and transpiration rates), leaf morphology, and wind speed (Crawford and Wolfe, 1999). Furthermore, warming will impact crop growth and hasten development, and these effects could counteract increased transpiration to some extent. Finally, as is discussed in detail in Chapters 7 and 8, soil warming could have significant effects on root growth and function, thereby impacting water relations. In general, at the high end of the optimal temperature range, high soil temperatures increase root hydraulic conductivity and thus water flux compared to cooler soil (Bowen, 1991).

Elevated Atmospheric Ozone Concentration

Atmospheric O_3 enters plant leaves through open stomates. Therefore, the effects of ozone on plants are linked to stomatal conductance values. In general, environmental factors that stimulate stomatal opening increase ozone susceptibility and factors that induce stomatal closing decrease susceptibility. This explains why water-stressed plants (low stomatal conductances) are typically more resistant to O_3 pollution compared with crops grown with frequent irrigation (high stomatal conductances) (Ashmore and Marshall, 1999). It also explains why CO_2 enrichment sometimes decreases the severity of O_3 damage to crops. Similarly, species or cultivars with inherently high stomatal conductance are often more sensitive than those with low stomatal conductance (Darrall, 1989).

The influence of O_3 pollution on stomatal function (and water relations) is complex and depends on the duration and severity of exposure. As a general rule, short-term exposure to a high $[O_3]$ results in stomatal closure while long-term exposure causes sluggish stomatal movements (McAinsh et al., 2002). Ozone concentrations exceeding 200 ppb often lead to stomatal closure and reduced leaf conductance (Figure 4.9) while lower concentrations often lead to impaired stomatal closure (Mulholland et al., 1997b; Robinson, Heath, and Mansfield, 1998).

How O_3 exposure influences stomatal function is not well understood but it is likely that O_3 directly affects metabolism of subsidiary cells or guard cells, since these are the first leaf cells—at least of

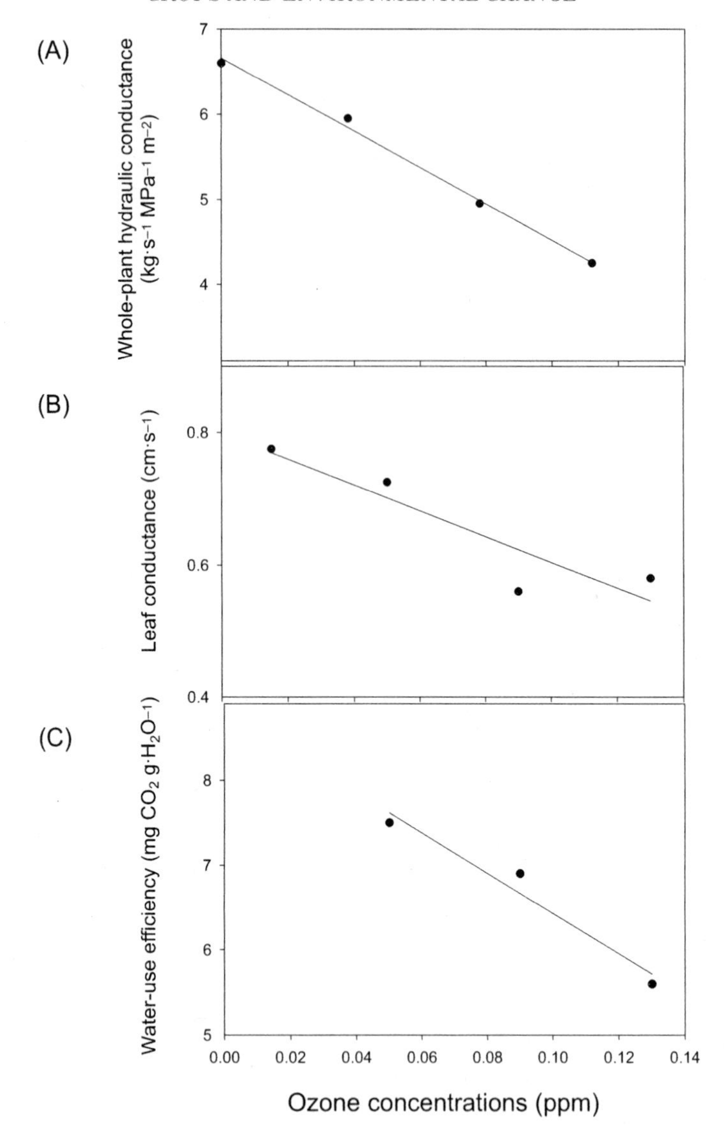

FIGURE 4.9. The effects of O_3 on (A) whole-plant hydraulic conductance of cotton plants expressed on a leaf-area basis (modified from Grantz and Yang, 1996a), (B) Stomatal conductance of soybean leaves measured with a leaf porometer, and (C) effects of O_3 on whole-plant WUE of irrigated soybean plants. *Source:* (A) Modified from Grantz and Yang (1996a); (B) and (C) modified from Reich, Schoettle, and Amundson (1985).

those unprotected by cuticular waxes—to come into contact with air pollutants (Robinson, Heath, and Mansfield, 1998). It was recently suggested that O_3 uncouples the linkage between guard cell and whole-plant function by interfering with the hormonal (i.e., ABA) and Ca^{2+}-dependent signaling mechanisms (McAinsh et al., 2002; Chapter 3). Disruption of signal perception and transduction could explain how pumping of molecules and ions back and forth across guard cell membranes, the process that causes the turgor changes that in turn drive stomatal opening and closing, is upset by O_3 pollution (McAinsh et al., 2002).

O_3 might also indirectly influence stomatal conductance. It is clear, for instance, that O_3 damages the photosynthetic apparatus leading to lower photosynthesis rates (Chapter 5). Less C-fixation in leaves would tend to elevate internal [CO_2]s, eventually leading to stomatal closure (Reich, Schoettle, and Amundson, 1985).

Atmospheric O_3 pollution could also influence water relations by affecting aspects of plant structure and function that are somehow involved in obtaining water from soil or distributing it throughout the plant. For instance, root growth is decreased (Chapter 8) and mycorrhizal associations are fewer (Chapter 10) in crops exposed to high atmospheric [O_3] and these changes could significantly influence the acquisition of soil water (Grantz and Young, 1996a). In addition, O_3 reduces the hydraulic conductivity of roots and whole plants, at least in cotton, which could limit water uptake as well as nutrient acquisition (Grantz and Young, 1996b; Figure 4.9).

Salinity

Whether you are a mariner or a wheat crop, there is a threshold of salinity above which water becomes unusable. Crops growing in saline soils are sometimes unable to acquire enough water to meet their needs, even when soil-water contents are high (Table 4.2). This is because water conduction through a plant depends on a favorable water-potential gradient between soil and plant, root and shoot, and plant and atmosphere (SPAC). An excess of dissolved salts can lower water potential values of the soil solution, thereby making the water potential gradient from soil to root unfavorable. This represents a major dilemma, one that, along with the direct toxic effects of specific salts on cellular metabolism may colimit crop success in salinity-affected regions (Table 4.2).

Often, initial growth inhibitions in salinized plants are dominated by water-stress effects, while long-term growth reductions are caused by ion toxicity (Chapter 3). Evidently, in the long-term (weeks to months), most plants compensate for the water-stress aspect of salt stress through osmotic adjustment of their cells (Munns, 1993; Katerji et al., 1997; Zhu, Hasegawa, and Bressan, 1997; Meneguzzo, Navari-Izzo, and Izzo, 2000). Osmotic adjustment is the lowering of cellular water-potential values through the accumulation of compatible organic solutes (also called osmolytes or osmoprotectants) such as proline, mannitol, fructans, glycine betaine, or ions such as K^+, Na^+, or Cl^-. Osmotic adjustment serves several functions: (1) protects cellular constituents from dehydration damage (osmoprotection), (2) maintains water uptake in soils of low water potentials, and (3) enables cells to maintain turgor. Apparently, the ability to compartmentalize salts within vacuoles and to balance these solutes with compatible organic solutes in cytoplasm appears to be an important adaptation to drought and saline environments for some plants (Munns, 2002). There is currently a great deal of interest in enhancing osmotic adjustment in crop plants to improve tolerance to salinity as well as drought.

TABLE 4.2. Generalized effects of salinization on salt-sensitive crops over various time scales.

Time	Water-stress effects	Salt-specific effects
Minutes	Instant reduction in leaf and root elongation rate then rapid partial recovery	
Hours	Steady but reduced rate of leaf and root elongation	
Days	Leaf growth more affected than root growth; reduced rate of leaf emergence	Older leaves visibly damaged
Weeks	Reduced final leaf size and/or number of lateral roots	Older leaves senesce
Months	Altered flowering time, reduced seed production	Younger leaves dead, whole-plant mortality

Source: Table modified from Munns (2002).

Notes: Effects attributed to both water-stress and salt-specific effects are shown.

It is important to note, however, that osmotic adjustment is metabolically expensive (see Chapter 5) and, although it may enhance *survival* under severe conditions, it also inhibits growth and reduces yield (Quisenberry, Cartwright, and McMichael, 1984). A review of the literature recently revealed that during drought-induced osmotic stress, positive correlations between osmotic adjustment and yields were usually obtained only when osmotic stress was so severe that yields were too low to be of economic value (Serraj and Sinclair, 2002). Salinity tolerance is probably linked more closely with the ability of a crop to exclude salts coupled with efficient compartmentalization of toxic ions (Chapter 3), not osmotic adjustment.

There is clearly significant disagreement regarding the importance of osmotic adjustment in conferring resistance to salt and drought stress. Nevertheless, the fact remains that crops growing in saline soil must at least adjust the water-potential values of their root cells to levels below that of the soil solution in order to maintain water uptake, and such a response must last throughout the duration of a crop. Regardless of the controversy over its value for leaves, osmotic adjustment is probably important in root cells, where it enables them to maintain growth and therefore access to soil water (Hall, 2001; Serraj and Sinclair, 2002). The significance of any genetic variability that exists with respect to this trait, and the extent to which that variability contributes to overall salinity tolerance, remains to be established (Neumann, 1997).

Other effects of salinity influence crop-water relations. Apparently, roots are able to detect low water-potential values (which are characteristic of drying and salinized soils), and respond by signaling leaves to close stomatal pores in order to conserve water (Munns, 1993). ABA is the most likely signal, as it is known to affect stomatal function and has been found to accumulate in the xylem of plants exposed to salinity (Zhao, Munns, and King, 1991). Salinized plants also develop smaller but thicker leaves, have fewer stomates, and produce thicker cuticular and epicuticular wax (i.e., they become more succulent; Rao and Rao, 1982). Salinity-induced increases in leaf-surface waxes have been reported in peanuts, pigeonpea, and gingelly (Rao and Rao, 1982). Reduced stomatal conductance likely contributes to salinity-induced growth reductions (by limiting CO_2 uptake), but may also protect the plant from accumulating lethal levels of toxic ions in leaves via the transpiration stream (Neumann, 1997).

Salinity also retards root growth and decreases root hydraulic conductivity (Joly, 1989; Evlagon, Ravina, and Neumann, 1992), which could further diminish a crop's water uptake. These effects, along with reduced total leaf area, combine to significantly reduce transpiration in salt-stressed crops. Finally, at the canopy level, salinity delays canopy closure, which may increase evaporation and decrease soil water (Neumann, 1997).

High temperatures can sometimes exacerbate salt toxicity in crops by stimulating transpiration and hastening the accumulation of salt ions in leaves. Conversely, high atmospheric humidity can decrease salt stress, perhaps by slowing transpirational water use (Bassil and Kaffka, 2002). Elevated CO_2 may boost salt tolerance by enhancing WUE and increasing the availability of carbon compounds for osmoregulation and growth (Schwarz and Gale, 1984; Munns, Cramer, and Ball, 1999).

Salinization: Avoidance and Management

Most soil salinity problems occur in arid and semiarid agricultural regions. Unfortunately, these are also the regions where irrigation provides the greatest benefit to crops. On dry or arid lands, soils experience little if any leaching and, as a result, salts tend to accumulate there. Accumulation of salt is often further aggravated by irrigation with poor quality water and improper irrigation practices (i.e., mistimed, too little, or too much). Excessive fertilization can also contribute to soil salinization.

The keys to sustainable crop production in areas most susceptible to salinization include minimizing irrigation and maximizing drainage so that salts do not accumulate (e.g., by installing perforated drainage pipes 1 to 2 m deep). In many areas, the efficiency of irrigation practices could be increased substantially simply by converting from inefficient sprinkler type (e.g., center-pivot) to more efficient subterranean (drip-type) irrigation systems. Much work also remains to be done in the area of irrigation scheduling (i.e., how much water does a crop really need, and when does it need it?).

The possibility of timing the application of poor-quality (i.e., brackish) water to coincide with resistant plant growth stages has been suggested many times (Shennan et al., 1995). This is based on research showing that crops are susceptible to salinity during early vegetative growth but are relatively resistant during reproductive stages (Botella, Cerdá, and Lips, 1993). Based on this principle, it should be possible to use a mixture of good- and poor-quality irrigation water, properly timed throughout the growing season, to produce a crop with acceptable yield. Of course, this strategy is not possible where good-quality irrigation water is not available during salt-sensitive growth stages. And in areas where salinity problems are associated with poor drainage or shallow groundwater, this might not be feasible either.

Another management option is to match the salinity level of the soil or irrigation water with the salinity tolerance of the crop, but switching to more salt-tolerant crops as salinization becomes more severe, although a common practice, is not a viable long-term solution. The best solution is to install irrigation systems with proper drainage so that salt is not allowed to accumulate in the first place.

SUMMARY

Worldwide, lack of fresh water limits crop production more than any other resource limitation. This shortage may get worse because irrigation practices are currently on the rise, leading to groundwater depletion and salinization of many important crop-production regions. Furthermore, changes in global hydrology accompanying climatic change are likely to affect water resources in unpredictable ways. Availability of water is crucial because growth and yield are often linearly related to the total amount of water consumed by a crop during a growing season.

Structural and functional properties of leaves, stems, and roots control water uptake from soil and transpiration from leaves. Underground, root length and architecture determine how much soil water a plant can access and root hydraulic conductivity controls how fast that water can flow through roots. Aboveground, stomates are especially important because they control stomatal conductance and transpiration, which in turn sets the demand for water from roots. Stomatal opening and closing are influenced by environmental conditions including light levels, humidity, temperature, [CO_2], plant water status, and air pollution levels.

It is difficult or impossible to predict water consumption by an entire field crop (total ET) from leaf-level properties such as stomatal conductance because of a number of canopy and regional feedbacks. For this reason, much of the research conducted to date regarding the influence of environmental changes on crops should be interpreted with caution, at least until many of these results can be validated with more field-scale measurements.

Rising CO_2 levels might lessen crop dependence on irrigation to some extent. This is because the amount of water needed to produce a given quantity of plant mass decreases as atmospheric [CO_2] goes up. At the crop canopy level, water consumption goes down only slightly, if at all, because of greater total leaf area and higher canopy temperatures. Although warming generally will increase ET and reduce WUE, it will also hasten plant development, which could counteract those effects to some extent.

In general, environmental factors that stimulate stomatal opening such as irrigation, humid air, and high light levels, increase O_3 uptake, and factors that induce stomatal closing such as drought and elevated [CO_2] decrease uptake. Ozone pollution affects plant-water relations by reducing partitioning of photosynthate to roots and by reducing stomatal conductance and WUE. Exposure of crops to O_3 pollution is expected to decrease water use by crops because it causes stomates to close, reduces plant leaf area, inhibits rooting, and lowers the root:shoot ratio.

Crops growing in saline soils are sometimes unable to acquire enough water to meet their needs because dissolved salts lower water-potential values of the soil solution, resulting in an unfavorable water-potential gradient from soil to root. Most crops overcome this unfavorable gradient by accumulating organic solutes and mineral ions in a process called osmoregulation. The importance of osmoregulation to whole-plant salinity tolerance, however, remains controversial. Salinity affects plant-water relations in other ways as well; for example, salinity often lowers stomatal conductance, hydraulic conductivity, and leaf area, which all contribute to reduced total water consumption.

The overall effects of environmental changes on water demand and consumption by a field crop are difficult to predict. They will depend to a large extent on whether stomatal closure and reduced ET driven by rising [CO_2] will be more or less important than the increases in canopy ET caused by global warming. Regionally, this balance will also be influenced by [O_3], which should decrease water consumption but also yields. Effects of elevated [CO_2] and warming on stomates will influence the extent of crop damage from air pollutants. Water stress associated with salinized soils could be aggravated by warmer temperatures but alleviated by elevated [CO_2].

Chapter 5

Photosynthesis, Respiration, and Biosynthesis

Two important facts of biology are: (1) the dry matter of organisms, including crops, is composed of a ubiquitous and relatively small number of macromolecules (compared to the much larger number of possible compounds), and (2) carbon is the major atomic component and provides the backbone of those macromolecules (Morowitz, 1968). The immediate source of carbon found in crops is atmospheric CO_2, so not only is atmospheric CO_2 an important greenhouse gas, it is also the key substrate of plant growth. Photosynthesis is the process plants use to assimilate atmospheric CO_2 (*assimilate* is both a verb, "to take in and appropriate as nourishment" or "absorb into the system," and a noun, "something that is assimilated"; herein, the noun refers to organic compounds formed during photosynthesis). The assimilated carbon, mainly carbohydrates, is then processed by respiration to supply usable energy, reducing power, and carbon-skeleton intermediates required for growth of new biomass and maintenance of existing biomass.

PHOTOSYNTHESIS

An overall summary equation of photosynthesis is:

$$CO_2 + H_2O \xrightarrow{\text{light}} \{CH_2O\} + O_2$$

where $\{CH_2O\}$ represents a carbohydrate. This summary equation greatly simplifies photosynthesis in plants, which involves dozens of physical and chemical reactions, most of them inside chloroplasts.

It is convenient to group the many individual reactions of photosynthesis into three sets: (1) physical and chemical processes that transport CO_2 from the atmosphere to the inside of chloroplasts; (2) "light reactions" that depend on the absorption of photons and result in the formation of a reductant (nicotinamide adenine dinucleotide phosphate [reduced] [NADPH]) and usable energy (adenosine triphosphate [ATP]) inside chloroplasts; and (3) "dark reactions" that use the products of the light reactions (i.e., NADPH and ATP) to assimilate (i.e., chemically reduce, or add electrons to) CO_2, resulting in the formation of sugars and other organic compounds.

Carbon Dioxide Transport

Photosynthetic CO_2 assimilation occurs mainly inside leaves, though stem tissues and many reproductive organs are also capable of photosynthesis. The main source of CO_2 assimilated in photosynthesis is the bulk atmosphere. Before CO_2 can be assimilated, it must diffuse from the atmosphere into chloroplasts within photosynthetic cells. The amount of CO_2 transported relative to the amount above a crop can be large. For example, a healthy crop might assimilate, in one day, the equivalent of all the CO_2 in the 100 meters of air above the crop (Loomis and Connor, 1992, p. 153). A much smaller fraction of CO_2 assimilated is derived from the soil beneath the crop (Moss, Musgrave, and Lemon, 1961; Monteith, Szeicz, and Yabuki, 1964), and much of the CO_2 diffusing out of soil is derived from recent photosynthate, either channeled through the respiration of living roots or the oxidation of dead roots and carbon compounds exuded from the crop root system.

The main stages of CO_2 transport are turbulent transport from the bulk atmosphere to the air immediately above the crop, transport through the crop "boundary layer" to the outside of individual leaf boundary layers, diffusive transport through leaf boundary layers to leaf surfaces, diffusive transport through stomatal pores in leaf surfaces (a small fraction of CO_2 may move directly through leaf cuticles) to substomatal cavities or air spaces, dissolution of CO_2 in water in the walls of mesophyll cells, diffusive transport of dissolved CO_2 through the walls and plasmalemmas of mesophyll cells, and diffusive transport through cytosol and chloroplast membranes into chloroplasts in mesophyll cells, where it is available for photosynthetic assimilation (C_4-photosynthesis uses HCO_3^- as inorganic carbon substrate whereas C_3-photosynthesis uses CO_2 directly; HCO_3^- forms spontaneously during CO_2 dissolution, but that reaction is catalyzed in plants by carbonic anhydrase). Chloroplasts are often located in (nearly) direct contact with plasma membranes, so transport through cytosol can be short and rapid. In this chain of transport processes stomata can be especially important because they are a major "bottleneck," are under short-term control by the crop, and respond rapidly and strongly to many environmental factors and changes (discussed in Chapter 4). The drawdown of [CO_2] from substomatal air spaces to sites of carboxylation in chloroplasts in crop leaves can be on the order of 50 percent (Evans et al., 1986; Loreto et al., 1994). (Concentration values given for dissolved CO_2, such as the CO_2 inside living cells, refer to the [CO_2] in air that would be in equilibrium with the dissolved concentration.) Further details of CO_2 transport processes were given in, e.g., Nobel (1999).

Light Reactions of Photosynthesis

The light reactions of photosynthesis absorb photons, split water into free oxygen (O_2) and protons, pump protons across thylakoid membranes inside chloroplasts (establishing a proton gradient across the thylakoid membrane that can be used for energy-requiring reactions), transport electrons through various macromolecular complexes associated with thylakoid membranes, and produce the reductant NADPH (from nicotinamide adenine dinucleotide phosphate [oxidized form] [$NADP^+$] and protons) and the "energy currency" ATP (from adenosine diphosphate [ADP] and unorganic orthophosphate [P_i]). Photons that are most important for the light reactions of photosynthesis have wavelengths between about 400 and 700 nm. This wave band is often called photosynthetically active radiation (PAR) (McCree, 1981). The PAR wave band corresponds approximately to electromagnetic radiation visible to humans.

Absorption of PAR driving the light reactions takes place in so-called light-harvesting antenna that transfer excitation energy between pigment molecules (mainly chlorophylls and carotenoids), eventually passing the energy to reaction centers within one of two different protein complexes (photosystems I and II) integral to thylakoid membranes. Photosystem II uses the energy of absorbed photons to split water molecules into oxygen and protons in the inside of a thylakoid. At the same time, photosystem II reduces plastoquinone molecules within the thylakoid membrane using electrons from the split water. The reduced plastoquinone is then oxidized during the reduction of the cytochrome b_6f complex within the thylakoid membrane. The reduced cytochrome b_6f complex passes electrons to plastocyanin while pumping protons from outside to inside the thylakoid. Photosystem I in turn uses the energy of other absorbed photons to transfer electrons from reduced plastocyanin to ferredoxin. The reduced ferredoxin is oxidized by ferredoxin-NADP reductase, producing NADPH from $NADP^+$ and protons outside the thylakoid. Protons previously pumped into the thylakoid by the cytochrome b_6f complex are used by the ATP synthase embedded in the thylakoid membrane to produce ATP from ADP and P_i. Perhaps four protons must pass through the ATP synthase, moving from inside to outside the thylakoid, per ADP phosphorylated. The light reactions are most effective when operation of both photosystems I and II are coordinated, as they generally are in healthy (unstressed) crops.

Dark Reactions of Photosynthesis

Plants use the NADPH and ATP produced by the light reactions of photosynthesis to assimilate CO_2 in the dark reactions. There are two different metabolic cycles used to photosynthetically assimilate CO_2 in most crops: (1) the C_3 carbon cycle and (2) the C_4 carbon cycle. The C_3 designation arises because the first product of C_3 photosynthesis is a three-carbon compound, whereas the C_4 label is used because the first product of CO_2 assimilated in the C_4 cycle is a four-carbon compound. (A third cycle, called crassulacean acid metabolism, or CAM, occurs in some higher plants, but not in the major crop species.) Plants using the C_3 cycle are called C_3 plants, and those using the C_4 cycle are called C_4 plants. Most major crops are C_3 species, though maize, sorghum, and sugarcane are C_4. The C_4 cycle includes the C_3 cycle as a subset of its reactions, so it is really a C_4 cycle coupled to the C_3 cycle.

Leaves of C_3 plants generally have one main type of photosynthetic cell, called mesophyll. C_4 leaves have two distinct cell types important to the dark reactions of photosynthesis: mesophyll and bundle-sheath cells. Bundle-sheath cells are found directly surrounding vascular tissue within C_4 leaves. Mesophyll and bundle-sheath cells are both needed for C_4 photosynthesis; the two cell types "cooperate." Mesophyll cells in C_4 leaves are typically no more than two or three cells distant from one or more bundle-sheath cells. An extensive network of plasmodesmata connect mesophyll and bundle sheath cells in C_4 leaves. This network facilitates the cooperative transfer of metabolites between the two cell types.

C_3 Photosynthesis

Carbon dioxide diffusing into the stroma of chloroplasts can be incorporated into carbohydrates through the reactions of the photosynthetic reductive pentose phosphate pathway (RPPP), also called the photosynthetic carbon reduction cycle (PCR). The RPPP requires inputs of NADPH and ATP, which are produced by the light reactions.

The CO_2 assimilation reaction of C_3 photosynthesis (i.e., the carboxylation reaction) involves the addition of one CO_2 (and one H_2O) molecule to a molecule of ribulose 1,5-P_2 (rubP; a five-carbon compound) within the stromal fraction of a chloroplast (Figure 5.1). The reaction results in the production of two molecules of 3-phosphoglycerate (a three-carbon compound). The enzyme that catalyzes the carboxylation reaction is rubP carboxylase-oxygenase, typically abbreviated "rubisco." Rubisco is commonly regarded as the most abundant enzyme on earth.

The 3-phosphoglycerate formed from CO_2 and rubP begins a cycle of metabolism that regenerates the rubP (so that more CO_2 can be assimilated) and produces triose-phosphates (triose-Ps) that exit the cycle and that are used to produce compounds such as sucrose and starch (i.e., the "final" products of photosynthesis). For every three CO_2 molecules assimilated in the RPPP, one net triose-P is formed and nine ATP and six NADPH are used (producing nine ADP, eight P_i, and six $NADP^+$) (Figure 5.1). Continued operation of the RPPP requires nearly simultaneous operation of the light reactions of photosynthesis (i.e., the light and dark reactions are coordinated).

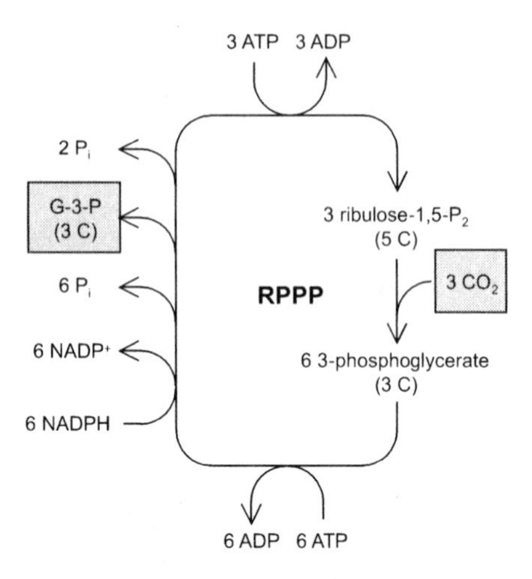

FIGURE 5.1. Simplified schematic representation of the photosynthetic reductive pentose phosphate pathway (or RPPP) in C_3 leaf mesophyll cells. Most individual reactions are omitted to highlight the inputs and outputs of the pathway (or cycle). The carboxylation reaction (i.e., reaction assimilating CO_2) joins ribulose-1,5-P_2 (a five-carbon compound) and CO_2 in the formation of two 3-phosphoglycerate (a three-carbon compound). The carboxylation is catalyzed by ribulose 1,5-bisphosphate carboxylase-oxygenase (rubisco). The pathway (cycle) is drawn to account for the production of one glyceraldehyde 3-P (G-3-P) molecule (a three-carbon compound) from three CO_2 molecules (the rubisco-catalyzed reaction involves a single ribulose-1,5-P_2 and a single CO_2, so the schematic applies to the net operation of three carboxylation reactions). Five glyceraldehyde 3-P molecules are used to regenerate three ribulose-1,5-P_2 molecules to maintain operation of the pathway (cycle). The sixth glyceraldehyde 3-P molecule exits the cycle and is available for other uses (e.g., production of sucrose or starch).

Photorespiration

In addition to being a carboxylase, rubisco also catalyzes the oxygenation of rubP in a process called *photorespiration* (because the process depends on the light reactions of photosynthesis, hence "photo," and because CO_2 is released by the cycle, hence "respiration"). Oxygenation of rubP results in the formation of one 2-phosphoglycolate molecule and one 3-phosphoglycerate molecule (Figure 5.2). Both CO_2 and O_2 compete for reaction with rubP, so photosynthesis and photorespiration occur simultaneously in C_3 plants during the day.

The 3-phosphoglycerate formed by oxygenation of rubP enters the RPPP and contributes to the regeneration of rubP. The 2-phosphoglycolate formed by rubP oxygenation eventually contributes to production of additional 3-phosphoglycerate, but only after releasing CO_2 in a complex set of reactions taking place in three organelles. The flow of carbon from 2-phosphoglycolate to 3-phosphoglycerate involves movement from the chloroplast to an adjacent peroxisome; an oxygenation reaction in the peroxisome (forming the highly reactive H_2O_2) followed by an amination reaction; transport of the resulting glycine from the peroxisome to a nearby mitochondrion in which two glycine molecules are combined to form one serine, one CO_2 (the "respiration" reaction of photorespiration), and one ammonia (which is reassimilated in a chloroplast); transport of the serine back to a peroxisome, where a deamination reaction occurs followed by a reduction reaction forming glycerate; and the transport of glycerate back to a chloroplast, where it is phosphorylated in the formation of 3-phosphoglycerate, which enters the RPPP. Thus, 25 percent of the carbon in 2-phosphoglycolate formed by oxygenation of rubP is released as CO_2 and the remaining 75 percent can contribute to rubP regeneration. Some of the serine formed in photorespiration might be removed from the photorespiratory cycle to support other areas of metabolism, including protein synthesis.

Although rubisco has a greater affinity for CO_2 than for O_2 (about 50 to 100 times greater), atmospheric $[O_2]$ is much greater (50 times) than atmospheric $[CO_2]$. Part of the discrepancy in atmospheric $[O_2]$ and $[CO_2]$ is removed inside chloroplasts because CO_2 is more soluble in water than O_2 and rubisco activity occurs in solution inside chloroplasts. Nonetheless, there is still enough O_2 inside C_3 chloroplasts to cause significant photorespiration. The ratio of rubP carboxylation to rubP oxygenation is temperature sensitive, such that oxygenation is relatively more rapid as temperature increases. This is due to an increase in the ratio of dissolved O_2 relative to dissolved CO_2 at higher temperature and to the kinetic properties of rubisco, which result in an increase in oxygenation relative to carboxylation at higher temperature. As a rule of thumb, the ratio of photosynthetic

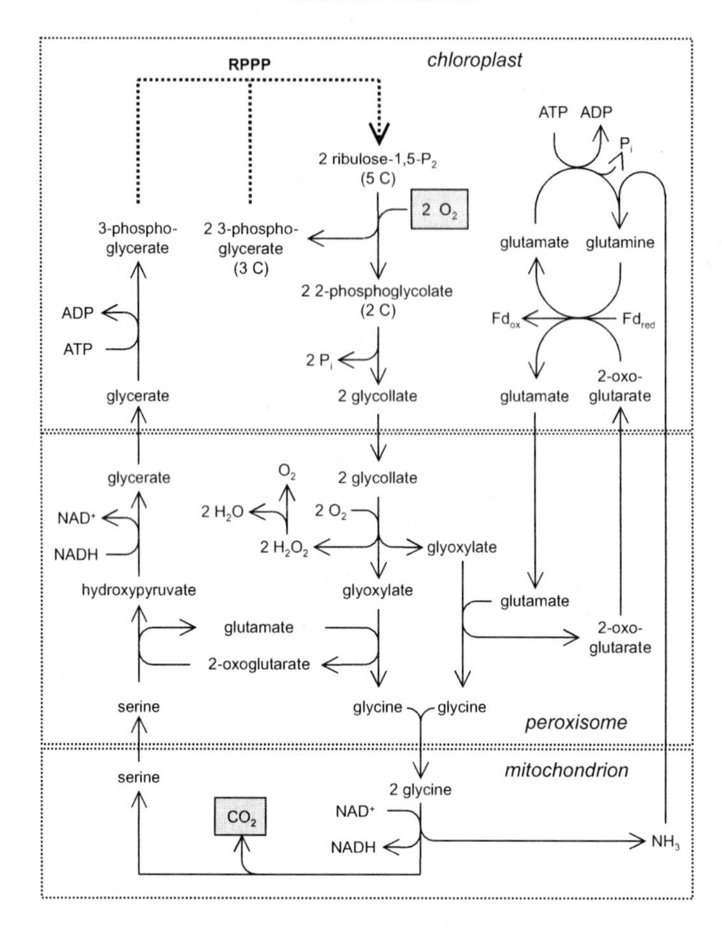

FIGURE 5.2. Path of carbon flow, and NH_3 turnover and assimilation, in photorespiration. Oxygenation of two ribulose 1,5-P_2 molecules in a chloroplast results in the release of one CO_2 molecule in a mitochondrion. Membrane transporters are omitted for simplicity. The NADH formed during glycine decarboxylation in a mitochondrion can be used to reduce hydroxypyruvate in a peroxisome through a malate/oxaloacetate shuttle between the two organelles (not shown). The 3-phosphoglycerate formed from both serine and ribulose 1,5-P_2 is used to regenerate ribulose 1,5-P_2 in the photosynthetic reductive pentose phosphate pathway (RPPP), which is needed for continued activity of photorespiration and photosynthesis. In most leaves, chloroplasts, peroxisomes, and mitochondria are in close proximity, which facilitates metabolite transport between the three organelle types.

CO_2 assimilation to photorespiratory CO_2 release is about five to one in the present atmosphere at 25°C.

Photorespiration appears to waste assimilate and reductant (NADH) and ATP, so the "purpose" of photorespiration is often questioned. The oxygenase activity of rubisco could be without benefit, at least under many circumstances. Photorespiration may, however, be of benefit to a plant in some cases. For example, when [CO_2] inside leaves is low (as might occur when stomates close to conserve water during drought) high light could produce excess ATP and reducing power because lack of CO_2 would limit the dark reactions of photosynthesis. Continued operation of the light reactions could then become damaging to organelles and cells because the energy absorbed by the photosynthetic pigments cannot be dissipated. In that case, photorespiration might function as a useful, and safe, sink for the products of the light reactions, but a larger physiological role (if any) of photorespiration remains unclear.

C_4 Photosynthesis

Some plants of tropical origin, including maize and sorghum, have evolved a C_4 photosynthetic strategy that minimizes (or completely eliminates) photorespiration. The first step of the C_4 photosynthetic pathway involves the carboxylation of phosphoenolpyruvate (PEP) by PEP carboxylase in the cytosol of a mesophyll cell (Figure 5.3). PEP carboxylase does not function as an oxygenase and also has a much higher affinity for CO_2 than does rubisco and therefore is saturated at [CO_2] of 150 to 200 ppm (Young and Long, 2000). The product of this initial reaction is a four-carbon acid. The four-carbon acids produced in mesophyll cells diffuse through plasmodesmata into bundle-sheath cells, where they are decarboxylated in chloroplasts producing CO_2 and the three-carbon compound pyruvate. The CO_2 molecule released enters the "normal" C_3 RPPP within bundle-sheath chloroplasts. Pyruvate is transported back into mesophyll cells, where it is converted into PEP to begin the cycle again.

Plants using the C_4 pathway concentrate CO_2 near rubisco in bundle-sheath chloroplasts; the [CO_2] near rubisco in C_4 leaves may be 20 times higher than ambient atmospheric [CO_2] (Young and Long, 2000). Such a high ratio of CO_2 to O_2 almost entirely suppresses

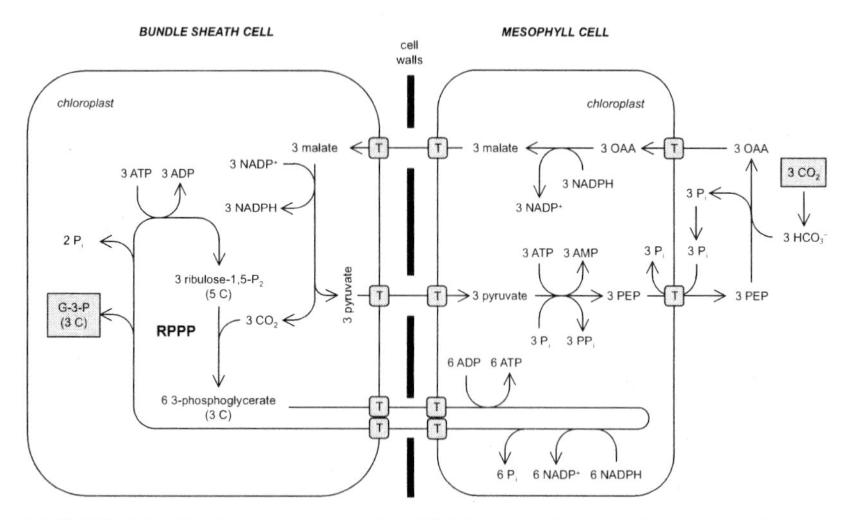

FIGURE 5.3. The four stages of the NADP-dependent malic enzyme (NADP-ME) type C4 photosynthesis, which occurs in maize, sorghum, and sugar cane: (1) assimilation of CO_2 (via HCO_3^-) in a mesophyll cell by cytosolic carboxylation of phosphoenolpyruvate (PEP) forming oxaloacetate (OAA), a four-carbon compound, which is transported to a mesophyll cell chloroplast and then reduced to malate, another four-carbon compound; (2) diffusive transport of the malate to a bundle-sheath cell chloroplast; (3) decarboxylation of the malate releasing CO_2 (and forming pyruvate) which is then assimilated by the action of ribulose 1,5-bisphosphate carboxylase-oxygenase and the rest of the reductive pentose phosphate pathway (RPPP; shown in abbreviated form), forming glyceraldehyde 3-P (G-3-P; a three-carbon compound); and (4) transport of the pyruvate to a mesophyll cell chloroplast where it is phosphorylated, forming PEP, which is then released to the cytosol to begin the cycle again. As drawn, two reactions of the RPPP (i.e., phosphorylation of 3-phosphoglycerate, forming 1,3-bisphosphoglycerate, and reduction of the 1,3-bisphosphoglycerate, forming G-3-P) occur in the mesophyll, following transport of 3-phosphoglycerate from the bundle-sheath to a mesophyll cell chloroplast. Triose-phosphate (e.g., G-3-P) is then transported back to bundle-sheath cells. The pathway is drawn to account for the production of one G-3-P molecule in the bundle-sheath cell from three CO_2 molecules in the mesophyll cell. Transport of malate and pyruvate between mesophyll and bundle-sheath cells is by diffusion through plasmodesmata. Transport of malate, pyruvate, OAA, PEP, and RPPP intermediates across chloroplast membranes is probably facilitated by transporters (T).

photorespiration (CO_2 out-competes O_2). The C_4 CO_2-concentrating mechanism also results in a relative insensitivity of C_4 photosynthesis to elevated atmospheric [CO_2], especially when compared to C_3 photosynthesis (Figure 5.4).

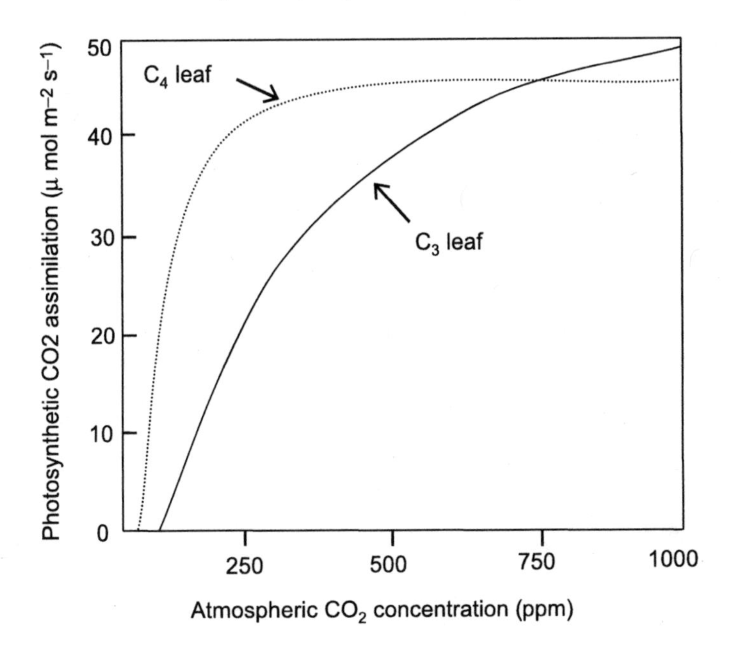

FIGURE 5.4. Photosynthesis of C_3 and C_4 leaves as a function of atmospheric [CO_2]. Because C_4 plants are already CO_2-saturated by present atmospheric [CO_2], increased [CO_2] has little effect on C_4 photosynthesis. Photosynthesis in C_3 plants, on the other hand, responds strongly to an increase in atmospheric [CO_2] because of the low affinity of rubisco for CO_2 coupled with a depression of photorespiration caused by increasing [CO_2]/[O_2] ratio. *Source:* Figure adapted from Berry and Downton (1982).

The transport of each CO_2 from mesophyll cells into the chloroplasts of bundle-sheath cells requires two ATP molecules. This energy requirement may explain why the C_4 pathway evolved primarily in hot climates, where C_3 plants suffer most from photorespiration. In cool climates, the extra cost of concentrating CO_2 in bundle sheaths exceeds the negative costs associated with photorespiration in C_3 species (i.e., five ATP and three NADPH per oxygenation reaction).

It is often important that C_4 plants assimilate more atmospheric CO_2 per unit of water transpired than do C_3 plants. Details concerning WUE of C_3 and C_4 plants, how the ratio of carbon fixation to transpiration might be affected by environmental changes, and the implications for crop water consumption, were outlined in Chapter 4.

In addition to improved WUE, there is some evidence that C_4 crops also make more efficient use of nitrogen in photosynthesis. It was calculated that C_4 plants require only 13 to 42 percent of the rubisco found in C_3 plants to synthesize equivalent quantities of carbohydrate (Watling, Press, and Quick, 2000). This represents a significant nitrogen savings because of the relatively large amount of nitrogen contained in the rubisco in C_3 leaves.

Sucrose and Starch Synthesis

The two major products of the overall process of photosynthesis, in both C_3 and C_4 plants, are sucrose and starch. Sucrose (a disaccharide, $C_{12}H_{22}O_{11}$) is formed in the cytosol from the triose-P exported from chloroplasts. It is the main form of carbon translocated out of leaves (Chapter 6) and into other parts of a crop (e.g., roots, stems, tubers, and grain) as well as being an important substrate for respiration throughout the plant. Starch (a polysaccharide, $(C_6H_{12}O_6)_n$) is formed from triose-P inside chloroplasts, where large amounts can accumulate in immobile starch grains. Starch stored in chloroplasts is later (e.g., at night) broken down, resulting in production of sugars in the cytosol that can in turn be translocated out of the leaf or used within the leaf for respiration or growth.

The main factor that determines whether triose-P is used to produce sucrose or starch (and not to regenerate rubP) is the fraction of triose-P exported from chloroplasts into the surrounding cytosol. That export depends on a supply of P_i in the cytosol because P_i is exchanged for triose-P at the chloroplast membrane (antiporter). The key question is, how does metabolism and growth outside the leaf control the cytosolic $[P_i]$ and therefore relative production of sucrose that is available for immediate transport out of photosynthesizing cells (Herold, 1980)? Although the answer remains uncertain, it is clear that whole-plant metabolism affects processes in photosynthesizing leaves, as outlined in Chapter 6. Moreover, species differ in their relative production of sucrose and starch during photosynthesis.

Field-Scale Considerations of Photosynthesis

Measurements of photosynthesis in single leaves provide some understanding of inherent variation in photosynthesis; for example, when comparing carbon assimilation rates in different crop varieties

(Hall, 2001). By linking measurements of photosynthesis in single leaves with analyses of photosynthetic enzymes, stomatal function, and leaf anatomy, it is often possible to determine to what extent photosynthetic variation is explained by various leaf attributes. Leaf- or plant-level analysis of photosynthesis may also be of use for seedlings or young, widely spaced plants in a crop (i.e., before plants begin to shade their own leaves and before plants have grown large enough to begin influencing neighboring plants).

The value of the leaf-level approach to understanding crop photosynthesis declines as a crop matures. Within a more or less fully developed crop canopy, different leaves neither have equal access to light, nor do they synthesize the same amount of carbohydrate per photon of light absorbed. Leaves that develop in full sun, for instance, have higher rates of maximum photosynthesis, are typically thicker, and contain more rubisco compared to leaves that develop (and are adapted to function best) in shade. Although shade leaves have low photosynthetic capacity, they also have low maintenance respiration requirements, which enables them to maintain a positive carbon balance even at very low light levels (Figure 5.5).

The variation in development and functioning of leaves that occupy different locations within a canopy partially explains why individual leaves may become saturated or may even experience photodamage at high light levels whereas photosynthesis at the canopy level is typically not saturated even on the sunniest days (Figure 5.5). Transmission of light through a canopy, and reflection and scattering of light as it passes through and between adjacent plants, are important considerations that are difficult to account for when measuring photosynthesis of single leaves.

In some respects, limitations of the leaf-level approach for understanding carbon assimilation by crops are analogous to those explained earlier for crop water relations because water loss and CO_2 uptake are both linked to stomatal conductance (discussed in Chapter 4; "The Crop Canopy"). By conducting canopy-scale measurements throughout the day it is possible to account for canopy-scale feedbacks on stomatal function. Canopy-scale measurements of photosynthesis also account for the fact that leaves occupying different microenvironments differ from one another in important ways and that available PAR decreases from the top of the canopy to the bottom. Hence, some of the most important studies will include mea-

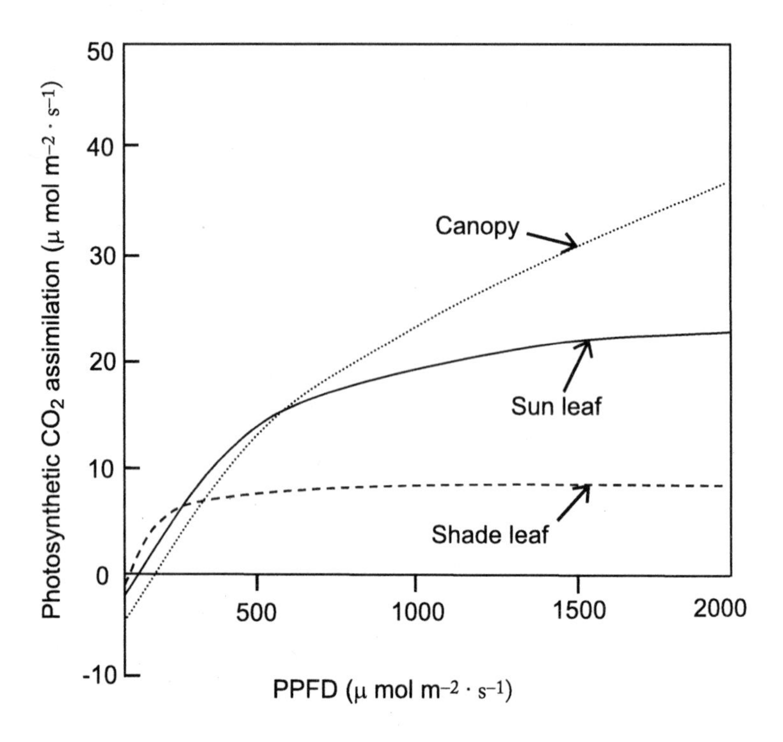

FIGURE 5.5. Photosynthesis of crop canopies (dotted line) and individual leaves adapted for full sun (solid line) or shade (dashed line) as a function of incident photosynthetic photon flux density (PPFD), i.e., light level. As a result of the complex, three-dimensional structure of crop canopies, in which different leaves are exposed to different PPFDs, crop canopies are typically not saturated with light even on the sunniest of days. The magnitude of the response to a given PPFD can vary considerably between different crop species.

surements of whole-canopy photosynthesis over multiple diurnal cycles rather than focusing on single leaves at single points of time (Hall, 2001).

Carbon acquisition is also affected by the rate of leaf-area development per unit of ground area and how long leaves remain phytosynthetically active once they are grown. Change in the longevity of leaves/canopies can affect photosynthesis integrated over a cropping season, which can in turn affect yield.

EFFECTS OF ENVIRONMENTAL CONDITIONS AND CHANGES ON PHOTOSYNTHESIS

Elevated Atmospheric Carbon Dioxide Concentration

Effects on C_3 Plants

Hundreds of experiments over the past 50 years have demonstrated, for nearly all C_3 crops, that photosynthesis increases as atmospheric [CO_2] is raised (Figure 5.4). Elevated [CO_2] (i.e., ~700 ppm) stimulates leaf-level photosynthesis on the order of 20 to 60 percent (Bowes, 1991, 1993; Drake, Gonzàlez-Meler, and Long, 1997), although this response varies considerably among crop species and with experimental methods (Table 5.1; Chapter 7).

Elevated atmospheric [CO_2] stimulates photosynthesis through several mechanisms: (1) rubisco is CO_2-limited at current atmospheric [CO_2] because it has such a low affinity for CO_2 (C_3 plants); (2) elevated [CO_2] leads to higher [CO_2]:[O_2] within chloroplasts, which reduces photorespiration (C_3 plants); (3) elevated [CO_2] appears to decrease the nonphotochemical dissipation of light energy within chloroplasts of well-fertilized plants, thereby alleviating photoinhibition (and boosting light-use efficiency and decreasing oxidative stress; see Chapter 3) (Hymus, Baker, and Long, 2001); and (4) elevated [CO_2] increases WUE and decreases water stress, which can contribute to higher rates of photosynthesis (C_3 and C_4 plants) over time. On the other hand, elevated [CO_2] reduces stomatal conductance (Figure 4.5), which places a limit on CO_2 transport from the atmosphere to chloroplasts, but the increase in [CO_2] per se usually more than fully compensates for this (Lawlor and Mitchell, 2000).

Elevated [CO_2] (i.e., about 700 ppm in many experiments) typically stimulates photosynthesis 50 percent or more over periods of minutes to hours, but over longer periods the magnitude of the stimulation can decline (Sims, Luo, and Seemann, 1998; Moore et al., 1999; Hymus, Baker, and Long, 2001; Gesch et al., 2002). Acclimation of photosynthesis to elevated [CO_2] has been attributed to several mechanisms. Perhaps the most widely held view is that acclimation is linked to carbohydrate accumulation and its effects on expression of genes encoding photosynthetic enzymes. Presumably, carbohydrate production in CO_2-enriched atmospheres often exceeds

TABLE 5.1. Proportional increases in photosynthesis of crops grown with CO_2-enrichment.

Species	Experimental unit	Ratio[a]	Experimental apparatus	Experimental [CO2]	Source
Wheat	canopy	1.40	field chambers	750	A
	canopy[b]	1.15	GC	680	B
	canopy	1.19	FACE	550	C
Faba bean	canopy	1.59	field chambers	750	A
Cotton	canopy	1.32	FACE	550	D
	canopy	1.25	SPAR	720	E
Rice	canopy	1.40	SPAR	500	F
	canopy	1.09	field chambers	650	G
Soybean	variable	1.23	variable	2× ambient	H
Sorghum	upper leaf	1.09	FACE	550	I
Wheat	upper leaf[c]	1.21	FACE	550	J
	flag leaf	1.26	FACE	550	K
	eighth leaf	1.69	FACE	550	K
45 species large rv	variable	1.58	variable	variable[d]	L
28 species small rv	variable	1.28	variable	variable	L
8 species high N	variable	1.57	variable	variable	L
8 species low N	variable	1.23	variable	variable	L

Sources: (A) Dijkstra et al. (1996); (B) Mitchell, Mitchell, and Lawler (2001); (C) Brooks et al. (2001); (D) Hileman et al. (1994); (E) Reddy, Hodges, and Kimball (2000); (F) Baker et al. (1996); (G) Sakai et al. (2001); (H) Allen and Boote (2000); (I) Brooks et al. (2001); (J) Garcia et al. (1998); (K) Osborne et al. (1998); (L) Drake, Gonzàlez-Meler, and Long (1997).

[a]Ratio represents photosynthesis rate of plants grown in high CO_2 relative to low CO_2 (unitless)

[b]Stands of wheat were grown in large boxes with 100 L rooting volume.

[c]This value represents the seasonal average of daily integrals of photosynthesis.

[d]Experimental CO_2 concentrations varied from 550 ppm to 1000 ppm, but most were ~700 ppm.

Abbreviations: FACE, free-air CO_2 enrichment; GC, growth chamber; SPAR, soil-plant-atmosphere research units.; rv, rooting volume was considered large (>10 L) or small (<10 L).

carbohydrate use by plants, and this imbalance between supply and use leads to a buildup of carbohydrates in the form of starch (chloroplasts) or sucrose (cytosol) in source leaves. Carbohydrate accumulation in turn either inhibits expression of genes coding for enzymes of photosynthesis or leads to inactivation of existing enzymes (Vu, Allen, and Bowes, 1983; Van Oosten and Besford, 1994; Paul and Pellny, 2003). For instance, the rubisco amount decreased on average 15 percent and activity was reduced 25 percent when plants were grown in CO_2-enriched atmospheres for prolonged periods of time (Drake, Gonzàlez-Meler, and Long, 1997). The concentrations of total soluble leaf proteins, carotenoids, and chlorophylls *a* and *b* also commonly decrease in plants grown in high [CO_2] (Vu et al., 1997; Surano et al., 1986; Polle et al., 1993; Pritchard et al., 2000).

If indeed photosynthetic acclimation is driven by accumulation of carbohydrates, then factors that limit sink activity, such as growing plants in small pots constraining roots (Arp, 1991) or growing plants with insufficient nutrients to support new growth, should exacerbate the acclimation response. This is precisely the case, as less photosynthetic acclimation occurs when plants are grown in large containers with plentiful nutrients compared to plants growing in small containers with limited nutrients (Drake, Gonzàlez-Meler, and Long, 1997). This may mean that much of the scientific literature on photosynthetic down-regulation in elevated [CO_2] is of limited significance to crops because they are not grown in pots, and many (though not all) crops are fertilized to reduce nutrient limitations.

Acclimation to high [CO_2], however, is probably not completely explained by down-regulation of photosynthetic genes in response to inadequate sink strength (Sims, Luo, and Seemann, 1998; Moore et al., 1999; Harmens et al., 2000). Photosynthetic acclimation to elevated [CO_2] has even been observed in plants with very large sink capacity, such as potato (Schapendonk et al., 2000). Effects of CO_2 enrichment on processing of carbohydrates in source leaves themselves can also lead to carbohydrate accumulation independent of sink activity. Starch accumulation, for instance, is sometimes a direct result of depressed photorespiration or may also follow from up-regulation of key enzymes in the pathway of starch synthesis, such as ADP-glucose pyrophosphorylase (Sims, Luo, and Seemann, 1998; Hendriks et al., 2003). Furthermore, a buildup of carbohydrates in source leaves could also follow from changes in export of triose-Ps from chloroplasts or from effects on sucrose synthesis, phloem loading, phloem transport, or phloem unloading. These possibilities have not been well explored, especially in crops.

Effects on C_4 Plants

Because the photosynthetic machinery of C_4 plants is saturated at relatively low concentrations of atmospheric CO_2, photosynthesis in C_4 crops such as maize, sugarcane, and sorghum is less affected by elevated atmospheric [CO_2] compared to C_3 plants (Figure 5.4). It was even suggested that rising [CO_2]s might decrease the efficiency of C_4 photosynthesis if the production of C_4 acids exceeds the capacity of rubisco to catalyze the carboxylation reaction. Under such conditions, a greater proportion of CO_2, generated via decarboxylation of C_4 acids within bundle-sheath cells, may leak out into the mesophyll (24 percent and 33 percent of CO_2 leaked out of bundle-sheath cells of sorghum grown at 350 and 700 ppm CO_2, respectively; Watling, Press, and Quick, 2000).

Despite predictions that C_4 photosynthesis should be relatively insensitive to CO_2-enrichment, C_4 crops often realize significant benefit from elevated [CO_2] (Cousins et al., 2001). There are several possible explanations for this. For example, C_4 leaves, before they have fully developed bundle-sheath anatomy, may largely function as C_3 leaves so elevated [CO_2] could stimulate early photosynthesis, and therefore growth, of C_4 leaves beyond what would be expected from the effect of elevated [CO_2] on C_4 photosynthesis (Poorter, Roumet, and Campbell, 1996; Cousins et al., 2001). Early stimulation of leaf growth would accelerate canopy development and consequently increase integrated seasonal PAR absorption.

Greater-than-expected productivity of C_4 crops in CO_2-enriched atmospheres might also be caused by increased WUE and alleviation of water stress. Stomates of C_4 plants are sensitive to atmospheric [CO_2]s in a fashion analogous to C_3 plants; as [CO_2] increases, stomatal conductance decreases, which decreases the amount of transpiration per molecule of CO_2 assimilated. This effect could improve photosynthesis in C_4 plants during periods of drought independent of any direct effect of atmospheric [CO_2] on photosynthetic carbon metabolism (Young and Long, 2000). Available evidence indicates that the stimulation of photosynthesis and growth in C_4 plants is usually linked to a decrease in water consumption.

Warming

The temperature at which photosynthesis is most rapid varies considerably from crop to crop, and therefore generalizations concerning the effects of global warming on photosynthesis itself are hard to make; but grouping crops according to the climates of their native habitats, seasonal "preferences," or photosynthetic pathways is a useful starting place. Photosynthesis of crops originating in cool climates is greater at lower temperature ranges compared to crops that have their geographical origins in warmer climates (see Table 8.1 for a list of crops adapted for cool and warm seasons/climates). Crops with the C_4 photosynthetic pathway typically have higher temperature optima than C_3 crops. Photosynthesis of wheat (C_3), for instance, functions best at temperatures between 15 and 30°C while photosynthesis of maize (C_4) is most rapid when temperatures are between 30 and 40°C (Figure 5.6).

Temperature preference of C_3 and C_4 plants can be explained in part by differences in quantum yield (amount of CO_2 assimilated per

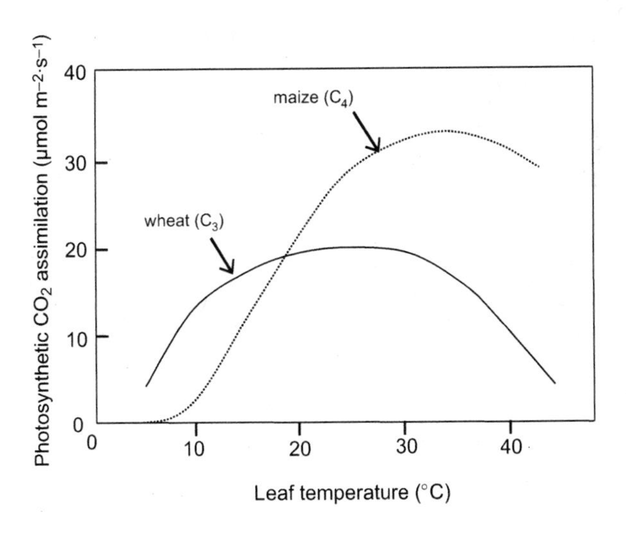

FIGURE 5.6. Leaf photosynthesis of wheat and maize as a function of temperature. *Source:* Adapted from Stone (2001).

photon absorbed) of C_3 and C_4 plants at high and low temperatures (Figure 5.7). At low temperatures, quantum yield for C_3 plants exceeds that for C_4 plants because of the extra metabolic costs (2 ATP) associated with transporting CO_2 from mesophyll cells into bundle-sheath cells in C_4 plants (Ehleringer and Björkman, 1977). At higher temperatures, however, photorespiratory costs in C_3 plants eventually exceed the metabolic costs of the CO_2-concentrating mechanisms in C_4 plants and C_4 photosynthesis becomes "more efficient." The superiority of C_4 photosynthesis may typically occur at and above 25°C (Young and Long, 2000).

As a consequence of differential effects of temperatures on quantum yields of C_3 and C_4 crops, C_4 crops could be at an advantage compared to C_3 plants as temperatures rise. The photosynthetic advantage granted C_4 plants by global warming is not likely to be as great as the photosynthetic advantage that C_3 crops could enjoy if at-

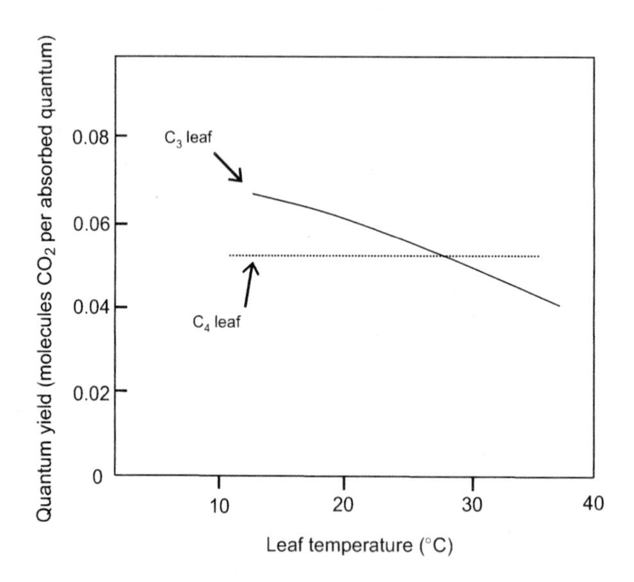

FIGURE 5.7. Quantum yield of individual leaves as a function of temperature for a C_3 and a C_4 species. The C_3 plant requires more light per molecule of CO_2 assimilated as temperature rises (i.e., quantum yield decreases) because photorespiration releases relatively more CO_2 (and consumes relatively more ATP and NADPH) as temperature increases. At lower temperatures, however, quantum yield efficiency is higher in C_3 plants compared to C_4. *Source:* Adapted from Ehleringer and Björkman (1977).

mospheric [CO_2] doubles, however. Interactions between warming and rising CO_2 will almost certainly be important for future photosynthesis.

When temperature exceeds 40°C, components of the photosynthetic apparatus are damaged and plants can suffer long-term, or even permanent, reduction in photosynthetic capacity (Berry and Björkman, 1980). Apparently, PSII within chloroplast thylakoids is particularly sensitive to high temperature (Al-Khatib and Paulsen, 1999), perhaps because high temperature inhibits thylakoid stacking (PSII is located primarily within the granal thylakoids) (Dekov, Tsonev, and Yordanov, 2000). Other components of the light reactions (i.e., PSI, cytochrome b_6f, and light-harvesting complexes) are less sensitive to high temperature. Changes in the activation state of rubisco have also been correlated with decreased photosynthesis at high temperatures (Law and Crafts-Brandner, 1999, 2001).

Many crops are capable of substantial photosynthetic acclimation to high temperatures when they are imposed gradually. This could be an important aspect of crop responses to warming considered over the course of a cropping season. Acclimation may be linked to protection of temperature-sensitive photosynthetic components by heat-shock proteins, altered expression of genes encoding components of the photosynthetic machinery, or through homeoviscous acclimation of chloroplast membrane systems (Chapter 3).

Interactive Effects of Warming and Rising [CO_2]

The effect of temperature on photosynthetic rate is greater at elevated [CO_2] (Berry and Björkman, 1980) (Figure 5.8). This is presumably related to a decrease in the ratio of photorespiration to photo-synthesis at high [CO_2] (Figure 5.9). Under ambient (nonsaturating) [CO_2] increased photorespiration negates a potential increase in photosynthesis that might be expected based on the Q_{10} of the carboxylation reaction, but at high [CO_2] photorespiration is inhibited and the temperature response of photosynthesis increases. Based on a biochemical model, a 300 ppm increase in [CO_2] causes a 5°C increase in the temperature optima of C_3 photosynthesis (Long, 1991).

Therefore, at least theoretically, the stimulation of photosynthesis by elevated atmospheric [CO_2] will be enhanced by warming. Fur-

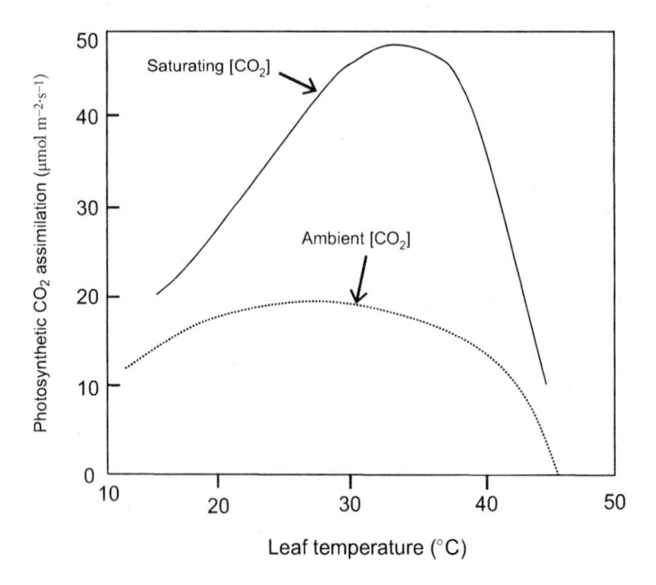

FIGURE 5.8. Photosynthesis as a function of temperature for plants grown at ambient and saturating concentrations of CO_2. *Source:* Adapted from Berry and Björkman (1980).

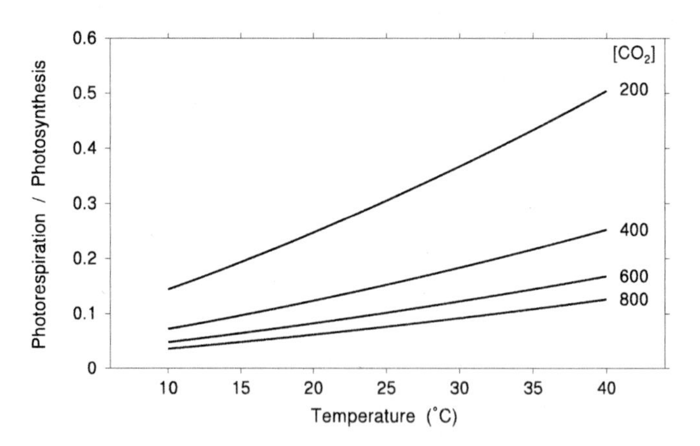

FIGURE 5.9. Ratio of photorespiratory CO_2 release (i.e., glycine decarboxylation in mitochondria; see Figure 5.2) to photosynthetic CO_2 uptake (i.e., ribulose 1,5-P_2 carboxylation in chloroplasts; see Figure 5.1) in C_3 leaves as a function of temperature and atmospheric $[CO_2]$ (based on Brooks and Farquhar, 1985). In this figure it is assumed that the $[CO_2]$ in substomatal cavities is 70 percent of the value indicated for atmospheric $[CO_2]$, a useful rule of thumb for many healthy crops in moderate to high light.

thermore, in elevated [CO_2], leaves should require a smaller pool of rubisco to maintain the same rate of photosynthesis as temperature rises (Drake, Gonzàlez-Meler, and Long, 1997). This effect could increase the efficiency of crop nitrogen use, leading to production of more canopy leaf area. Such an effect could further boost canopy carbon assimilation at high temperature and elevated [CO_2], leading to higher yields.

Unfortunately, theory and reality often do not meet. In the words of Morison and Lawlor (1999):

> Although there are sound theoretical reasons for expecting a larger stimulation of net CO_2 assimilation rates by increasing [CO_2] at higher temperatures, this does not necessarily mean that the pattern of biomass and yield responses to increasing [CO_2] and temperature is determined by this response. (p. 659)

In their review, these authors concluded that: "there is little unequivocal evidence for large differences in response to [CO_2] at different temperatures, as studies are confounded by the different responses of species adapted and acclimated to different temperatures, and the interspecific differences in growth form and development patterns" (p. 659). We take a similar view: leaf-level photosynthetic responses to short-term changes in temperature and/or [CO_2] can be negated by factors operating at field scales integrated over entire cropping seasons.

Elevated Atmospheric Ozone Concentration

Elevated atmospheric O_3 reduces photosynthesis significantly in sensitive crops. A recent review of 53 studies of soybean, for instance, reported that leaf-level photosynthesis rates were reduced 20 percent on average by exposure to 70 ppb O_3 compared to carbon-filtered air (Morgan, Ainsworth, and Long, 2003). Similarly, photosynthesis of barley exposed to 180 ppb O_3 was lowered by 17 percent compared to 10 ppb (Plazek, Rapacz, and Skoczowski, 2000). These effects are normally attributed to decreased stomatal conductance (Chapter 4, Figure 4.9; Figure 5.10), direct oxidative damage to chloroplasts, premature leaf senescence, or a combination of these three mechanisms (Pell, Schlagnhaufer, and Arteca, 1997; Ojanperä, Pätsikkä, and Yläranta, 1998; Zheng, Shimizu, and Barnes, 2002).

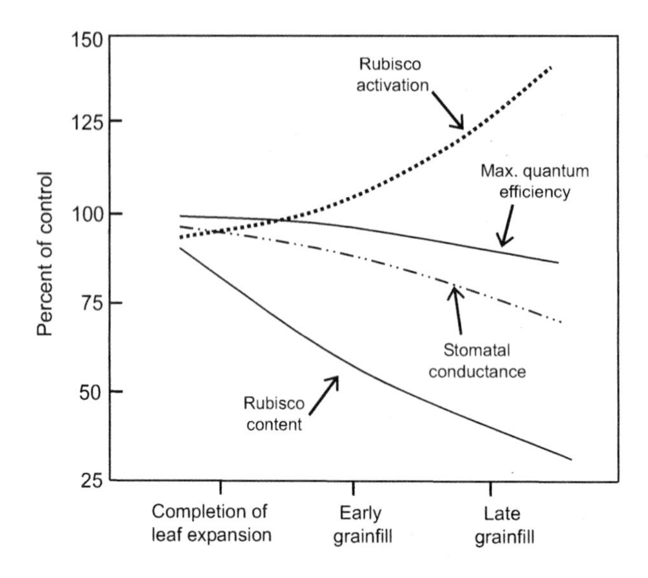

FIGURE 5.10. Change in photosynthetic properties of wheat flag leaves over 40 days of exposure to 70 ppb O_3 compared to O_3-free air. Rubisco activation is the percentage of the rubisco present that is in its active form. Maximum quantum efficiency is the ratio of photosynthesis per unit of absorbed light in a low-light environment. *Sources:* Long and Naidu (2002).

Studies of the effects of elevated [O_3] on crop-canopy photosynthesis are rare, but reductions at the field scale are probably greater than leaf-level responses due to combined effects of less CO_2 assimilated per unit leaf area, reduction in total leaf area, and shorter leaf-area durations (premature senescence). The relative contributions that stomatal and nonstomatal changes resulting from elevated [O_3] make to photosynthetic reductions is not known and probably varies among crop species and with the timing or duration of ozone exposure (Meyer et al., 1997; Carrasco-Rodriguez and del Valle-Tascon, 2001). Whether effects of elevated [O_3] on stomatal conductance can be attributed to direct damage to the stomatal complex itself or indirect effects mediated through changes in photosynthesis (and thus intercellular [CO_2]), is unknown.

The direct effects of elevated [O_3] on photosynthesis have been linked to decreased amounts of rubisco in a number of crops, including pea, radish, potato, clover, and oat (reviewed in Zheng, Shimizu, and Barnes, 2002). Elevated [O_3] has also been shown to cause breakdown of photosynthetic pigments, including chlorophyll (Meyer et al., 2000; Donnelly et al., 2001a; Morgan, Ainsworth, and Long, 2003) which can lead to a decrease in the efficiency of light harvesting and thylakoid electron transport, and thus availability of NADPH and ATP to drive the dark reactions of photosynthesis (Krupa et al., 2000). Apparently, O_3 (or more likely its toxic derivatives) directly damages photosynthetic enzymes and pigments and/or the genes which code for them. Alternatively, O_3 often leads to accumulation of carbohydrates in leaf tissue, which in turn could suppress photosynthesis (see also Chapter 6). A third possibility is that O_3 leads to up-regulation of the hormone ethylene, which in turn influences expression of photosynthetic genes, including those that code for chlorophyll *a/b* protein, glyceraldehyde-3-phosphate dehydrogenase, and the small subunit of rubisco (Figure 5.3; Glick et al., 1995; Pell, Schlagnhaufer, and Arteca, 1997). In many studies, observed drops in photosynthesis of crops fumigated with O_3 are probably associated with premature leaf senescence (see Chapter 3), not necessarily specific effects on the photosynthetic process itself.

Rising [CO_2] has the potential to alleviate negative effects of rising [O_3] on photosynthesis by reducing stomatal conductance and limiting the amount of O_3 taken up into mesophyll tissue. Elevated [CO_2] increased leaf longevity in both potato (Donnelly et al., 2001a) and wheat (Mulholland et al., 1997a; Ommen et al., 1999) exposed to high [O_3] and also mitigated against chlorophyll breakdown. Although elevated [CO_2] can lessen the negative effects of O_3 on photosynthesis, it probably will not eliminate them completely in the future if both [O_3] and [CO_2] increase at recent rates.

Salinity

Salinity stress has little effect on leaf-level photosynthesis (Lazof and Bernstein, 1999). Reduced plant growth generally occurs independently of any change in photosynthesis when plants are subjected to salt stress (see Chapter 8). In fact, plants grown in saline substrates probably assimilate more CO_2 than they can use, and carbohydrates accumulate as a result (Lazof and Bernstein, 1999).

Photosynthesis has been reported to be suppressed by salinity (for example, Seemann and Sharkey, 1986; Plaut, Grieve, and Maas, 1990; Khan, Silberbush, and Lips, 1994), but in these cases, it is likely that reduced growth *caused* the decrease in photosynthesis, rather than reduced photosynthesis limiting growth (Munns, 1993). A likely chain of events leading up to reduced photosynthesis in salt-stressed crops is: salt inhibits cell expansion in growing sinks → carbohydrate use decreases → starch and soluble sugars accumulate → photosynthesis is down-regulated through feedback inhibition. In addition, reduced growth caused by salinity will limit leaf area and therefore PAR absorption and whole-crop photosynthesis. Premature leaf senescence caused by salt toxicity can further reduce seasonal integrated photosynthesis.

RESPIRATION

There are several definitions of respiration (sometimes called "dark" respiration) in plants. One physiologically based definition is CO_2 release (in the dark for whole plants or photosynthetic organs). Another definition is O_2 uptake (again, in the dark for whole plants or photosynthetic organs). The relationship between these two is called the respiratory quotient (RQ), defined as CO_2 release/O_2 uptake. Most studies of crop respiration are physiological, i.e., they involve measurements of CO_2 release from the crop or crop organs. Unfortunately, respiration cannot be easily separated from concurrent photosynthesis during the day, and in any case, the database on crop respiration is limited.

Respiration can also be defined in biochemical terms. Perhaps the most useful biochemical definition is the activity of glycolysis, the oxidative pentose phosphate pathway (OPPP), the tricarboxylic acid (TCA) cycle, and mitochondrial electron transport and related oxidative phosphorylation. These biochemical reactions and pathways release CO_2 at several points (i.e., in the early stages of the OPPP and in the TCA cycle) and consume O_2 (i.e., they reduce O_2 in the formation of H_2O) as terminal steps in mitochondrial electron transport. Other nonphotosynthetic and nonphotorespiratory reactions also release CO_2 and/or consume O_2 (e.g., some biosynthetic pathways), so a strict correspondence between the physiological and biochemical definitions of respiration does not exist. Nonetheless, most nonphotosynthetic and nonphotorespiratory CO_2 and O_2 exchanges in crops are associated with the biochemical reactions of respiration.

A summary equation for respiration can be written as:

$$\{CH_2O\} + O_2 \longrightarrow CO_2 + H_2O + x\text{ATP} + y\text{NAD(P)H} + \text{heat}$$

As with the summary equation for photosynthesis, which is essentially the reverse of the present equation, this summary equation simplifies dozens of chemical reactions and intracellular transport processes.

The function of respiration in crops (defined biochemically) is to produce usable energy in the form of ATP, reducing power in the forms of NADH and/or NADPH (i.e., NAD[P]H), and carbon skeleton intermediates used as the building blocks of biosynthesis (i.e., the foundation of dry-matter growth). These products of respiration (i.e., ATP, NAD[P]H, and carbon skeletons) are also needed for maintenance of existing dry matter, various intra- and intercellular transport processes, and assimilation of nutrients from the soil solution, especially nitrogen.

The percentage of the carbon assimilation during photosynthesis that is later released as CO_2 through the processes of respiration ranges from 40 to 60 percent in crops (Table 5.2). This percentage might be even higher in tropical cropping regions because of high rates of respiration during warm nights. Whereas photosynthesis occurs predominantly in green leaf tissue, all plant tissues respire, and they do so 24 hours per day. Respiration is therefore important to the overall carbon balance of a crop, and it influences the formation of yield. Effects of environmental changes on respiration could be important to yield for a number of reasons, including changes in the fraction of assimilate available for growth of seeds or tubers.

Respiration is often divided into two functional components: growth respiration and maintenance respiration. Physiologically, growth respiration is the CO_2 generated by growth processes and may consume roughly one-quarter of total photosynthate produced in some crops (Amthor, 1989). The ratio of growth respiration to growth seems to be largely independent of environmental conditions, unless the environment significantly influences the biochemical composition of growing tissues (see "Biosynthesis" section). The amount of

TABLE 5.2. Estimated fraction of C assimilated in photosynthesis that is eventually released as CO_2 through respiration.

Crop	Fraction of C assimilated that is later released through respiration[a]	Source
Alfalfa	0.33-0.49	Thomas and Hill (1949)
Barley	0.48	Biscoe, Scott, and Monteith (1975)
Bean	0.35-0.62	Tanaka and Osaki (1983)
Maize	0.25-0.47	Tanaka and Osaki (1983)
	0.38	Yamaguchi (1978)
Potato	0.28-0.40	Tanaka and Osaki (1983)
	0.26-0.68	Gawronska et al. (1984)
Rice	0.17-0.57	Tanaka and Osaki (1983)
	0.45	Yamaguchi (1978)
	0.30-0.97	Tanaka and Yamaguchi (1968)
	0.34-0.57	Lian and Tanaka (1967)
Soybean	0.45-0.70	Tanaka and Osaki (1983)
	0.57	Yamaguchi (1978)
Sugarbeet	0.29-0.33	Thomas and Hill (1949)
	0.25-0.40	Tanaka and Osaki (1983)
Wheat	0.32-0.37	Tanaka and Osaki (1983)
	0.30-0.40	Austin et al. (1977)
	0.50-0.63	Ruckenbauer (1975)

Source: Table adapted from Amthor (1989).

Notes: All estimates of respiratory C released are from field-grown crops. It should be noted that values are from crops at various stages of development and that the proportion of C assimilated that is released through respiration increases as a crop matures.

[a]mol C (respired) per mol C (assimilated)

growth respiration is expected to change in proportion to any changes in growth caused by environmental changes.

Maintenance respiration is the CO_2 generated in the processes of maintaining plant cells in a functional state and is proportional to the amount of plant mass times its specific cost of upkeep. Maintenance processes include (1) replacement of RNA, labile enzymes, and membrane lipids that turn over during normal metabolism or as a result of environmental changes, (2) establishment and maintenance of favorable ion gradients between different plant compartments, and (3) cellular processes involved in alleviating environmental stress (Penning de Vries, 1975; Amthor, 1994a,b). Respiration supporting maintenance processes can account for more than 50 percent of total respiration under some conditions, and maintenance processes (and therefore maintenance respiration) are expected to be affected by many environmental conditions such as temperature. Changes in plant size brought about by environmental changes are expected to cause proportional changes in maintenance respiration. Furthermore, environmental changes might speed up or slow down crop development, which in turn could influence the proportion of photoassimilates released as CO_2 during respiration. (Early in crop development, the majority of CO_2 released is associated with growth respiration. As plants mature and relative growth rates slow, more CO_2 released in respiration is attributable to maintenance respiration.)

Respiration in crops is presumably controlled by the demand for (i.e., use of) its products, mainly ATP and carbon skeletons. The availability of carbohydrates (substrates of respiration and growth) and the amount of respiratory enzymes place limits on respiratory rate, but it is the regeneration of ADP (from ATP) and use of carbon skeletons (i.e., elimination of negative feedbacks on respiration imposed by the build-up of respiratory products) by the growth and maintenance processes that control respiration rate in the short term.

Whether environmental changes alter respiration rate is not as important a question, in most cases, as is how environmental changes might alter the relationships (i.e., stoichiometries) between respiration, growth, transport, and maintenance processes. For instance, if environmental change stimulates respiration as a result of faster growth (i.e., faster growth respiration), and if that growth stimulation leads to higher yields, then an increase in respiration is necessary and desirable. However, if an environmental change increases respiration as a result of greater maintenance costs (i.e., faster maintenance respiration) in the absence of more growth, then that increase in respiration would be undesirable to the extent that it represented a drain on substrates that might otherwise be used for growth. It is important that production of some valuable crop constituents such as proteins and lipids are metabolically more expensive to produce (and therefore require more respiration) compared to other biochemical components such as reserve and structural carbohydrates.

EFFECTS OF ENVIRONMENTAL CONDITIONS AND CHANGES ON RESPIRATION

Elevated Atmospheric Carbon Dioxide Concentration

Direct Effects on Respiration

There are several reports that elevated [CO_2] has a direct, inhibitory effect on a plant's respiration rate. A *direct* effect of CO_2 on respiration is an instantaneous effect of CO_2 per se rather than a respiratory response to an effect of [CO_2] on another process, such as photosynthesis or growth. In many cases, a doubling of [CO_2] in the dark was reported to directly inhibit leaf, shoot, or whole-plant respiration rates by 10 to 50 percent (e.g., Amthor, 1997; Reuveni, Gale, and Zeroni, 1997). Other experimental evidence has been contradictory, however, and it may be that elevated [CO_2] has no, or only a limited, direct effect on respiration (Table 5.3). For example, in a recent study involving over 600 measurements, including soybean and maize, no direct effect of [CO_2] on respiration was observed (Davey et al., 2004; and see also Bouma et al., 1997; Amthor, 2000a; Amthor et al., 2001; Jahnke, 2001; Bruhn, Mikkelsen, and Atkin, 2002; Burton and Pregitzer, 2002; Jahnke and Krewitt, 2002).

If elevated [CO_2] does directly inhibit respiration (and this is presently unclear), several effects on growth and yield might result. If the elevated [CO_2] mainly affected an inefficient or unnecessary component of respiration, the reduced rate of respiration could result in a savings of assimilate that could then be used for additional useful components of respiration and for additional growth. In that case, a reduced respiration rate would benefit the crop. Such an effect of elevated [CO_2] would be counter to notions that respiration is under tight control by the use of its products, or it would mean that the use of some respiratory products is for unnecessary metabolism.

An alternative outcome would result if elevated [CO_2] directly inhibited a useful component of respiration (or a useful component of metabolism using the products of respiration). In that case, elevated

TABLE 5.3. Some reported short-term (i.e., direct) effects of [CO_2] on respiration rate in crop species.

Crop and organ	[CO_2]s used (ppm)	Effect of [CO_2]	Source
Alfalfa, leaf	360, 730	ne, +	A
root or shoot	50, 350, 750, 950	–	B
Bean, leaf	200-4200	ne	C
Castor bean, leaf	100-800	ne	D
Clover (white), leaf	340, 680	ne	E
Lettuce, seedling	0, 1000	–	F
Maize, leaf	0, 310	–	G
Pea, shoot	300, 520	+	H
root + nodule	0, 300, 3000	ne	I
Rice, whole crop	350-700	–	J
Ryegrass (perennial)	340, 680	ne	K
Soybean, leaf	350, 750	ne, –	L
	500, 1000	ne	M
	490, 850	–	N
	350, 700	–	O
	10, 800	ne	D
	220, 370, 700, 1400	–	P
plant + soil	350/370, 550, 700/740	ne, –	L,Q
Sorghum, leaf	500, 1000	ne	M
Tobacco, leaf	0-1200	ne	R
	350, 2000	ne	S
Tomato, leaf	350, 700	–	L,T
plant + soil	350, 700	–	L
Wheat, ear	75, 176, 282	ne	U
	350, 1000	–	V
vegetative shoot	400, 600	ne	W
plant + soil	370, 710	ne	X
etiolated seedling	0, 660	–	Y

Sources: (A) Ziska and Bunce (1994); (B) Reuveni and Gale (1985); (C) Jahnke (2001); (D) Amthor et al. (2001); (E) Ryle et al. (1992); (F) Reuveni, Gale, and Mayer (1993); (G) Cornic and Jarvis (1972); (H) Hellmuth (1971), measure-

TABLE 5.3 *(continued)*

ments at 22 and 27 °C; (I) Mahon (1979); (J) Baker et al. (2000); (K) Ryle, Powell, and Tewson (1992); (L) Bunce (1990); (M) Byrd (1992), from drawn curves; (N) Thomas and Griffin (1994); (O) Bunce (1995a); (P) Bunce (2002); (Q) Bunce and Ziska (1996); (R) Ruuska et al. (1998); (S) Jahnke and Krewitt (2002); (T) Van Oosten, Wilkins, and Besford (1995); (U) Stoy (1965); (V) Reuveni and Bugbee (1997); (W) Amthor (2001); (X) Gifford (1995), after 72 h in dark; (Y) Nátr, Driscoll, and Lawlor (1996).

Symbols: –, inhibition by higher [CO_2]; ne, no effect of [CO_2]; +, stimulation by higher [CO_2].

[CO_2] might slow processes responsible for crop growth and health such as biosynthesis, maintenance, transport, or nutrient assimilation. For example, Bunce (2002) reported that the rate of nighttime translocation of assimilates out of soybean leaves was inhibited to the same degree that the respiration rate in those leaves was inhibited. This might lead to reduced seed growth if the assimilate supply to pods was significantly reduced.

Indirect Effects on Respiration

It is relatively easy to speculate on how (and why) elevated atmospheric [CO_2] should affect the respiration rate independently of any direct effects of CO_2 on respiration (Amthor, 1991). It is well-known that elevated CO_2 enhances photosynthesis and plant growth (at least in C_3 plants). Increased photosynthesis and growth also stimulate translocation. Elevated [CO_2] should, therefore, result in greater whole-plant respiration supporting growth and translocation as well as respiration supporting ion uptake and nitrogen assimilation (assuming that bigger plants contain more minerals and proteins, which they usually do). The resulting increase in plant size should in turn stimulate whole-plant maintenance respiration. Finally, elevated [CO_2] often results in a higher proportion of nonstructural carbohydrates (i.e., reserve materials), and this might enhance respiration associated with wastage (e.g., Azcón-Bieto and Osmond, 1983; Tjoelker, Reich, and Oleksyn, 1999)—that is, "substrate-induced respiration" of Warren Wilson (1967)—though it must be kept in mind that elevated nonstructural carbohydrate concentrations in source leaves may also stimulate respiration through increased phloem loading and translocation. Thus, because elevated [CO_2] stimulates photosynthesis, translocation, growth, and nonstructural carbohydrate levels, it is expected that rising atmospheric [CO_2] will increase whole-plant respiration, and there is evidence for this response in elevated-CO_2 experiments (Amthor, 1997; and see Davey et al., 2004).

In addition to increased growth, elevated CO_2 can also cause lower protein concentrations, perhaps in part through "dilution" by increased nonstructural carbohydrate levels. This response might be expected to reduce both growth respiration and maintenance respiration per unit biomass, though not necessarily respiration on a whole-plant basis, and there is evidence supporting these responses in several experiments (Amthor, 1997).

Thus, elevated atmospheric [CO_2] is expected to affect respiration to the extent that it alters: (1) rates of processes supported by respiration; (2) stoichiometries between respiration and the processes it supports; and (3) rates of futile cycling, alternative pathway activity, and other forms of wastage (Amthor, 2000b). Unfortunately, the experimental and observational database is limited, so these generalizations are mainly based on simple correlations. There are too few simultaneous measurements of respiration and the processes it supports to draw firm conclusions or explanations.

The expected effects of elevated [CO_2] on respiration brought about through changes in photosynthesis, translocation, growth, plant size, and/or plant composition are termed "indirect" (Amthor, 1997) because the same respiratory responses would be expected if any other environmental factor (e.g., temperature, nutrient availability) caused the same changes in photosynthesis, translocation, growth, plant size, and/or plant composition.

Elevated Atmospheric Ozone Concentration

Direct Effects on Respiration

If O_3 (or its products) damages respiratory enzymes or mitochondrial membranes, it may alter the rate of respiration. In most cases, this would probably slow respiratory reactions, but if damage to mitochondrial membranes reduced or eliminated effective coupling of electron transport and oxidative phosphorylation and/or reduction of O_2, the rate of respiration might increase, albeit with a loss of ADP phosphorylation per CO_2 released. Effects of O_3 (or its products) on respiratory enzymes or membranes can be called *direct effects* of O_3 on respiration.

High levels of O_3 (e.g., greater than 200 ppb for more than an hour), which were commonly used in laboratory experiments in the 1960s and 1970s, can cause extensive damage to leaf cells, and such damage almost certainly includes direct effects on respiration. Whether more modest levels of O_3 directly affect respiration is unclear—available experimental data have little to say about direct effects of modest levels of O_3 on respiration—but even more modest O_3 levels may damage cells and probably directly affect respiration, though such effects may only slightly alter respiration rate.

Indirect Effects on Respiration

Because elevated atmospheric [O_3] can slow photosynthesis in crops, elevated [O_3] is likely to indirectly affect respiration rate (*sensu* the direct effect of O_3 is on some nonrespiratory process, but that nonrespiratory process in turn affects respiration rate). A slowing of CO_2 assimilation in leaves will typically slow the use of carbohydrates in respiration and the processes supported by respiration. This might be expressed as a reduced respiration rate in leaves or as a reduced respiration rate in other organs dependent on leaves for a supply of carbohydrate (in addition to reducing photosynthesis, elevated [O_3] may more directly interfere with translocation of carbohydrates from leaves to other organs [Chapter 6]). Such a slowing of respiration would be a symptom of, rather than a cause of, altered carbohydrate metabolism resulting from O_3 pollution. For example, Edwards (1991) observed slower respiration in roots of loblolly pine seedlings exposed to twice-ambient [O_3]; a similar indirect effect of elevated [O_3] on root respiration might occur in crops as well.

Elevated atmospheric [O_3] might also indirectly stimulate respiration rate, at least in leaves. A main effect of O_3 (and its products) is disruption of membrane structure (Chapter 3), and it might also damage other macromolecules. If such damage is not lethal to a cell, that cell may respond by repairing or replacing the damaged structure(s). That repair or replacement would elicit respiration for provision of carbon skeletons, reductant, and/or ATP. Additional active transport would also be needed to reestablish metabolite gradients across repaired membranes. These repair activities can be interpreted as an increase in maintenance processes and associated maintenance respiration. Such repair activities can include the synthesis of new lignin and related compounds (Dizengremel, 2001) which would entail additional respiration. Experiments have indicated increased leaf respiration in response to elevated [O_3], and those increases in respiration rate can dissipate if the [O_3] is reduced (Amthor, 1994a).

As with other effects of environmental change on respiration, it will be necessary to quantify concomitant changes in respiration and the processes supported by respiration (e.g., translocation, growth, and maintenance) before clear explanations of effects of increasing atmospheric [O_3] on respiration will be possible. Little quantification of such changes has occurred for crops.

Warming

A short-term (seconds to hours) temperature increase (over the physiologically relevant range) stimulates respiration rate, often with a Q_{10} of about 2.0 to 2.5, but over the long term (days to years), respiration may acclimate and/or adapt to temperature change (e.g., Amthor, 1994a). Short-term changes in temperature probably affect respiration mainly through kinetic effects on the processes using respiratory products. Whether, and to what extent, processes supported by respiration acclimate and adapt to temperature probably determines the effects of long-term temperature change on respiration. That is, in the long term, temperature probably affects respiration through its effects on growth and maintenance processes and developmental state rather than through changes in respiratory capacity or kinetics per se, though respiratory capacity may also be affected by long-term temperature change. The study by Marcelis and Baan Hofman-Eijer (1995) indicated that cucumber fruit maintenance respiration did not acclimate to temperature (and growth respiration was independent of temperature in those studies), but there are too few data available to make generalizations about temperature acclimation (if any) of maintenance respiration in crops.

Because of acclimation and/or adaptation, short-term responses of respiration to temperature need not reflect long-term responses. Stated another way, the "long-term Q_{10}" of respiration will generally be smaller than the "short-term Q_{10}" because of some degree of acclimation and/or adaptation. Perhaps the most important issue is how growth will respond to warming. If warming enhances growth and plant size (for whatever reasons), it is likely that both growth respiration and maintenance respiration will be enhanced as well.

In the end, understanding the effects of long-term warming on respiration will depend on knowledge of how warming affects: (1) rates of processes that require respiration as a source of C-skeletons, ATP, and/or NAD(P)H; (2) specific respiratory costs of those processes; and (3) the amount of ATP produced per unit of substrate respired and the extent of any wastage respiration. Unfortunately, such knowledge is presently limited.

Salinity

Salinity generally increases respiration when it is expressed on the basis of a unit of leaf, stem, or root tissue. For instance, it was calculated that a root growing in a medium containing 500 $mol \cdot m^{-3}$ NaCl (~50 $dS \cdot m^{-1}$), would require 50 to 70 percent more ATP compared to a root growing at the same rate in a nonsaline soil (Munns and Termaat, 1986). Furthermore, the halophytes *Salicornia fructosa* and *Suaeda maritama* growing in saline soils would expend 10 to 20 percent of their total ATP, maintaining cellular ionic balances through compartmentation and osmoregulation (Yeo, 1983). In an example more relevant for agriculture, sorghum respiration was increased 6 percent when grown in media with an osmotic potential of –0.56 MPa (~15.6 $dS \cdot m^{-1}$; McCree, 1986). Increased respiration is driven by extra metabolic costs associated with compartmentation of Na^+ and Cl^- in addition to production of organic osmotica such as proline, glycine-betaine, pinitol, mannitol, and sorbitol (reviewed in Volkmar, Hu, and Steppuhn, 1998).

While maintenance needs almost invariably increase in salt-stressed plants, growth respiration decreases in proportion to reductions in biomass accumulation (Chapter 8). Respiration at the whole-plant level may therefore decline in response to salt stress because growth respiration decreases more than maintenance respiration increases (McCree, 1986).

BIOSYNTHESIS

Crop biomass is largely composed of fractions associated with a few classes of organic macromolecules: carbohydrates, proteins, lipids, lignins, and organic acids, in addition to minerals. The "cost" of producing these fractions can vary among compounds within a class, but even more so between classes. Quantifying the cost (assimilate requirement) to produce a unit of each major fraction (class of compound) of biomass was pioneered by Penning de Vries, Brunsting, and van Laar (1974). It has been reevaluated and updated several times (e.g., Zerihun, McKenzie, and Morton [1998] for protein; Amthor [2003] for lignin).

It takes more assimilate to construct a unit of dry matter containing a high concentration of protein and/or lipid compared to a low concentration of protein and lipid (Penning de Vries, Brunsting, and van Laav, 1974; Table 5.4). Hence, it is the composition of a plant or tissue, not just its mass, that dictates the assimilate requirements for growth. In terms of the energy content of a crop, there is less difference in the amount of photosynthate that is used to grow crops of different composition because the energy content of proteins and lipids is higher than that of carbohydrates, for example.

Effects of environmental changes on crop chemical composition are treated elsewhere in more detail (mineral nutrient content, Chapter 7, Table 7.3; composition of grain, Chapter 9; production of sec-

TABLE 5.4. Theoretical growth costs of various crop organs.

Crop organ type	**Organ growth possible (in grams carbon) per gram carbon available for growth**
Tubers and beets (mainly carbohydrates)	0.88-0.89
Low-lipid grain (wheat, rice, sorghum, maize, millet, chickpea, field bean, cowpea)	0.83-0.86
High-lipid organs (sunflower seed, soybean seed, cotton seed, coconut, peanut)	0.70-0.76

Source: Amthor (2000b).

Notes: These values exclude both maintenance and transport costs

ondary metabolites, Chapter 10), but as an example, elevated atmospheric [CO_2] can increase the carbohydrate concentration of plant tissues at the expense of proteins and lipids. In such a case, more biomass can be grown from a unit of assimilate. Such changes are probably too small, however, to greatly alter the construction costs of those tissues; estimates based on sporadic data indicate that 700 ppm CO_2 may decrease construction costs of C_3 plants by 3 percent (or less) compared to 350 ppm CO_2 (Poorter et al., 1997; Amthor et al., 1994; Amthor, 1997). Warming, on the other hand, tends to increase the proportion of protein relative to carbohydrate, and this effect might be associated with increased construction costs. The same could be said for crops exposed to high [O_3] that also sometimes increase (grain) protein content. Whether warming, salinization, or elevated [O_3] will influence crop composition enough to significantly alter construction cost has not been well studied, but effects of environmental change on construction cost are likely to be less important than effects on assimilation, maintenance costs, and growth rate, so construction cost changes (if any) caused by environmental change will probably be of limited significance to yield.

SUMMARY

Carbon derived from CO_2 in the atmosphere is the key substrate of crop growth. In the process of photosynthesis, pigments within chloroplasts in leaf mesophyll cells absorb PAR and use some of the associated energy to produce NADPH (from $NADP^+$) and ATP (from ADP and P_2). The energy stored within the bonds of these molecules is used to assimilate CO_2.

Two types of photosynthesis occur in crops, the C_3 system (wheat, rice, soybean, potato, bean, sugarbeet, cotton) and the C_4 system (maize, sorghum, sugarcane). The C_3 system includes photorespiration (which results in loss of some carbon as CO_2 nearly immediately after it is assimilated) and is sensitive to atmospheric [CO_2]. Photorespiration is temperature sensitive, such that oxygenation is relatively more rapid than carboxylation at high temperature. The C_4 system concentrates CO_2 near the enzyme rubisco, which effectively eliminates photorespiration and also reduces the sensitivity of photosynthesis to atmospheric [CO_2]. The final product of both C_3 and C_4 photosynthesis is triose phosphates, three-carbon compounds that are later converted into more complex carbohydrates such as sucrose and starch.

Elevated atmospheric [CO_2] generally enhances photosynthesis in C_3 crops and can enhance photosynthesis in C_4 crops over the longer term by enhancing WUE. Growth in high [CO_2], however, often leads to starch and soluble sugar accumulation in the leaves of C_3 crops. An abundance of sugar and starch has been linked to down-regulation of photosynthetic rates through repression of genes encoding photosynthetic enzymes. Furthermore, growth of crops in high [CO_2] also causes partial stomatal closure, which reduces the potential for higher photosynthesis in CO_2-enriched environments. These feedbacks notwithstanding, 700, compared to 360, ppm CO_2 increases photosynthetic rates from 20 to 60 percent in most species.

Elevated atmospheric [O_3] can reduce photosynthesis by several mechanisms, including decreased stomatal conductance, direct oxidative damage to chloroplasts, and/or premature leaf senescence. Future research will be required to sort out stomatal versus nonstomatal effects of O_3 pollution on photosynthesis.

Effects of warming on photosynthesis are complicated. Photosynthetic response to warming depends on the climatic conditions to which a crop is adapted and its photosynthetic pathway. Because photorespiration in C_3 crops increases as temperature increases, C_3 crops generally have lower temperature optima for photosynthesis compared to C_4 crops. Therefore, warming could favor C_4 crops. The potential for photosynthetic acclimation to higher temperatures in crops the effects of long-term warming on seasonal (canopy) photosynthesis remain poorly understood.

Salinity reduces photosynthesis, but this is an indirect effect. Salinity directly inhibits growth; slower growth causes carbohydrates to accumulate, which in turn leads to lower rates of leaf-level photosynthesis. At the canopy level, photosynthesis integrated over an entire cropping season declines with salinity as a result of less leaf area and premature leaf senescence (shorter leaf-area duration).

Respiration breaks down assimilate and couples this to production of usable energy in the form of ATP, reducing power in the forms of NADH and NADPH, and carbon skeleton intermediates used as building blocks of biosynthesis (i.e., the foundation of dry-matter growth). Respiration can be directly affected by several environmental changes, but it is likely that indirect effects of environmental change on respiration will be more important than direct effects. Factors that enhance growth and/or plant size tend to increase whole-plant respiration. Conversely, factors that limit growth typically reduce respiration. When crops must actively respond to environmental stresses such as elevated atmospheric [O_3] or soil salinity, they may increase their respiration rate for various repair or detoxification mechanisms. This may represent a loss of assimilate that could otherwise be used for growth.

Chapter 6

Partitioning of Photosynthate

Photosynthesis assimilates the carbon that forms the backbone of plant matter (biomass). Partitioning (Box 6.1) of that assimilated carbon influences crop growth and yield by dividing: (1) translocated carbon among leaves, stems, roots, tubers, seeds, and so on, and (2) carbon within an organ or tissue among various organic molecules such as lipids, proteins, storage carbohydrates, structural carbohydrates, and lignins. Thus, partitioning determines how photosynthesis contributes to desired products, such as seeds or tubers, and the biochemical quality of those products.

SOURCE AND SINK

Partitioning (sometimes called allocation; Box 6.1) is a function of the activities of, and interactions between, sources of assimilate and sinks for assimilate. Movement of carbon from sources to sinks involves translocation in (through) phloem. Clearcut distinctions between sources and sinks are sometimes elusive, though it is clear that crop plants have many sources and many sinks. Some sinks are small but critical, such as meristems. All sources were once sinks, and some organs can simultaneously be both sources and sinks. For example, the mature portion of a leaf is mainly a source, while a younger part of the same leaf may be a sink. This can also apply to other plant parts, such as an inflorescence, in which the growing seeds are sinks and the surrounding tissues carry out photosynthesis and function as sources.

Some assimilate is used within the source for its own growth, maintenance, and respiration. Other assimilate in a source is stored in various "reserve" compounds (e.g., starch) for various periods. The remainder of the assimilate (including that mobilized from storage) is

BOX 6.1. Definitions

Allocate v: To apportion for a specific purpose or to specific persons or things or to set apart or earmark. Allocate (allocation) is sometimes used to describe source-sink relationships with respect to assimilate partitioning, but source organs never apportion assimilate among sinks or for specific purposes, and assimilate is not earmarked for particular destinations. Thus, plants do not allocate assimilate. Farmers, on the other hand, allocate seeds, fertilizers, and pesticides to different fields.
Assimilate v: To take in or appropriate as nourishment, to absorb into the system.
Assimilate n: Something that is assimilated. Herein will refer to organic compounds formed from (or being) the products of photosynthesis, most typically, various sugars.
Harvest index: The useful fraction of aboveground biomass of a crop. Typically seed mass per unit total shoot mass. Abbreviated HI.
Partition v: to divide into parts or shares. Through a combination of processes in sources, conducting cells (e.g., phloem), and sinks, assimilate is partitioned among plant organs (and to soil through exudation, etc.). Within organs and cells, assimilate is further partitioned to various structural and reserve compounds, some of which are oxidized to CO_2, water, and heat.
Sink n: An organ or tissue that uses or consumes the assimilate produced in sources. Any growing tissue or storage organ is a sink.
Source n: Any organ or tissue that assimilates CO_2, forming various organic molecules, most notably sugars. Structures (usually mature leaves) that provide assimilate needed for production of above- and belowground plant biomass.
Translocate v: To move assimilates and other molecules through vascular tissue from one location to another.

loaded into phloem cells in the source and then transported through the phloem complex throughout the plant to sinks (Figure 6.1).

The question is frequently asked whether yield is limited by source activity or sink capacity. This question has added importance herein because environmental changes may have differential effects on source and sink activities. Probably, both source activity and sink capacity limit yield under different circumstances. Sources and sinks can also partly regulate each other (see, e.g., Paul and Foyer, 2001), a fact that led Evans (1993, p. 184) to conclude that "because the source activity feeds forward to determine the sink capacity while the sink activity feeds back to modulate the source, they are not inde-

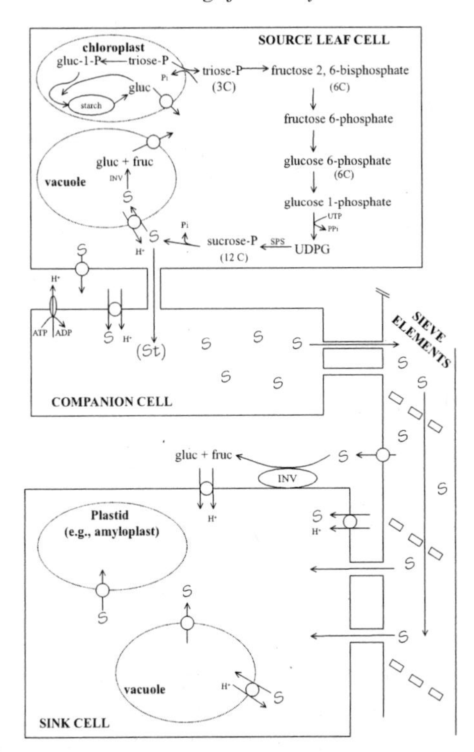

FIGURE 6.1. The major components of assimilate partitioning within source-leaf cells, mechanisms of phloem loading (both apoplastic and symplastic), long-distance transport of sucrose (S) in phloem sieve tubes, unloading of assimilates, and transport into sink cells. Carbohydrate synthesized in chloroplasts of source leaves can remain within chloroplasts in the form of starch or may enter the cytosol, where most is converted into sucrose. Some sucrose is used locally to support growth, maintenance, or storage in source cells, but in mature leaves, most is eventually exported across the source-leaf cell membrane into the cell-wall region and then across the companion-cell membrane into the companion-cell cytoplasm (called apoplasmic phloem loading). This is an energy-consuming process (driven by pumping of H^+ ions by a membrane-bound ATPase into cell walls). Conversely, in some tropical crops, the sucrose may pass through plasmodesmata into companion cells, where it is converted into stachyose (St). Stachyose or sucrose can then be transported through the phloem sieve elements to sink cells. Carbohydrate in the phloem must be unloaded into sinks by either apoplastic or symplastic routes. In developing seeds, sucrose enters the cell-wall region, where it is typically hydrolyzed to glucose (gluc) and fructose (fruc) subunits in a reaction catalyzed by the enzyme invertase (INV). A hexose transporter may then shuttle these hexose sugars into sink cells for use in storage or metabolism. In other sinks, such as meristems and potato tubers, sucrose may travel through plasmodesmata, connecting sieve elements with adjacent sink cells.

pendent determinants of yield and may both be limiting to it." A corollary is that if an environmental change has a direct effect on either source activity or sink capacity (or activity), it is also likely to have indirect effects on the other part of the source-sink balance.

HARVEST INDEX

For crops, a critical outcome of partitioning is the harvest index (HI). Harvest index generally refers to the desired fraction (e.g., grain) of aboveground biomass of a crop at the time of harvest (Donald, 1962; Donald and Hamblin, 1976). In addition to biomass per se, HI can apply to individual elements such as nitrogen (Evans, 1993). Measurements or estimates of HI usually exclude abscised leaves, so they exaggerate the fraction of total aboveground production found in grain. Emphasis on aboveground, as opposed to total, biomass is a practical matter; it is difficult to measure root biomass, so whole-plant biomass estimates are relatively rare. In the future, HI may come to mean the useful fraction of the whole plant, and explicitly account for production of shed leaves and roots. For example, potato HI would be the fraction of whole-plant biomass production found in tubers. Another special case is sugarcane, for which HI may be thought of as the nonstructural carbohydrate (or sucrose) content of the stem divided by aboveground (or better, whole-plant) biomass. For now, however, HI usually refers to amount of grain divided by total shoot biomass present during harvest.

Changes in HI (i.e., partitioning to grain) played a large role in yield increase in many crops during past decades. For example, between 1880 and 1980, HI of several cereal crops increased about 50 percent (Evans, 1993), though changes in HI played only a minor role in maize-yield increase during the 1900s (Richards, 2000). In many cases, increased HI brought about by breeding has been the result of shorter plants, with relatively less stem mass. But whole-plant growth of crops per unit ground area, not just HI, increased as well because of increased applications of fertilizers, herbicides, and pesticides in addition to other agronomic practices (Loomis and Connor, 1992; Evans, 1993).

METHODS FOR STUDYING PARTITIONING

Two of the most important aspects of whole-plant partitioning are end-of-season HI and root/shoot ratio (though root/shoot ratio can be less important than HI for crops). Two of the most common methods of quantifying partitioning are to (1) weigh different parts of the plant or (2) label CO_2 made available for photosynthesis with an isotopic signature (^{14}C has been most often used) and then to measure the amount of that signature that later appears in different organs or compounds. In most cases, estimates of partitioning provide only snapshots of the state of a crop. A significant amount of carbon translocated to roots, for example, is secreted into the soil as exudates, or is lost through root death and decomposition before most measurements of root mass or isotopic signature are made. Similarly, leaves (or even whole tillers) may senesce and abscise before HI is estimated. In all cases, some carbon is lost during respiration, and this fraction is difficult to estimate.

To accurately describe partitioning, it is necessary to quantify rates of carbon gain (i.e., photosynthesis + import), as well as rates of carbon loss (respiration + export + leakage). Consideration of how the carbon is used within sinks (i.e., for growth, maintenance, storage, or respiration), and accounting for its chemical form, are also needed. Current understanding of the mechanisms controlling partitioning remain rudimentary. Although it is obvious that both genetic and environmental factors are involved in partitioning, only a few specific molecular control points have so far been identified.

CARBON PARTITIONING WITHIN SOURCE LEAVES

Triose phosphates, the initial products of photosynthesis (Chapter 5), are transported from chloroplasts into the cytosol where the majority are utilized to synthesize sucrose (Figure 6.1). Some of that sucrose (and other compounds) is loaded into phloem in sources, through which it is translocated to sinks. But not all newly assimilated carbon is immediately available for export. Some is used locally to supply the carbon skeletons needed for synthesis of amino acids, phenolic compounds, lipids, antioxidants, and a host of other compounds involved in source-cell structure and functioning. Another fraction of assimilate is oxidized by respiration in sources to supply ATP and reducing equivalents (mainly NADH and NADPH). During drought and salinity stress, some sugars may be retained in source leaves as part of osmotic adjustment (Chapter 4). Partitioning of assimilates to storage pools within the source (e.g., starch in chloroplasts and soluble sugars in vacuoles) accounts for the rest of newly assimilated carbon.

Carbohydrates in storage pools (reserves) in source organs are typically made available (i.e., mobilized) for export through phloem during the night, when photosynthesis is inactive. It is this process of storage and mobilization that maintains a continuous (24 hours per day) supply of assimilate to sinks even though assimilation occurs only during daylight.

Crops can also temporarily store assimilate in stem and root tissue, which functions as a transient stop along the way from source to ultimate sink. Relatively large amounts of sugars that accumulate in stem tissue during the vegetative growth phase of cereal crops may be mobilized and translocated to reproductive structures during grain filling. For example, organic compounds stored temporarily in stems may account for 10 to 30 percent of final grain yield in maize, rice, sorghum, sunflower, and wheat (Wardlaw, 1990; Evans, 1993). In sugarcane, sugars stored in stems (for which the crop is usually harvested) may typically account for about 20 percent of total biomass (Irvine, 1983).

PHLOEM LOADING, TRANSPORT, AND UNLOADING

To be translocated from source leaves to sinks, assimilates must be transported from mesophyll cells to leaf veins (vascular bundles). Although the complete pathway from mesophyll cells to phloem remains undefined for most crops, it is generally thought that sucrose moves through the symplasm from mesophyll cell to mesophyll cell (i.e., through the plasmodesmata) and then into bundle-sheath cells. Once in vascular bundles, assimilates are actively (i.e., consuming energy) loaded into phloem sieve tubes, through which they can be translocated to sink organs (Lalonde et al., 1999; Turgeon, 2000; Figure 6.1). In the absence of phloem loading in sources, the pressure gradient that drives transport through phloem from source to sink dissipates and translocation is greatly reduced or ceases.

The active loading of assimilates into phloem sieve tubes (i.e., into the sieve cell–companion cell complex) can occur by two pathways, symplastic or apoplastic. In the case of symplastic phloem loading, assimilates pass into companion cells through plasmodesmata without ever leaving the cytoplasm. Conversely, with apoplastic loading, assimilates leave the cytoplasm of bundle-sheath cells, enter the cell-wall compartment, and then return to the cytoplasm in neighboring companion cells (Turgeon, 2000). The majority of crop species rely mainly on apoplastic phloem loading.

Apoplastic phloem loading is made efficient in many species by highly specialized transfer-type companion cells (transfer cells have cell-wall ingrowths that presumably function by increasing the surface area, and thus the efficiency of transport, between cell walls and cytoplasm). Loading of sucrose from the cell-wall complex into the companion-cell cytoplasm is facilitated by cotransport with H^+ ions. A concentration gradient of H^+ ions between the cytoplasm and cell-wall regions of companion cells, established by an ATP-dependent proton pump, apparently facilitates this cotransport (Figure 6.1).

In symplastic phloem loaders, which include cucurbit species, cotton, and perhaps other tropical crops (A.J. van Bel, personal communication, 2002), assimilates enter leaf veins without ever leaving the cytoplasm. This is probably accomplished by converting sucrose, a disaccharide that readily passes through plasmodesmata, into raffinose (a trisaccharide) and stachyose (a tetrasaccharide) within companion cells (Figure 6.1). Conversion into raffinose and stachyose "traps" carbohydrates within veins because these molecules are too large to pass back through plasmodesmata into bundle-sheath cells. This is called the polymer-trap hypothesis and is widely accepted as the probable mechanism of symplastic phloem loading (Schulz, 1998).

Because many tropical and subtropical crops load phloem symplastically while temperate and boreal plants load phloem apoplastically, it has been suggested that these mechanisms may have evolved to function best within certain temperature ranges. Recent evidence, however, indicates that the geographic distributions of species exhibiting these phloem-loading mechanisms is not adequately explained by temperature effects on symplastic versus apoplastic phloem loading (Schrier, Hoffmann-Thoma, and van Bel, 2000). It has also been suggested that species which load phloem via a symplastic pathway may be less efficient at translocating the extra assimilates produced by plants growing in elevated atmospheric [CO_2]s. For example, symplastic phloem loaders accumulated 41 percent nonstructural carbohydrates in source leaves compared to 25 percent in apoplastic phloem loaders (Körner, Palaez-Riedl, and van Bel, 1995). Whether a phloem-loading mechanism makes a given crop more or less able to utilize extra carbohydrates that might be produced in a higher-CO_2 world, or better adapted to a warmer climate that might result from global warming, remains to be determined.

The high concentration of chloroplasts in the well-developed bundle-sheath cells characteristic of C_4 species—compared to the more diffuse chloroplast distribution throughout the mesophyll in C_3 leaves—means that the average path from chloroplast to phloem is shorter in C_4 crops, and this may be important to the rapid translocation associated with rapid C_4 photosynthesis in favorable conditions.

Metabolic Costs of Phloem Loading

The energetic requirements of phloem loading are imprecisely known, and may vary among species and perhaps environments, but clearly depend on the number of active processes involved (e.g., number of membranes that a sucrose molecule must cross to enter phloem) and the number of ATP (or other energy-source molecules) required for each active process. In photosynthetically active sources, that ATP might be supplied directly by photosynthesis, but at night, it will mainly come from respiration.

If one active step is involved in phloem loading (e.g., sucrose transport into a companion cell, as indicated in Figure 6.1), and if one ATP is required per sucrose for that step, and if respiration generates 60 ATP per sucrose completely oxidized to CO_2 (Amthor, 2000b), then for every 60 molecules of sucrose loaded into phloem, one additional sucrose molecule will be oxidized in the source to provide the energy for that loading. This cost (i.e., 1 ATP per sucrose, or 0.017 CO_2 released per C translocated in sucrose) is small, and may not reflect total respiratory cost of phloem loading in crops, at least at night. Specifically, the ATP cost of nighttime conversion of starch to sucrose (i.e., mobilization of reserves) may be two to three times larger than the ATP cost of loading that sucrose into phloem (Bouma et al., 1995). For source leaves deriving a significant fraction of the carbon they translocate from starch, the respiratory cost of mobilization *plus* phloem loading may therefore be three to four ATP per sucrose (or 0.050-0.067 CO_2 per C in sucrose). In experiments, the estimated respiratory cost of carbohydrate mobilization plus phloem loading was 0.22 CO_2 per C translocated from tomato leaves (Ho and Thornley, 1978), 0.1 CO_2 per C translocated from cotton leaves (Hendrix and Grange, 1991), and 0.058 CO_2 per C translocated from potato leaves (Bouma et al., 1995). These ratios indicate that a significant fraction of nighttime source-leaf respiration can be associated with phloem loading (and see Bouma et al., 1995).

Although environmental change is not expected to significantly affect the ratio of respiration to phloem loading, the rate of phloem loading, and therefore source-leaf respiration rate, might be affected by environmental change. Factors that stimulate translocation from sources should stimulate respiration in those sources; conversely, factors that slow translocation from sources should slow the accompanying respiration. For example, nighttime respiration of expanded cotton leaves was stimulated by daytime elevated atmospheric [CO_2] (Thomas et al., 1993), and this was explained (at least in part) in terms of additional respiration to support additional phloem loading (Amthor, 1997).

Transport Through Phloem

Phloem is made up of an interconnected network of cells called sieve elements and their associated companion cells. Other cells, including undifferentiated parenchyma and fibers, are also present in phloem, although they do not appear to play a direct role in transport. Openings in sieve-element end-walls (sieve plates) connect adjacent phloem cells, forming a network (sieve tubes) that ramifies throughout the plant body.

Assimilates are transported long distances through sieve tubes by bulk flow from sites of high to low pressures. The loading of solutes (mostly sucrose, but also other sugars, amino acids, and some minerals) into the phloem of source leaves lowers the solute potential inside sieve elements, which causes an influx of water from surrounding tissues. Hydrostatic pressure is thereby generated within sieve tubes in source leaves. Meanwhile, assimilates are unloaded into sinks throughout the plant (e.g., root tips, growing tubers, or growing seeds, as well as some young leaves). Removal of solutes in sinks reduces the solute potential of nearby phloem sieve cells, which causes movement of water out of those sieve elements into surrounding tissues. High pressure associated with phloem loading in sources coupled with lower pressure associated with phloem unloading in sinks results in a pressure gradient from source to sink. It is this osmotically generated pressure gradient that drives the bulk flow of water and the solutes it contains from sources to sinks. Active processes are not thought to be important to transport within phloem per se.

The physical capacity of phloem to carry assimilates probably does not constrain crop yields under most field conditions. This was illustrated effectively for wheat and sorghum when it was shown that grain growth was unaffected following damage to half the phloem tissue leading to the developing kernels (Wardlaw, 1990; and see Bancal and Soltani, 2002). For this reason, the influence of environmental changes on phloem structure, as it affects phloem functioning, has been largely ignored. It is much more likely that environmental changes will affect phloem function by altering source metabolism (e.g., photosynthetic rate) and sink activity (e.g., cell division or grain-filling rates).

Phloem Unloading

Sucrose is unloaded from vascular tissues into organs where it is used for growth, storage, and maintenance. The rate of unloading is probably related to the rate of metabolic activity in sinks. Transfer of assimilate from phloem into sinks may be passive or active, as illustrated in Figure 6.1.

REGULATION OF WHOLE-PLANT CARBON PARTITIONING

At the most general level, knowledge of carbon partitioning is based on understanding of the physics of phloem function. When the rate of carbohydrate use by sinks is rapid, turgor of associated sieve elements remains low, which favors bulk flow of assimilates into sinks. So, to some extent at least, the ability of a sink to metabolize imported sucrose defines its demand, and thereby influences the proportion of assimilates it will receive relative to competing sinks.

Thus, sinks with a high sucrose demand "outcompete" sinks with a low demand for sucrose. For example, symbiotic soil organisms such as nodulating bacteria and mycorrhizal fungi promote carbon flow to roots (compared to other organs) by stimulating demand by rapidly consuming sucrose.

The physical location of a sink relative to a source is also important. Other factors being equal, sinks near to a source preferentially receive photosynthate from that source relative to sinks located farther away. When young maize leaves (far away from roots) were labeled with ^{14}C, for instance, little of the labeled carbon was translocated to roots. But when the oldest leaf (close to roots) was labeled, the roots received more than half of the labeled carbon (Hofstra and Nelson, 1969).

Sink Control

Several key sucrose-metabolizing (catabolizing) enzymes are likely to play major roles in controlling carbon partitioning by dictating how fast sucrose is utilized by sinks. Spatial (interorgan) differences in the activities of these enzymes may determine which organs receive the most assimilate. In this way, assimilates can be preferentially partitioned to specific organs. The enzymes that hydrolyze sucrose, i.e., invertase (producing the hexoses glucose and fructose) and sucrose synthase (producing UDP-glucose and fructose), are both probably important to partitioning (Roitsch et al., 2000). Invertase activities, for example, are often highly correlated with the rate of sucrose import into sinks (Geiger, Koch, and Shieh, 1996) as well as the rate of growth by sinks (Wardlaw, 1990). Invertase isozymes are found in cell walls (i.e., extracellular invertase; Roitsch et al., 2000), cytosol, and vacuoles, and their activities are influenced by environmental conditions. Activities of sucrose-cleaving enzymes are generally high in active meristems and growing tissues, but decline rapidly as tissues mature and demand for photosynthate falls.

In the case of developing seeds, sucrose is unloaded from phloem tissue into the cell-wall region (apoplast) where it is cleaved by invertase (Figure 6.1). The resulting glucose and fructose must then be transported across the plasmalemma into sink cells, where they can support many aspects of sink function, especially seed growth and storage of carbohydrates, proteins, and/or lipids. By virtue of their role in facilitating movement of fructose and glucose from apoplast into sink symplast, hexose transporters might control sucrose utilization in sinks (i.e., sink demand). Furthermore, hexose transporters themselves are regulated by both exogenous and endogenous signals. This indicates that an understanding of how environmental change affects those transporters might lead to an understanding of how environmental change affects partitioning (Hellmann et al., 2000). Activity of those transporters is probably most important where carbohydrates must pass through membranes between phloem and sink cells; that is, where there is no direct connection between phloem and sink symplasm as is the case for developing seeds (sucrose enters root and shoot apical meristems, as well as storage organs such as potato tubers, without ever leaving the symplast).

Source Control

While sink activity is obviously important to partitioning among competing organs, recent evidence indicates that source activity may also exert control over carbon partitioning (Farrar and Jones, 2000; Sweetlove and Hill, 2000). In source leaves, control over export of carbohydrate versus storage in the source is important. Regulation of sucrose phosphate synthase (SPS) activity by reversible phosphorylation and fructose bisphosphatase by the regulatory molecule fructose 2,6-bisphosphate, appear to be critical control points for controlling partitioning between starch and sucrose (i.e., storage and export) in leaf tissue (Rufty, Huber, and Kerr, 1985; Geiger and Servaites, 1991; Mooney and Winner, 1991; Evans, 1993; Komor, 2000). These enzymes apparently are sensitive to environmental conditions and there is interest in modifying their expression (using recombinant DNA techniques) to increase a plant's capacity to synthesize and transport sucrose (Signora et al., 1998).

Control over carbon partitioning in source leaves does not end in mesophyll cells. There is a family of sugar transporters, active in source tissues, that facilitates the transmembrane uptake of sucrose from bundle sheath or mesophyll cells into the sieve cell–companion cell complex. These transporters, which control phloem loading and therefore the potential for sucrose transport to sinks, are regulated by hormones, sucrose concentration in leaf cells, and diurnal changes in carbohydrate pools (Hellmann et al., 2000). These sucrose transporters might represent another important control point that could be affected by environmental change.

A linkage between environmental conditions and molecular control of carbon partitioning is likely and, when better characterized, should provide insight into mechanisms of partitioning. The elucidation of this connection, however, has proven difficult. One factor is the confounding effect of plant developmental stage on expression of these same elements. For example, in sink tissues such as growing seeds, mRNAs for the most abundant proteins (including sugar transporters and sucrose metabolizing enzymes) increase as a sink develops and then disappear by maturity. Therefore, environmental effects on partitioning are superimposed on developmental and spatial regulation of gene expression. In order to tease apart these control networks, the signal molecules involved (if any) must be identified and the effects of the environment on those signal molecules must be understood.

Sucrose

Sucrose is perhaps the most critical signal molecule in the control of carbon partitioning, and its regulatory effects on gene expression are apparent throughout a plant. Sucrose signaling facilitates coordination of source and sink activity by communicating the status of one to the other. For example, when leaf photosynthesis is rapid and sucrose supply is ample, sucrose export increases and growth and metabolism in sinks is stimulated (Rolland, Moore, and Sheen, 2002; Table 6.1). Sucrose stimulates sinks by serving as a growth substrate and also by enhancing expression of genes controlling growth and storage (feed-forward regulation). Soluble sugars, for example, have been shown to induce tuber formation in potatoes and root storage in sweet potato (Simko, 1994). In this way, source activity (i.e., photosynthesis and assimilate export) influences not only the growth and metabolism in sinks but also the capacity of sinks to develop and grow.

Information concerning sink status is also transduced to source leaves via sucrose. When sink activity (mainly growth) does not keep up with sucrose supply, sucrose accumulates in phloem tissue. High sucrose concentrations in source phloem sap inhibits further phloem loading by lowering the activity of a transmembrane sucrose

TABLE 6.1. Genes up-regulated by sugar abundance.

Process genes/products	Plant and tissue	Effectors	Source
Starch synthesis			
starch phosphorylase	potato tuber	S	A
starch synthase	potato lvs detached	S, G, F	B
branching enzyme	cassava stms/lvs	S, G, F	C
Storage proteins			
sporamin (A & B)	sweetpotato lvs	S, G, F	D, E
β-amylase	sweetpotatoes/lvs	S, G, F	E
patatin class I	potato tubers and lvs	S	F
proteinase inhibitor	potato lvs	S, G, F	G
Respiration			
PGAL-de.	*Arabidopsis* lvs	S	H
β-isopropylmalate de.	potato, tomato	S	I
PP-F-6-P PT	spinach lvs	G	J
Sucrose metabolism			
invertase	maize root tips,	S, G, F	B, K
sucrose synthase	maize	S, G, F	L
	rice embryos	S, G, F	M
	faba bean cotyledon	S	N
SPS	sugarbeet petioles	G	O
	transgenic potato	sol. sugars	P

Sources: (A) St. Pierre and Brisson (1995); (B) Kossmann et al. (1991); (C) Salehuzzaman, Jacobsen, and Visser (1994); (D) Ishiguro and Nakamura (1994); (E) Nakamura et al. (1991); (F) Kim, May, and Park (1994); (G) Johnson and Ryan (1990); (H) Yang et al. (1993); (I) Jackson, Sonnewald, and Willmitzer (1993); (J) Krapp and Stitt (1994); (K) Xu et al. (1995); (L) Koch et al. (1995); (M) Karrer and Rodriguez (1992); (N) Heim et al. (1993); (O) Hesse, Sonnewald, and Willmitzer (1995); (P) Müller-Röber, Sonnewald, and Willmitzer (1992).

Notes: These examples are from Koch (1996), Table 2. Abreviations: lvs, leaves; rt, root; S, sucrose; G, glucose; F, fructose; PGAL-de., phosphoglyceraldehyde dehydrogenase; β-isopropylmalate de., β-isopropylmalate dehydrogenase; PP-F-6-P PT, pyrophosphate:fructose-6-phosphate-phosphate-phototransferase; SPS, sucrose phosphate synthase; sol., soluble.

transporter (Chiou and Bush, 1998). This leads to the accumulation of sucrose in source-leaf mesophyll cells, which eventually causes down-regulation of photosynthetic genes and/or up-regulation of genes encoding carbohydrate storage enzymes (Table 6.2; Koch, 1996). However, when sucrose concentrations are kept low in phloem by high sink activity, the activity of sucrose transporters remains high, sucrose levels in leaves remain low, and photosynthetic rates can remain rapid. This system effectively links sink activity, phloem loading, leaf sucrose concentration, and whole-plant partitioning of assimilate (Farrar and Gunn, 1996; Chiou and Bush, 1998).

An environmental change that affects any aspect of source or sink activity has the potential to alter the full chain of events, from photosynthesis in sources to growth and storage in sinks. Warming, for example, is likely to impact partitioning by directly stimulating sink

TABLE 6.2. Genes up-regulated by sugar depletion.

Process genes/products	Plant and tissue	Effectors	Source
Photosynthesis			
Rubisco S-subunit	tobacco and potato	S, G	A
triose-P transporter	tobacco lvs	S	B
Mobilization			
isocitrate lyase	cucumber cotyledons	S, G, F	C, D
malate synthase	cucumber cotyledons	S, G, F	C, D
proteases	maize rt tips	G	E
asparagine synthetase	*Arabidopsis* shoots	S	F
Sucrose metabolism			
acid invertase	maize rt tips	G, S, F	G, H
sucrose synthase	maize rt tips	S, G, F	I
	bean seeds	F, G	J
SPS	sugarbeet lvs/rts	S	K

Sources: (A) Krapp et al. (1993); (B) Knight and Gray (1994); (C) Graham, Baker, and Leaver (1994); (D) Graham, Denby, and Leaver (1994); (E) Brouquisse et al. (1991); (F) Lam, Peng, and Coruzzi (1994); (G) Koch et al. (1995); (H) Xu et al. (1995); (I) Koch et al. (1992); (J) Heim et al. (1993); (K) Hesse, Sonnewald, and Willmitzer (1995).

Notes: These examples are from Koch (1996), Table 1. *Abbreviations:* lvs, leaves; rt, root; S, sucrose; G, glucose; F, Fructose.

growth and respiration, whereas rising [CO_2] is more apt to directly stimulate source activity. Ozone pollution can inhibit photosynthesis as well as the phloem loading processes.

Hormones

Hormones affect nearly all plant developmental and physiological processes (Table 6.3). Auxins, cytokinins, gibberellins, and ABA can all be found in phloem tissue, where they are thought to influence as-

TABLE 6.3. Effects of hormones on developmental and physiological processes in crops.

	Phytohormones and their effects on crops[a]				
Plant response	**IAA**	**GA**	**CK**	**C = C**	**ABA**
Nucleic acid synthesis	+	+	+	–	–
Cell division	+	(0)	+	–	–
Cell enlargement	+	+	+	(–)	(–)
Cell-wall growth	+	0	0	0	0
Membrane permeability	+	+	+	+	+
Cambial activity	+	+	+	(–)	(–)
Cell differentiation	+	+	+	(–)	(–)
Shoot elongation	+	+	+	–	–
Apical dominance	+	–	–	(0)	(0)
Leaf growth	+	+	+	–	–
Root initiation	+	0	0	(–)	(–)
Bolting	0	+	0	0	–
Flowering	+	+	0	+	–
Floral development	+	+	(+)	(+)	(–)
Fruit development	+	+	+	(0)	(0)
Senescence	–	–	–	+	+
Abscission	–	–	–	+	+
Nutrient distribution	+	0	+	0	0

Source: Table modified from Nilsen and Orcutt (1996)

[a] +, promotes; –, inhibits; 0, has no effect; ( ), effect is questionable but likely; IAA, indoleacetic acid (an auxin); GA, gibberellin; CK, cytokinin; C = C, ethylene; ABA, abscisic acid.

similate partitioning by modifying phloem loading, unloading, and/or sink activity (Hendrix, 1995; but see also Beveridge, 2000). We take the view that hormones are critical, but that they are mainly messengers rather than ultimate causes of the effects of environmental change on partitioning. This is an easy view to adopt because so little is known of the mechanisms linking environmental change to hormone production, transport, and degradation, or the mechanisms linking specific hormones to source and sink activities.

In sources, gibberellins stimulate phloem loading and sucrose export (Lalonde et al., 1999), while auxins and cytokinins affect phloem transport rates. Photosynthesis is also subject to hormonal control.

In sinks, gibberellins and auxins regulate the rate of cell enlargement and cytokinins stimulate cell division. Effects on growth are important because they dictate sink strength and assimilate demand (Offler, Thorpe, and Patrick, 2000). Cytokinins have also been shown to induce activities of a cell-wall invertase and a transmembrane glucose transporter in shoots, thereby stimulating shoot sink activity (Baker, 2000). Conversely, cytokinins repress root growth (van der Werf and Nagel, 1996). A link between environmental conditions, hormone dynamics, and carbon partitioning is well illustrated in plants grown in nitrogen-depleted soil; soil nitrogen limitations lead to decreased cytokinin production by roots (root tips are important sources of cytokinins) and decreased cytokinin transport to shoots. This chain of events is probably linked to shoot growth limitations in crops growing in low-nitrogen soil (Aiken and Smucker, 1996; Stitt and Scheible, 1998).

Several factors prevent generalizations about the role of hormones in linking environmental changes to partitioning: (1) a few hormones may control many developmental events, (2) effects of one hormone can be modified by another/other hormone(s), (3) hormones often act in an organ-specific manner, and (4) minute quantities of hormones, which are difficult to measure, often have large effects on growth.

ENVIRONMENTAL CHANGES

Warming

Temperature affects all aspects of carbon partitioning, e.g., source activity, phloem sap viscosity, and sink activity. This is reflected in changes in the relative sizes of crop organs when temperature fluctuates. For example, the ratio of leaf to stem mass sometimes increases as temperature increases, even though individual leaves may be thinner (Farrar, 1988, 1991).

Warming generally decreases phloem sap viscosity and, assuming an equivalent pressure gradient from sources to sinks, this effect should increase mass flow rates through the phloem (but see Bancal and Soltani, 2002). In addition, rates of solute diffusion from mesophyll cells to phloem, and from phloem to sink cells, will increase with warming (Johnson and Thornley, 1985). In other words, the physical capability for both short- and long-distance transport of assimilates may be enhanced by warming (Bowen, 1991).

Warming increases enzyme activity, at least until a temperature threshold is reached beyond which enzymes become damaged (Chapter 3). Different enzymes have different temperature sensitivities, however, so it is difficult to speculate about how different aspects of source and sink metabolism might be affected by warming (Table 6.4). For instance, a reported Q_{10} value for fructan-fructan fructosyl transferase was 1.6 compared to a value of 3.9 for adenosine diphosphate glucose starch synthase (see Table 6.4 for other examples).

Probably the most important effects of warming on carbon partitioning will follow from changes in sink activity (metabolism and growth). Warming generally stimulates sink metabolism and reduces carbohydrate pools in most plant tissues, and as discussed previously, this could influence the expression patterns of many genes involved in partitioning (Rowland-Bamford et al., 1996). There is another significant consequence of enhanced metabolism for partitioning: since growing shoot organs have greater access to photosynthate than roots (because they are closer to sources) root/shoot ratio sometimes decreases with warming (Batts et al., 1998), but because warming accelerates development, it can be difficult to judge whether warming directly affects partitioning or merely shifts ontogony or both (Farrar and Gunn, 1996). Hormones may also influence relative root and shoot activities with warming, and affect root/shoot partitioning (Bowen, 1991). In any case, it may be that warming of the magnitude of 2 to 5°C will have only small effects on root-shoot partitioning (Farrar and Gunn, 1996). Decreased partitioning to roots has implications for access to soil resources and could be most significant for yield of root and tuber crops.

TABLE 6.4. Q_{10} values for enzymes and processes important to carbon partitioning in some crops.

Process	Q_{10}	Species	Source
Enzyme activity			
SPS	1.7 (2-10)[a]	*Lolium temulentum*	A
SST	1.4 (2-10)	*Lolium temulentum*	B
FFT	1.6 (2-10)	*Lolium temulentum*	B
invertase	2.0 (2-10)	*Lolium temulentum*	B
invertase	2.0	barley	C
ADPG PP	2.0 (2-10)	*Lolium temulentum*	A
ADPG SS	3.9 (2-10)	*Lolium temulentum*	A
Sucrose import			
sugar beets	3.0	sugar beet discs	D
roots	1.6 (5-30)	barley	E
Membrane transport	1.6-3.0	generalized value	E
Translocation			
diffusion	2.0		F
velocity in phloem	1.1 (10-25)	bean	G
mass transfer rate	1.3 (10-25)	bean	G
Phloem sap fluidity	1.3 (10-25)	castor bean	G

Sources: (A) Pollock (1986a); (B) Pollock (1986b); (C) Wagner, Keller, and Wiemken (1983); (D) Saftner, Daie, and Wyse (1983); (E) Farrar (1988); (F) Johnson and Thornley (1985); (G) Giaquinta and Geiger (1973).

[a]Temperature range used to estimate Q_{10}

Notes: SPS, sucrose phosphate synthase (responsible for sucrose synthesis); SST, sucrose-sucrose fructosyl transferase (enzyme involved in fructan synthesis); FFT, fructan-fructan fructosyl transferase (also involved in fructan synthesis); ADPG PP, adenosine diphosphate glucose pyrophosphorylase (enzyme involved in starch synthesis); ADPG SS, adenosine diphosphate glucose starch synthase (also involved in starch synthesis). The numbers in parentheses indicate the temperature range in which the Q_{10} is applicable (where reported).

Harvest index of grain crops may decrease with, or be unaffected by, warming (Rawson, 1995; Wheeler et al., 1996; Batts et al., 1998; Van Oijen et al., 1999). Although a general effect of modest warming is more rapid substrate utilization, which in turn can induce changes in gene expression that lead to greater source activity, this may not produce greater grain yield. In fact, grain yield is commonly reduced with warming because the *duration* of grain filling is shortened (Chapter 9). Furthermore, high temperature (5 to 15°C above optimal) around the time of anthesis can inhibit the transport of carbohydrates into flowers and developing fruits (for pepper; Aloni, Pashkar, and Karni, 1991). This effect is probably related to the effects of temperature on the function of enzymes in those sinks.

Effects of warming on partitioning among biochemical compounds in grain (i.e., grain quality) are outlined in Chapter 9.

Elevated Atmospheric Carbon Dioxide Concentration

Elevated atmospheric [CO_2] directly stimulates source activity (i.e., photosynthesis) in C_3 crops. This can lead to greater carbon assimilate supply to all sinks, and has the potential to stimulate sink growth capacity. An increase in root/shoot ratio is common in elevated-[CO_2] experiments, although decreases are observed too (Figure 6.2; Rogers et al., 1996). The effect of elevated [CO_2] on HI is also variable. In a survey of 112 experiments with wheat, HI declined with elevated [CO_2] in 31 experiments and increased in 55 experiments, indicating a tendency toward a modest stimulation of HI by elevated [CO_2] (Amthor, 2001). Effects of elevated [CO_2] on the HI of other crops, such as soybean, are similarly inconsistent (Farrar and Williams, 1991; Allen, Baker, and Boote, 1996; Polley, 2002).

Variable effects of elevated [CO_2] on partitioning in crops may be related to differences in growth habits (e.g., determinate versus indeterminate), soil nutrient and water availability, and other environmental factors (Rogers et al., 1996). For instance, HI was sometimes positively related to [CO_2] in wheat subjected to water stress and elevated [O_3], but this was not always the case (Amthor, 2001). Temperature did not interact with [CO_2] to affect wheat HI (Amthor, 2001). On the whole, relatively few data are available to assess effects of [CO_2] in combination with other factors on HI, and intra- and interexperiment variability in measured HI is large.

Interactive effects of elevated [CO_2] with soil nitrogen availability appear to be especially important. Wheat HI was negatively related to [CO_2] with nutrient limitations in some cases, but not in others (Amthor, 2001). Root/shoot ratio goes up in plants growing under nitrogen-limited conditions; this adaptation presumably enables plants

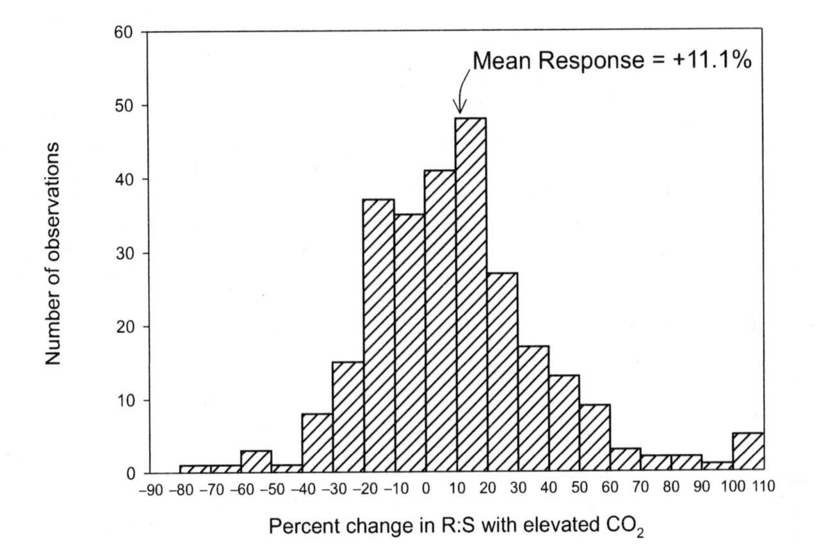

FIGURE 6.2. Histogram showing the percent change in root/shoot ratio for crops growing in CO_2-enriched atmospheres. A total of 264 observations on crops are included. Data from Rogers et al. (1996).

to partly overcome nitrogen limitations through greater soil exploration by more or larger roots for a given total plant size. This might explain why the increase in root/shoot ratios in crops growing in CO_2-enriched atmospheres is usually more marked when they are grown with low compared to high nutrient supplies (Read and Morgan, 1996). Changes in root/shoot ratio may be expected when crops are nutrient-limited (e.g., in many low-input cropping systems).

Variation in sink growth capacity among crop species also contributes to differing effects of elevated [CO_2] on partitioning because different crops tend to be more or less "sink limited," depending on their inherent genetic potential for sink growth. Determinate crops are expected to experience smaller effects of elevated [CO_2] on partitioning than indeterminate crops. Potatoes and other crops with large storage organs may have adequate sinks for "extra" assimilate, whereas crops with weaker sinks, such as rice, may not.

Some partitioning shifts are almost universally observed in plants grown in elevated [CO_2]; crop leaves are often thicker, but plants produce less leaf mass and leaf area relative to root, stem, and branch mass when grown in elevated [CO_2] (Pritchard et al., 1999; Poorter and Nagel, 2000). This is easily interpreted in terms of enhanced source (leaf) activity per unit source organ in elevated [CO_2], reducing the "need" for leaf compared to other organs.

The capacity for enhanced export rates in plants growing in elevated [CO_2] is poorly known. Some have questioned whether the transport system has the potential to accommodate rates of carbon assimilation much higher than occur today (Grimmer and Komor, 1999). But (C_3) crops are generally larger when grown in elevated [CO_2], so they clearly can support some additional translocation supporting additional growth.

Increased carbohydrate export notwithstanding, leaf carbohydrates, especially starch, accumulate in some CO_2-enriched plants, indicating that more carbon is assimilated than is used for export or local metabolism (Grodzinski, Jiao, and Leonardos, 1998; Baxter and Farrar, 1999; Grimmer, Bachfischer and Komor, 1999). To the extent that crops growing in the field in elevated [CO_2] accumulate carbohydrates in source leaves (by the end of the night, after opportunities for nighttime translocation) and/or downregulate photosynthesis (Chapter 5), source activity (capacity) presumably exceeds sink activity (capacity). This issue is important to overall growth, but how it might affect partitioning per se is unclear. As illustrated in Table 6.1, large soluble-carbohydrate concentrations may reduce photosynthetic capacity by changing gene expression. Furthermore, accumulation of large starch grains in mesophyll chloroplasts might reduce rates of photosynthesis even further by physically damaging the photosynthetic machinery (Stitt, 1991) or may hinder the diffusion of CO_2 from intercellular spaces to chloroplast stroma (Sawada et al., 2001). Photosynthetic acclimation to elevated [CO_2] may have implications for long-term carbon partitioning processes in crops by reducing canopy-level carbon assimilation and therefore production of biomass and yield (pod filling in legumes, grain filling in cereals, and tuber filling in root crops). If this occurs in field crops, which is presently unclear, and if the limiting process can be identified, that limiting process would be a target for breeding to take fuller advantage of increasing atmospheric [CO_2].

Elevated Atmospheric Ozone Concentration

Three effects of elevated tropospheric [O_3] on crops are especially important to translocation and partitioning: (1) reduced instantaneous rate of leaf photosynthesis (Darall, 1989; Andersen, 2003); (2) reduced leaf longevity and therefore time-integrated amount of photosynthesis (Gelang et al., 2001); and (3) direct inhibition of phloem loading and/or translocation (Grantz and Yang, 2000; McKee and Long, 2001). In addition, elevated [O_3] may increase leaf respiration (e.g., Amthor, 1988; Amthor and Cumming, 1988; Aben, Janssen-Jurkovicová, and Adema, 1990), perhaps reducing the fraction of assimilate available for translocation. But in spite of reduced photosynthesis (and enhanced respiration) in leaves, carbohydrates accumulate in leaves, indicating that the effects on phloem loading and/or translocation can be quantitatively the most important (Andersen, 2003; Figure 6.3). The accumulation of carbohydrates in source leaves might lead to downregulation of photosynthesis.

How O_3 actually interferes with phloem loading and/or transport is unclear. Several researchers have suggested that O_3 might damage sugar carriers in the plasmalemma of companion or sieve cells (Ru-

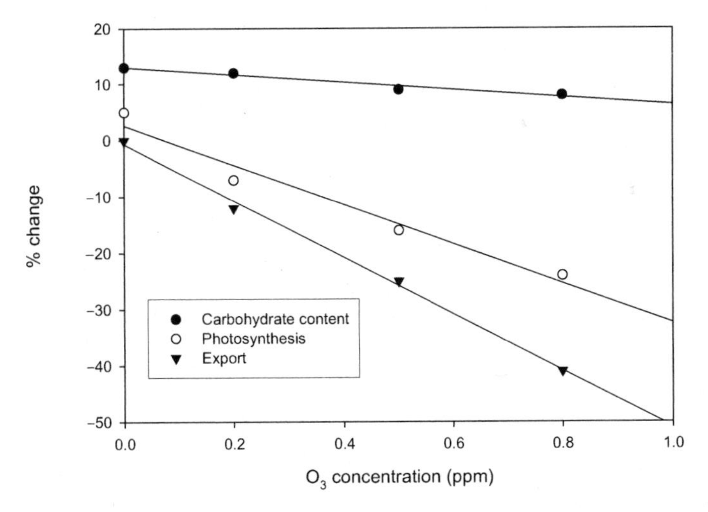

FIGURE 6.3. Effect of exposure to O_3 for 0.75 h on carbohydrate content, photosynthesis, and carbohydrate export in Pima cotton (relative to values prior to exposure). Export and carbohydrate contents were measured 1.3 h after onset of O_3 exposure. *Source:* Data from Grantz and Farrar (2000).

neckles and Chevone, 1992; Mortensen and Engvild, 1995; Meyer et al., 1997) but this remains unproven. Nonetheless, a disturbance in phloem loading or export would explain why carbohydrates can accumulate in sources.

Effects of elevated [O_3] on HI are variable and depend on the growth stage of the crop at the time of exposure (Gelang et al., 2001). In their early review of the literature, Cooley and Manning (1987) found that elevated [O_3] caused an increase in HI in maize grown for feed grain, peanut, soybean, and tomato, but a decrease in bean, sweet corn (maize grown for direct human consumption), cotton, pepper, millet, and wheat. Harvest index is more likely to decline when plants are exposed to elevated [O_3] just before, or during, flowering because of direct damage to reproductive tissues (Black et al., 2000; see Chapter 9).

A common consequence of elevated [O_3] is reduced root/shoot ratio (Cooley and Manning, 1987). Diminished phloem transport capacity could explain preferential partitioning to shoots over roots, but this idea has received little study. Reduced carbon partitioning to root systems can be particularly problematic over long time periods because it leads to poor root development and low hydraulic conductivity. Both these factors can negatively affect the water balance of crops, which in turn can lead to long-term reductions in photosynthesis (Grantz, Zhang, and Carlson, 1999). Thus, changes attributable to altered carbon partitioning can drive photosynthetic rates lower—lower even than would be expected from combined effects of oxidative damage to the photosynthetic apparatus (Rennenberg, Herschbach, and Polle, 1996) and end-product inhibition of photosynthesis caused by the accumulation of carbohydrates in leaves.

Salinity

Availability of photosynthate does not play a major role in limiting crop growth and yield in salt-stressed crops (Aslam et al., 1986). Nonetheless, salinity can affect partitioning, even though direct effects on export or transport of carbohydrates through the phloem (Munns and Termaat, 1986), or import of assimilate into sinks (Hu, Schnyder, and Schmidhalter, 2000), are unlikely. Instead, the primary effect of salinity that can affect partitioning is inhibition of growth, particularly of leaves (because this is where salts accumulate). Reduced leaf growth can cause carbohydrates to accumulate in leaves (for barley, rice, soybean, and cotton, Aslam et al., 1986). Accumulation of soluble carbohydrates such as mannitol might promote salt tolerance by contributing to a cell solute balance that is favorable for metabolism (Everard et al., 1994), but carbohydrate accumulation might also lead to down-regulation of photosynthesis (Balibrea et al., 2000). Although accumulation of carbohydrates in leaves is often interpreted as an adaptive partitioning response (i.e., osmotic adjustment) that confers salt tolerance, mannitol accumulation may be an indirect effect of salinity-induced growth retardation rather than an adaptation to NaCl per se.

An almost universal effect of salinity is an increase in the root/shoot ratio (Munns and Termaat, 1986; Brugnoli and Björkman, 1992). Indeed, with mild salinity, growth of roots may actually increase even while shoot growth decreases (Chapter 8). This indicates that transport of carbohydrates belowground is unaffected, or even stimulated, by salinity. An increased root/shoot ratio is commonly interpreted as an adaptive response for extracting water from dry soil or from soil of low solute potential. However, higher root/shoot ratios in salt-stressed plants could also be a simple effect of reduced shoot (leaf) growth, with some of the "extra" assimilate arising from reduced leaf growth being made available for enhanced root growth. Thus, as with sugar accumulation in leaves, increased root/shoot ratio may be a consequence of salt stress, not an adaptation to it.

Some effects of salinity on partitioning among biochemical components of grain are considered in Chapter 9.

SUMMARY

Partitioning of photoassimilate to crop organs (e.g., grain in wheat, rice, and maize; tubers in potato and cassava) and biochemical compounds (e.g., sugar in sugarcane stems or sugarbeet roots; oil in rapeseed grain; protein in soybean grain) is critical to how well crops serve humanity. Assimilate partitioning is a complicated process that can be (or is) controlled simultaneously by sources and sinks. The key components of assimilate partitioning are:

1. carbon assimilation in sources,
2. addition of assimilate to storage or export pools in sources,

3. short-distance transport of assimilate to phloem veins,
4. loading of assimilate into phloem sieve elements,
5. long-distance (interorgan) transport of assimilate through phloem sieve tubes,
6. unloading of assimilate into sinks, and
7. use of assimilate in sinks for biosynthesis, respiration, and storage processes.

Plants can acclimate and adapt to environmental change by modifying one or more components of partitioning. For example, crops growing without adequate soil nutrients or water often partition photosynthate, preferentially to roots, thereby maintaining a balance between processes required in roots (e.g., water and nutrient uptake) and those required in shoots (e.g., photosynthesis). Similarly, plants exposed to high soil salinity, O_3 pollution, elevated [CO_2], or warming must be able to reconcile the many physiological effects of these changes with the requirement to maintain a balance among different activities in different organs. This will often involve a change in the way photosynthate is partitioned in crops to different organs or compounds. These shifts may either be adaptive, enabling crops to enhance or maintain growth and yield, or nonadaptive, being accompanied by yield reductions or mortality.

Warming will stimulate most components of photosynthate partitioning, including carbohydrate export, phloem transport rates, unloading, and sink activity. However, the concurrent decrease in the duration of crop development, especially grain filling (Chapter 9), will probably negate (and might reverse) the stimulatory effects warming will have on those processes. Warming can limit HI in many crops, and if global warming brings with it significantly higher temperature extremes around the time of anthesis, HI could drop precipitously.

In general, elevated atmospheric [CO_2] appears to have only small effects on HI. It usually increases root/shoot ratio, but this apparently depends on the availability of soil nutrients. Most tissues in C_3 crops, especially in leaves, contain more nonstructural carbohydrates when the plants they are part of are grown in elevated [CO_2]. As a result, greater quantities of carbohydrate export from leaves can be expected, perhaps mainly during the night (Grimmer and Komor, 1999). During the course of C_3 crop development, carbohydrates may accumulate in leaves, leading to photosynthetic acclimation (down-regulation) to elevated [CO_2]. Acclimation may be weaker (or more slowly developing) in crops with large storage organs or in those crops able to grow more branches, tillers, and fruits (i.e., indeterminate crops such as cotton). Those indeterminate C_3 crops are the ones in which the greatest yield increases are to be expected as a result of elevated [CO_2]. A key result of greater photosynthesis caused by elevated [CO_2] is increased partitioning of assimilate to growing seeds and storage organs. If other factors are unchanged, this should enhance the yield of C_3 crops.

Elevated tropospheric [O_3] can impact photosynthate partitioning by lowering rates of photosynthesis, shortening leaf-area duration, and directly interfering with carbon export from leaves. Reduced ability to export assimilate from leaves may be most important. Direct damage to sugar carriers in the plasmalemma of companion or sieve cells may be the cause of carbohydrate retention in source leaves exposed to elevated [O_3]. Regardless of the mechanism, less carbohydrate available for belowground growth leads to lower root/shoot ratios, which can compromise a crop's capacity to acquire soil resources.

Salinity does not greatly affect photosynthate partitioning processes. It does, however, restrict growth of shoots more than roots, and this commonly leads to an increase in the root/shoot ratio. Carbohydrates often accumulate in leaves of salt-stressed plants, but it remains unclear whether they are retained there for osmotic adjustment or if carbon is simply assimilated more quickly than it can be exported.

Chapter 7

Mineral Nutrition

Some time postgermination, after a seedling has depleted minerals, fats, proteins, and carbohydrates stored in the seed, it must actively and selectively acquire nutrients from both soil and air. Plants require nine nutrients in relatively large quantities (macronutrients; H, C, O, N, K, Ca, Mg, P, and S) and about eight in small quantities (micronutrients; Cl, Fe, B, Mn, Zn, Cu, Ni, and Mo, and perhaps Na), each supporting specific functions within the plant (Table 7.1). Dry mass of crop plants is typically about 6 percent H, 40 to 45 percent C, 42 percent O, and 1 to 2 percent N, with the remaining 5 to 11 percent comprised of other, mostly essential, elements (Wignarajah, 1995). Nearly all C in plants is obtained from atmospheric CO_2 during photosynthesis, and the H and O are obtained mainly from soil water. The other nutrients are acquired from the soil solution, where they are "mined" by roots, with the exception of some N in N_2-fixing plants (Chapter 10), which does not require the soil solution. We call these mineral nutrients.

Plant chemical composition differs substantially from the chemical compositions of atmosphere and soil solution. Most nutrients are needed in plant tissues in much higher concentrations than found in soil solution, while a few are required at lower concentrations. Plant success depends on assimilation and maintenance of nutrient levels between well-defined upper (the toxicity threshold) and lower (the deficiency threshold) limits (Figure 7.1). The range of soil nutrient concentrations required for plants to maintain maximal growth (the "luxury consumption zone") can slide in either direction along the x-axis of Figure 7.1, depending on soil and atmospheric conditions, biotic or abiotic stress levels, and tissue concentrations of other nutrients.

Mineral nutrients are needed for fixation, partitioning, and use of C (Cakmak and Engels, 1999). Any environmental factor that

TABLE 7.1. Essential plant nutrients, their mobility in the plant, functions, and deficiency symptoms.

Nutrient assimilated	Form assimilated	Plant mobility	Key plant functions	Key deficiency symptoms
Nitrogen	NH_4^+ (pH > 6.8) NO_3^- (pH < 6.8)	high	Structural component of proteins Component of heterocyclic molecules and alkaloids	Chlorosis Increased root:shoot Slow growth Roots less branched Early flowering Shorter life cycle Purple stems/leaves
Phosphorus	HPO_4^{2-} (pH > 7.0)	high	Component of nucleic acids Component of ATP, ADP, AMP, PPi Backbone of DNA, RNA Component of sugar phosphates Membrane constituent	Dark green leaves Stunted growth Maturity delayed
Potassium	$K+$	high	Enzyme activator Osmoregulation Maintains cation/anion balance	Stomatal closure Susceptibility to frost Stunted growth Nectrotic lesions Plants lodge easily
Magnesium	Mg^{2+}	high	Component of chlorophyll Enzyme cofactor/activator Regulates cellular pH Maintains cation/anion balance Important for ribosome structure	Interveinal chlorosis Mottled leaves

Sulfur	SO_4^{2-}	intermediate	Forms disulfide bonds in proteins Constituent of thiamine, biotin, ferredoxin, and 2' metabolites	Increased root:shoot Decreased chlorophyll Symptoms similar to N deficiencies
Calcium	Ca^{2+}	low	Component of middle lamellae Essential for membrane function Second messenger (enzyme activator)	Reduced cell division and elongation Crinkled leaves
Iron	Fe^{3+} $Fe(OH)^{2+}$ $Fe(OH)_2^+$	low	Constituent of catalase, cytochromes, ferrodoxin, and phytoferritin	Interveinal chlorosis
Manganese	Mn^{2+} Mn^{3+} Mn^{4+}	low	Enzyme activator Found in chloroplast membranes Required for photolysis of H_2O	Interveinal chlorosis Reduced grain yield Low photosynthesis Reduced [chlorophyll]
Boron	H_3BO_3	low	Carbohydrate metabolism (?) Cell-wall synthesis (?) Phenolic/lignin metabolism (?) Respiration (?) Pollen tube elongation(?) RNA metabolism	Abnormal root growth Inhibited DNA/RNA synthesis Misshaped leaves Fruit/flower abortion
Chlorine	Cl^-	intermediate	Required for photolysis of H_2O Required for leaf expansion Osmoregulation	Slowed root growth and subapical swelling Leaf damage such as poor growth, wilting, chlorosis, and necrosis

TABLE 7.1 *(continued)*

Nutrient assimilated	Form assimilated	Plant mobility	Key plant functions	Key deficiency symptoms
Copper	Cu^{2+} (aerated soil) Cu^{+} (wet soils)	intermediate	Vital for electron transport Phenolic/lignin metabolism	Necrosis of apices Distorted new leaves Dark green leaves
Zinc	Zn^{2+} $ZnOH^{+}$ (high pH)	intermediate	Essential for enzyme function Essential for ribosome and membrane structure	Reduced leaf growth Shortened stem internodes
Molybdenum	MoO_4^{2-}	intermediate	Component of nitrate reductase Required for ABA synthesis (?) Needed for purine degradation	Interveinal chlorosis Twisted young leaves Decreased flowering or premature flower abscission

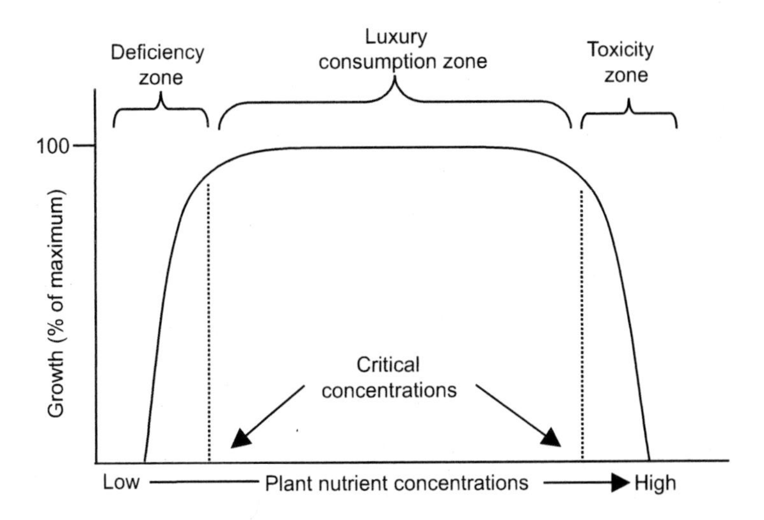

FIGURE 7.1. Generalized dose response curve describing crop growth responses to a range of tissue nutrient concentrations (Grattan and Grieve, 1994). The slopes of the tails of the dose response curve (i.e., above and below critical concentrations) is nutrient dependent (Marschner, 1995). For micronutrients the slope is steepest, and for N the slope is the flattest (assuming the nutrient supply is expressed in the same mass units).

contributes to soil nutrient imbalances (deficiencies or toxicities), inability to acquire or assimilate nutrients, or disrupted nutrient transport within the plant, will negatively affect crops. Vegetative growth, flower initiation and development, development and viability of pollen grains, and growth of storage organs such as tubers and seeds could all suffer (Cakmak and Engels, 1999). However, environmental factors that enhance nutrient uptake or improve nutrient use efficiency (amount of product or process per unit nutrient in tissue) will likely enhance crop yield.

Effects of environmental changes are difficult to predict because nutrient uptake is a multidimensional process, regulated at levels spanning from molecules to landscapes (Figure 7.2). For instance, environmental changes are likely to influence crop nutrition by affecting the membrane-bound transporter proteins that shuttle nutrients from soil into plant. At the same time, by influencing water consumption by an entire crop, environmental changes might also affect

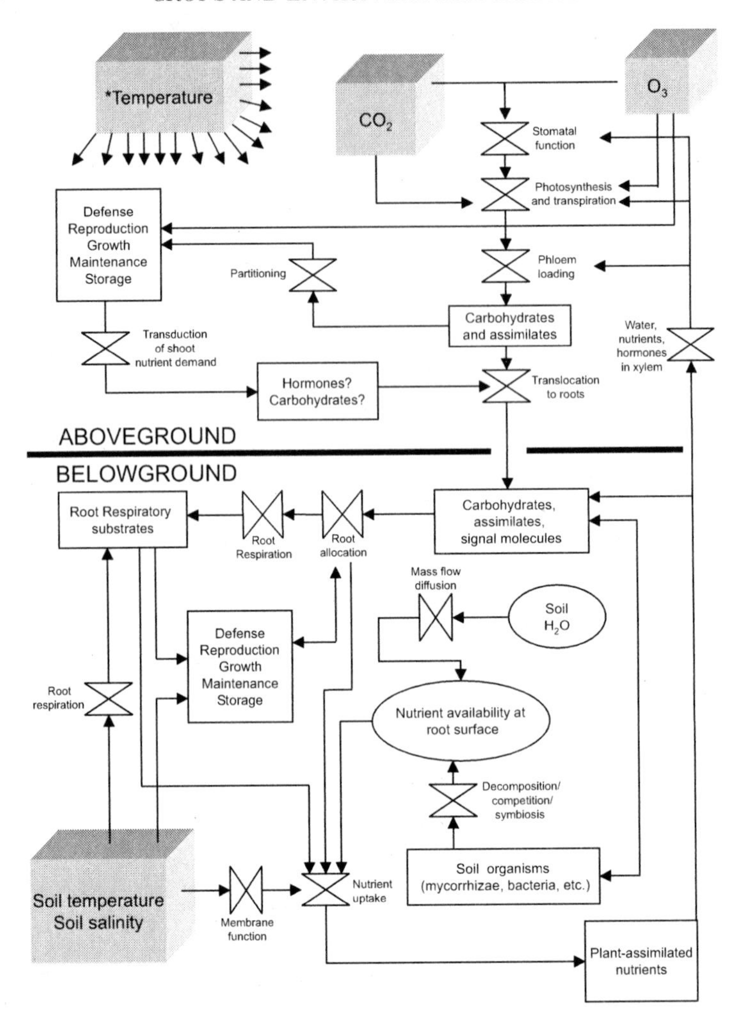

FIGURE 7.2. Conceptual model illustrating the processes that affect nutrition of crop plants, and the pools and processes most likely to be impacted by a changing environment. Shaded 3-D boxes indicate environmental changes covered in this book. Double triangles represent the processes that mediate flux or signaling between pools. Squares represent metabolic pools except for the boxes representing the combined processes of defense, reproduction, growth, maintenance, and storage (for both shoots and roots). Ovals represent soil resources. Root architectural and demographic considerations have been omitted from this model. *Changes in both aerial and root-zone temperatures will affect all aspects of plant function.

the flow of water (and all the nutrients dissolved in it) through soil and plant. Changes in branching patterns, depth, and lateral spread of roots are likely to affect mineral nutrition, as are shifting relationships between plants and soil microorganisms such as mycorrhizae and bacterial N_2 fixers (see Chapter 10).

NUTRIENT PASSAGE FROM SOIL TO ROOT AND BEYOND

Mineral nutrients must be present at the interface of root and soil before they can be taken up into root cells. Nutrients reach the root surface by three methods (Table 7.2):

1. *Root interception:* Roots growing through soil push aside soil particles and come into more or less direct contact with mineral nutrients. Root interception depends on structural aspects of roots, such as root surface area, length, and volume. Environmental changes that influence root growth are likely to affect uptake of nutrients by root interception.
2. *Mass flow:* Nutrients dissolved in the soil solution reach the root surface primarily by mass flow. Environmental changes that affect mass flow of water through the SPAC will affect this mechanism of nutrient uptake.
3. *Diffusion:* Nutrients move along their concentration gradient from soil to the root surface. Pure diffusion in solution occurs over only short distances (mm) and is the least important mechanism of nutrient uptake (except for potassium). Uptake by diffusion depends to a great extent on root structural characteristics.

Once at a root surface, nutrients must cross the epidermis, cortex, and endodermis before being loaded into xylem for transport to plant structures. (See Figure 3.6 for a diagram of the pathway water and dissolved nutrients must take to reach root xylem.) Two paths may be taken by nutrients on the journey from soil to stele (the conducting tissue composed of xylem and phloem elements surrounded by a layer of living cells called the pericycle): apoplastic and symplastic (Jacoby, 1995).

TABLE 7.2. The significance of root interception, mass flow, and diffusion in providing a maize crop with nutrients.

Nutrient	Requirement for yield of 150 bu acre maize crop	Amount (lb/acre) supplied by		
		Root interception	Mass flow	Diffusion
Nitrogen	170	2	168	0
Phosphorus	35	1	2	33
Potassium	175	4	35	136
Calcium	35	60	150	0
Magnesium	40	15	100	0
Sulfur	20	1	19	0
Copper	0.1	0.01	0.4	0
Zinc	0.3	0.1	0.1	0.1
Boron	0.2	0.02	0.7	0
Iron	1.9	0.2	1.0	0.7
Manganese	0.3	0.1	0.4	0
Molybdenum	0.01	0.001	0.02	0

Source: Adapted from Barber and Olson, 1968.

Apoplastic transport of water and solutes across the root involves rapid, passive, nonselective movement through the cell-wall/intercellular space continuum, without entering a cell. In the apoplast, nutrients can freely wash out of roots with water. Because of the presence of immobile negative charges in cell walls, significant cation exchange may also occur as a mineral ion makes its way to the stele (Marschner, 1995). Apoplastic passage is primarily confined to the epidermis and cortex because a ring of hydrophobic suberin, impregnated in radial walls of endodermal cells, forces water and solutes to enter the symplast before it can reach xylem. In order to pass from the apoplast to the symplast (the many interconnected protoplasts of a tissue) a membrane must be crossed, and it is here that selectivity is exerted.

Symplastic radial transport across roots involves slower, nonreversible passage of water and solutes from cell to cell through plasmodesmata, the pores linking together the cytoplasm of adjacent cells. Passage of solutes from apoplast to symplast can occur within several different cell types: through root hairs or other epidermal cells, exodermis (suberized cells between the epidermis and cortex found in some species, including maize and onion [Enstone and Peterson, 1992]), cortex, or endodermis. Regardless of where nutrients first enter the symplast, they must actively or passively cross a membrane barrier. The mechanics of solute uptake by plant cells through membranes is discussed in Chapter 3. Briefly, positively charged mineral ions are passively assimilated by root cells through uniporters while anions are actively taken up against an electrochemical gradient by protein carriers. In addition, passive uptake of some cation species is probably followed by their active transport back out of roots (or active transport into vacuoles for storage).

To summarize, soil mobile nutrients are carried to root surfaces primarily by bulk flow with water. Uptake of immobile nutrients is accomplished by root interception or diffusion. At the root surface, solutes pass (through apoplast and/or symplast) through the epidermis and cortex, are forced into the symplasm of the endodermis, and are then symplastically transported from cell to cell until reaching the xylem. In the stele (centermost cells of roots), solutes pass back across a membrane barrier and enter nonliving xylem tissue (apoplast) the pipes through which water, and all that is dissolved in it, is pulled into transpiring leaves. By forcing solutes from the apoplast to the symplast and then back again to the apoplast, two membrane barriers and thus two layers of selectivity are present.

REGULATION OF ION UPTAKE

Many nutrient transporters located in membranes of root cells are expressed constitutively (low-affinity transport systems or LATS), while others are induced by nutrient deficiencies (high-affinity transport systems or HATS). Inducible transporters enable plants to adjust uptake rates in response to environmental conditions and nutrient requirements. Adaptation likely involves either repression/induction or activation/deactivation of protein transporters in response to messages transduced from internal (e.g., hormones, [Ca^{2+}], carbohydrate levels) and external (environmental) conditions (Reid, 1999). How metabolic and structural nutrient demands are communicated to root cells, where nutrients are taken in, and the adaptations that occur there are unknown.

Lack of knowledge about how nutrient transporters function and are regulated limits our ability to accurately predict how plant nutrient relations will be affected by environmental changes. We are currently limited to studies of the effects of environmental conditions on the overall, or net, flux of nutrients into plant systems. Net nutrient uptake by crops can be quantified by measuring uptake kinetics (described later) or by measuring nutrient concentrations of plants grown under different experimental conditions.

KINETICS OF NUTRIENT UPTAKE

The pioneering work of Epstein and Hagen (1952) demonstrated that the rate of ion uptake by roots increased with increasing ion concentration only until a saturation rate was reached. When the concentration at which all transport proteins become saturated with their substrate is reached, uptake begins to behave independently of solute concentration in soil. Kinetics of nutrient uptake by roots therefore follow the typical enzyme/substrate relationships described by the Michaelis-Menton equation:

$$v = V_{max}\,[ion] / (K_m + [ion]) \qquad [7.1]$$

where:

v= ion uptake rate (for a specific ion species),
[ion]= ion concentration,
V_{max}= uptake rate when all available carriers are occupied, and
K_m= apparent affinity of the transporter for the ion (in units of ion concentration and equal to the ion concentration when $v = V_{max}/2$).

Methods used to solve Equation 7.1 involve uptake rate measurements across a range of ion concentrations. These provide information about ion uptake kinetics per se when gross uptake from soil by roots is known, but can also be used to describe net ion uptake (i.e., the balance of uptake and efflux through leaks or transporters) as long as uptake is significantly faster than efflux. Gross uptake rate can be estimated experimentally by measuring the short-term influx of labeled tracers (Reid, 1999).

Unfortunately, multiphasic uptake kinetics are often encountered when plants are supplied with a broad range of nutrient availabilities (Jacoby, 1995), limiting the power of Equation 7.1 for some nutrients under some conditions. The multiphasic nature of transporter systems may be attributed to a single transporter that undergoes a nutrient-concentration-dependent phase change, or from the simultaneous functioning of separate high- and low-affinity transport systems (Jacoby, 1995; BassiriRad, 2000). In spite of some limitations, and because alternative methods are lacking, Equation 7.1 is often used to describe ion uptake kinetics in crops.

ENVIRONMENTAL CHANGES

Elevated Atmospheric Carbon Dioxide Concentration

The general stimulation of growth by elevated [CO_2] is often accompanied by reduced specific activity of root physiology (e.g., nutrient uptake and respiration per unit of root length or mass decreases) despite greater carbohydrate availability in root tissues (Jongen et al., 1995; BassiriRad et al., 1996; Pritchard and Rogers, 2000). In fact, even when supplies of nutrients are freely and continuously available, their uptake does not always increase proportionally with growth and thus mineral concentrations in tissues decrease (although nutrient *content* does increase) (Reeves et al., 1994; Newbery et al., 1995; Hodge et al., 1998) (Table 7.3).

However, little evidence exists of consistent effects of growth in elevated [CO_2] on ion uptake kinetics, though the topic has received relatively little study (BassiriRad, 2000). For example, exposure to elevated CO_2 did not change the kinetics of NO_3^- or NH_4^+ uptake in soybean and sorghum (BassiriRad et al., 1999), carob seedlings (Cruz, Lips, and Martins-Luocâo, 1993), or bentgrass (Newbery et al., 1995). Most available information on effects of [CO_2] on crop nutrition comes from tissue nutrient analyses, not from uptake kinetic studies (Table 7.2).

TABLE 7.3. Effect of elevated atmospheric $[CO_2]$ on concentration (generally expressed as mg g^{-1} d wt) and content (mg plant) of mineral nutrients.

Crop	$[CO_2]$ (high/low)	Conc[a]	Cont[b]	Notes (nutrient)	Source
Wheat (L)[e]	550 / 350	0.94	—	FACE[c], ample irrigation (N)	A
		0.94	—	FACE[d], limited irrigation (N)	
		0.94	—	FACE[c], ample N (N)	
		0.75	—	FACE[d], low N (N)	
(G)	550 / 350	0.97	1.10	FACE[d], ample irrigation (N)	B
		0.96	1.17	FACE, limited irrigation (N)	
		0.97	1.10	FACE[d], ample N (N)	
		0.91	0.99	FACE, low N (N)	
(P)	700 / 350	0.96	1.00	Laboratory chamber (N)	C
		0.90	0.96	(K)	
		1.00	1.08	(P)	
Cotton (P)	550 / 370	0.89	1.21	FACE (N)	D
		1.00	1.32	(P)	
		0.93	1.26	(K)	
		0.86	1.17	(Ca)	
		0.94	1.28	(Mg)	
		0.91	1.23	(Cu)	
		0.94	1.28	(Fe)	
		0.97	1.31	(Zn)	
Rice (P)	589 / 389	—	1.13	FACE, very high N (N)	E
			1.03	FACE, ample N (N)	
	589 / 389	—	1.06	FACE, very high N (N)	F
		—	1.03	FACE, ample N (N)	
		—	0.99	FACE, low N (N)	
Soybean	Literature	0.94	—	28 observations (N)	G
Cotton	Literature	0.66	—	17 observations (N)	
Oat	Literature	0.78	—	3 observations (N)	G
Pea	Literature	0.91	—	1 observation (N)	

Crop	$[CO_2]$ (high/low)	Conc[a]	Cont[b]	Notes (nutrient)	Source
Sorghum	Literature	0.99	—	1 observation (N)	
Wheat	Literature	0.84	—	20 observations (N)	
Maize	Literature	0.86	—	4 observations (N)	

Sources: (A) Sinclair et al. (2000); (B) Kimball et al. (2001); (C) Van Vuuren et al. (1997); (D) Prior et al. (1998); (E) Kobayashi, Lieffering, and Kim (2001); (F) Kim et al. (2001); (G) Cotrufo, Ineson, and Scott (1998)

[a] Conc = concentration of nutrients in high $[CO_2]$ relative to low $[CO_2]$ (unitless).

[b] Cont = content of nutrients in high $[CO_2]$ relative to low $[CO_2]$ (unitless).

[c] Results averaged over four years.

[d] Results averaged over two years.

[e] Letters in parentheses after the crop indicate which part of the plant is being referred to: (L), leaves; (G), grain; (P), whole plant.

Aside from direct changes in nutrient transporters, changes in plant water use resulting from growth in high CO_2 might account for reduced concentrations of nutrients in plant tissues, especially those with high soil mobility (such as N and K) (Box 7.1). Reductions in stomatal conductance and transpiration are common in crops grown in CO_2-enriched atmospheres. Corresponding reductions in bulk flow of water through the SPAC could slow passage of mobile nutrients through the soil to the root surface. Fewer nutrients available at root surfaces could decrease uptake of solutes independent of direct changes in membrane transporters (Lambers, Stulen, and van der Werf, 1996; Van Vuuren et al., 1997).

Decreased water flow through the plant-soil system is unlikely to change uptake rates of immobile nutrients, however. For nutrients with low soil mobility, effects of elevated CO_2 on nutrient content or concentration are more likely attributable to changes in spatial rooting patterns. In fact, uptake of immobile elements such as P and Zn may be enhanced (relative to more mobile nutrients) by CO_2-enrichment because of more thorough soil exploration due to increased root length density (Chapter 8). However, greater root proliferation could increase the overlap of nutrient depletion zones of adjacent branches, sometimes leading to less uptake per unit of root length (i.e., greater root/root competition).

BOX 7.1

Hypothesized mechanisms (both negative and positive) whereby rising atmospheric [CO_2] is likely to affect crop mineral nutrition.

Potential explanations for commonly observed reductions in tissue nutrient concentrations:

1. Reduced uptake efficiency (less uptake per unit length of root)
   a. Changes in root membrane transporter function
   b. Reduced passage of nutrients to the root surface as a result of slower bulk flow of water through the SPAC
   c. Increased resistance to radial water flow through roots (i.e., reduced hydraulic conductance)
   d. Overlap of depletion zones by adjacent root branches (greater root/root competition)
   e. Increased competition of soil microbes versus roots for finite resources
2. Artifacts
   a. Nutrients are diluted by large pools of nonstructural carbohydrates (i.e., starch and soluble sugars)
   b. Available nutrients are used up by larger plants (probably only applies to container studies)

Potential compensatory mechanisms whereby increasing [CO_2] might increase total plant nutrient acquisition:

1. Changes in root architecture
   a. More roots exploring a given volume of soil more thoroughly
   b. Greater horizontal or vertical soil exploration by longer roots
2. Enhanced root exudation (e.g., organic acids, sugars, amino acids) could stimulate microorganisms and hence nutrient availability by affecting:
   a. Activity of organisms important for decomposition or nutrient mineralization
   b. N_2 fixation
   c. Mycorrhizae

Plant preference for different chemical forms of the same nutrient may also change in a higher CO_2 environment. For example, NH_4^+ uptake rate by six grasses was unaffected by elevated [CO_2] (ambient + 350 ppm), whereas NO_3^- uptake was slowed (Jackson and Reynolds, 1996). Conversely, uptake rates of both NH_4^+ and NO_3^- by soybean and grain sorghum were unaffected by elevated [CO_2] (BassiriRad et al., 1999). While it is becoming clear that plant preference for NO_3^- versus NH_4^+ can be affected by elevated [CO_2], the direction and magnitude of preference changes may be species specific, and the mechanisms driving such shifts are poorly understood (BassiriRad, 2000). Implications of a preference switch from NO_3^- to NH_4^+ (or vice versa) include the different costs of uptake and different costs of assimilation of the two N species. (A word about N_2 fixation: In N_2-fixing crops, elevated [CO_2] stimulates N_2 fixation, presumably because of greater photosynthesis and supply of respiratory substrate to roots and nodules; Chapter 10.)

Growth in High CO_2 May Reduce Plant Demand for N

Plants growing in CO_2-enriched environments may require less N to function than those growing in ambient CO_2 conditions. Under high CO_2, regeneration of ribulose-bisphosphate (rubP) and inorganic phosphate (P_i) tends to limit photosynthetic rates more than carboxylation by rubisco (i.e., fixation proper) (Chapter 5). This may enable the photosynthetic apparatus to function efficiently while maintaining only a fraction of the rubisco, which normally represents some 60 percent of total soluble leaf proteins. Growth in high CO_2 also alleviates photorespiration, and this might lead to a smaller diversion of N into enzymes of the photorespiratory pathway (Fangmeier et al., 2000). In other words, at least in the case of N, it could be demand rather than supply that most limits uptake and eventually leads to the decrease in plant [N] that almost invariably accompanies growth in supra-ambient CO_2 levels.

Nutrient Availability Mediates the Growth Response to High [CO_2]

In spite of the many compensatory mechanisms whereby exposure to elevated CO_2 may enable plants to compensate for soil nutrient limitations (Box 7.1), low productivity resulting from nutrient limitations usually is not remedied by growth in higher atmospheric [CO_2]s. Often, the CO_2 response is completely absent (or may even become negative) when nutrients are at very low concentrations. Rarely does the *relative* growth response to CO_2-enrichment under low-nutrient conditions exceed the relative growth enhancement observed in plants supplied with ample nutrients (although this is sometimes the case when atmospheric CO_2 concentrations are very high, much higher than we expect to experience within the next 100 years). The *absolute* yield response to elevated CO_2 is always maximal in adequately fertilized plants (Rogers et al., 1999; Stitt and Krapp, 1999).

Conclusions

In general, elevated [CO_2] stimulates crop growth. Although larger plants accumulate more total nutrients (i.e., nutrient content increases), the concentration of nutrients (g nutrient per g plant dry mass) ordinarily decreases. Exposure to high CO_2 leads to less nutrient uptake per unit of root length. Although the physiological basis is not understood, mineral nutrition is significantly influenced by elevated atmospheric [CO_2] (Sinclair, 1992). It is clear that conventions regarding the soil and plant nutrient levels required for maximal yields might need to be adjusted up as atmospheric [CO_2] rises.

Salinity

High NaCl concentrations in soil interfere with plant mineral nutrition. In fact, some researchers have suggested that the reduction in crop growth and productivity on saline soils results *primarily* from nutritional imbalances. Symptoms of salt stress can be lessened by supplying nutrients according to the severity of the salt stress (Figure 7.3). Under mild salinity stress, optimal nutrient availability may boost salt tolerance (+). At moderate salinity levels nutrient supply often has no impact on salt tolerance (0). When salt stress is severe, nutrient levels that would be optimal under nonsaline conditions can aggravate salinity stress (–). The nutritional status of crops therefore plays an important role in determining their sensitivity to salinity stress, and salt stress directly affects mineral nutrient uptake by the crop (Grattan and Grieve, 1994).

Na^+ and Cl^- Uptake

Sodium ions enter root cells passively. Passive uptake is made possible by negative membrane potentials inside cells coupled with a Na^+ chemical-potential gradient (Hasegawa et al., 2000). Flux of Na^+ into plant cells may, in turn, dissipate cell-membrane potential, facilitating passive Cl^- uptake via an anion channel. Otherwise, Cl^- uptake occurs via a Cl^-/H^+ symport process (secondary active transport).

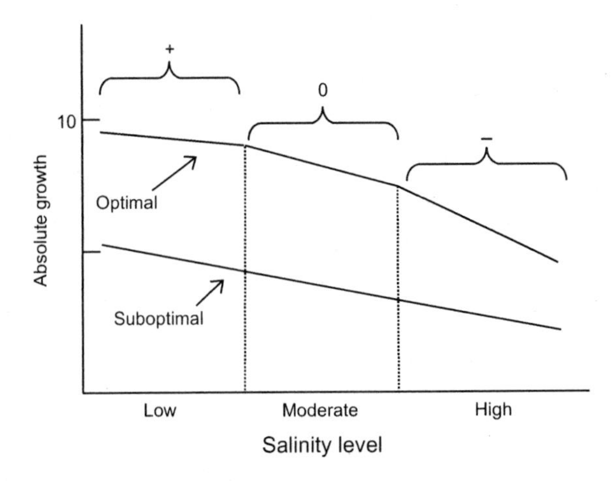

FIGURE 7.3. Theoretical representation of the interaction of salinity with nutrient availability (optimal or suboptimal). At low salinity, optimal fertility increases salt tolerance (+), at moderate salinity it has no effect on salt tolerance (0), and at high salinity it decreases salt tolerance (–). *Source:* Redrawn from Grattan and Grieve (1994).

Although the exact transport mechanism whereby Na^+ is taken into root cells is unknown, studies indicate that Na^+ competes with K^+ for uptake. Apparently these ions are transported by membrane proteins that lack the ability to distinguish one from the other (Hasegawa et al., 2000). The hydrated ionic radii of sodium and potassium are very similar, making discrimination by transport proteins difficult (Blumwald, Aharon, and Apse, 2000).

In some instances, accumulation of Na^+ in root cells might be advantageous, as this ion can serve as an inexpensive osmoticant (Yeo, 1983; Lazof and Bernstein, 1999). The advantages of accumulating Na^+ to maintain a favorable osmotic balance under drought or salinity stress might explain why so few plants exclude Na^+ from root cells (Yeo, 1983; Lazof and Bernstein, 1999; Hasegawa et al., 2000). Of course, *excessive* accumulation of Na^+ is a major problem for crops because it is toxic at cytosolic concentrations in excess of about 100 mM (Amtmann and Sanders, 1999).

Potassium

Competition for membrane carriers favors Na^+ over K^+ uptake as [NaCl]s increase, which can lead to K^+ deficiencies. This is important because K^+ is involved in turgor-pressure maintenance in young expanding tissues, stomatal movements, and enzyme function (Hawkins and Lewis, 1993a; Shabala, 2000). Furthermore, within cytoplasm, Na^+ can displace K^+ from enzyme activation sites (Amtmann and Sanders, 1999). Many crops exposed to high salinity accumulate Na^+ in shoot tissues at the expense of K^+, and this shift is associated with growth inhibition.

Although high cytosolic Na^+:K^+ ratios are generally considered incompatible with metabolism, some plants are able to substitute Na^+ for K^+ to some extent, allowing them to maintain normal cellular function under salt stress. Crops generally fit into one of four groups with respect to their propensity to substitute Na^+ for K^+ (Marschner, 1995).

Group A: Includes members of the Chenopodiaceae (beets, turnips, swiss chard) and some C_4 grasses. In these plants a high proportion of K^+ can be substituted for by Na^+ without a loss of yield. In fact, yield is sometimes increased by adding Na^+.

Group B: Includes cabbage, radish, cotton, pea, wheat, and spinach. A much smaller proportion of K^+ can be substituted for by Na^+ before a decrease in growth is observed (compared to Group A).

Group C: Includes barley, millet, rice, oat, tomato, and potato. Only minor substitution of K^+ by Na^+ is possible. Na^+ has no specific effects on growth at low to moderate concentrations.

Group D: Includes maize, rye, soybean, and common bean. No substitution of K^+ is possible.

Our most important crops are in Group C or Group D (except for wheat and cotton). These tend to be more salt sensitive (except for barley) than those in Groups A and B (Marschner, 1995). This phenomenon underscores the link between salt sensitivity and K^+ nutrition.

Lazof and Bernstein (1999) argue that too few studies have looked at the effects of salt stress on other nutrients to conclude that deviations in Na^+:K^+ ratios represent the chief mode of salt toxicity. They reviewed 50 articles (1980 to 1995) on plant response to salinization, and K^+ uptake and/or transport was examined in 72 percent whereas only 18 percent studied any other nutrient. They also point out that in spite of the large database on K^+/Na^+ interactions, rarely has the addition of extra K^+ to salt-stressed plants actually increased yield (Grattan and Grieve, 1994).

Interactions of soil salinity with several other nutrients have been implicated. Of these, Ca^{2+} has received the most attention. For example, Cramer, Läuchli, and Polito (1985) reported that excess NaCl leads not only to membrane depolarization and loss of K^+ but Ca^{2+} ions are also displaced from root plasma membranes (of cotton root hairs, in this case). Displacement of calcium ions might represent a key event in salinity-induced disruption of normal membrane permeability. In fact, in cases where both were measured, rarely has [K^+] decreased more than [Ca^{2+}] (Lazof and Bernstein, 1999).

Calcium

Increasing Ca^{2+} supply to crops often ameliorates symptoms of salt stress (Bressan, Hasegawa, and Pardo, 1998; Epstein, 1998). As an example, raising the Ca^{2+} content of nitrate-fed wheat from 2 to 12 mM in the presence of high Na^+ produced an 11 percent increase in growth (Lewis, Leidi, and Lips, 1989). Similar effects have been observed for many other crops as well.

Ca^{2+} may alleviate salinity stress by stabilizing membranes and consequently enhancing the selective absorption of other mineral nutrients into root cells (Cramer, Läuchli, and Polito, 1985; Lewis, Leidi, and Lips, 1989; Hawkins and Lewis, 1993a,b; Huang and Xu, 2000). The positive effects of calcium on membranes, which include maintaining membrane structure, permeability, and transporter function, are thought to counteract the negative effects of Na^+ on membranes, including depolarization, disruption of ion selectivity, and loss of transporter function (Epstein, 1998). Ca^{2+} also blocks voltage-independent, nonselective ion channels (VIC channels), major pathways for Na^+ uptake into cells under saline conditions (Shabala, 2000).

The positive relationship between salt tolerance and soil Ca^{2+} availability might also stem from the role of calcium as a second messenger. At least in some plants, the process of adapting to salt stress is initiated when Ca^{2+} from cell walls and vacuoles floods into cytoplasm, which in turn activates a Ca^{2+}-dependent signal transduction cascade leading to up-regulation of the SOS pathway (Chapter 3; Bressan, Hasegawa, and Pardo, 1998). The SOS system functions in sequestering Na^+ in vacuoles and cell walls, where it causes the least damage to metabolism.

Nitrogen

Salt-sensitive crops contain less tissue N when grown under saline conditions compared to the same genotypes grown under nonsaline conditions and salt-resistant genotypes grown under saline conditions. For instance, salinization of wheat reduced the N uptake rate 30 percent, while the V_{max} of NO_3^- uptake dropped 38 percent, and the N concentration of the plant fell 35 percent (Hawkins and Lewis, 1993b). NH_4^+-fed plants exposed to a similar salinity treatment experienced a 50 percent reduction in the V_{max} of NH_4^+ uptake and a 16 percent drop in total plant N content.

How salinity reduces N uptake is unknown, but it is probably linked to the negative effects Na^+ has on membrane permeability, as discussed earlier (Hawkins and Lewis, 1993a). Fortunately, lack of N caused by low to moderate salinity stress can be counteracted by adding extra N (beans, Lunin and Gallatin, 1965; Wagenet et al., 1983; maize and millet, Ravikovitch, 1973; tomato, Papadopoulos and Rendig, 1983; cotton and maize, Khalil, Amer, and Elgabaly, 1967). Under severe salinity stress, however, addition of extra N is apparently not beneficial (Grattan and Grieve, 1994; Figure 7.3).

NO_3^- appears to be a better source of N than NH_4^+ in salt-affected areas. Growth inhibition in salt-stressed plants is usually greater when plants are fed NH_4^+ compared to NO_3^- (Lazof and Bernstein, 1999).

Phosphorus

In contrast to K^+, Ca^{2+}, and N, which decrease in concentration in salt-stressed plants, salinization sometimes leads to *accumulation* of P in shoot tissues. Phosphorous concentrations commonly used in solution cultures often become toxic when the rooting media is amended with NaCl (Grattan and Maas, 1984). In fact, when grown in saline media, two soybean varieties were killed by 0.20 mmol L^{-1}

P (a level commonly used in nonsaline media), whereas no injury was observed on the same cultivars treated with 0.02 mmol L^{-1} P. Salt tolerance also decreased with increasing P concentrations in maize plants (Bernstein, Francois, and Clark, 1974).

Studies also showed that salinity stress led to decreased transport of ^{32}P to young tissues and shoot apical meristems (Martinez, Bernstein, and Läuchli, 1996). Lazof and Bernstein (1999) suggested that the increase in [P] of the shoot may be a byproduct of decreased ability to transport P to young growing leaves. If P transport mechanisms to sinks were disturbed by NaCl, they explained, this could result in "starvation" at some sites and toxicities at others. This may have been the case in two grain sorghum cultivars, in which soil salinity consistently decreased P concentration in leaves (Francois, Donovan, and Maas, 1984). It is possible that information regarding P starvation within actively growing tissues could be transduced to roots by signal molecules, resulting in up-regulation of root P assimilation, further exacerbating toxicities in some tissues. This idea remains speculatory.

Most studies reporting P accumulation in response to salt stress were conducted in either sand or solution cultures in the laboratory (Grattan and Grieve, 1994). For field-grown crops grown in soil, shoot P concentrations are more likely to decrease. Differences between the rooting media probably explain this variability. In soil, especially those high in salts, less P is generally *available* to plants (even at similar [P] to those used in solution culture experiments) because much of it is bound to soil particles or organic matter. Further research in the field will probably show that interactions between P nutrition and salinity tolerance reflect the model illustrated in Figure 7.3.

Magnesium

Even though Mg^{2+} is required for microtubule assembly during mitosis, and inhibition of cell division/expansion are typical symptoms of salinity stress, the effects of salt stress on Mg^{2+} nutrition have not been thoroughly studied. Perhaps this is because [Mg]s are often high in saline soils and therefore this line of research might prove largely academic. Nevertheless, shoot Mg^{2+} contents have been reported to decrease in salt-stressed sorghum (Francois, Donavan, and Maas, 1984), white lupin (Jeschke, Wolf, and Hartung, 1992), and

Atriplex amnicola (Aslam et al., 1986) plants but were unchanged by salinity in maize (Grieve and Maas, 1988). Studies to date are too few and contradictory to make any generalizations concerning the influence of soil salinity on Mg^{2+} nutrition in crops.

Other Nutrients

Effects of salinity on few other nutrients have been studied. In fact, few studies of effects of *any* environmental stress on plant micronutrient dynamics have been published, perhaps because micronutrients are rarely growth limiting in the field.

Elevated Atmospheric Ozone Concentration

Atmospheric O_3 does not penetrate the soil and therefore does not directly influence nutrient uptake by roots (Blum and Tingey, 1977). Exposure of shoots to high $[O_3]$ can indirectly affect plant mineral nutrition, however.

First, because O_3 is absorbed principally through stomates, nutritional factors that affect stomatal movement (e.g., K^+ concentrations), ontogeny, or distributions across the leaf surface (i.e., stomatal density and indices) may lead to nutrient-O_3 interactions. Second, nutritional factors that affect the cellular repair process (such as those involved in alleviating oxidative damage to enzymes), could also indirectly affect crop sensitivity to O_3 damage. Third, O_3 exposure can lead to up- or down-regulation of genes, including those that code for plant hormones (Figure 3.3). Hormones have wide-ranging effects throughout the plant which can affect root growth and other aspects of nutrition. Fourth, O_3 diminishes photosynthesis, growth, and perhaps also demand for nutrients. Fifth, mineral nutrition may affect the uptake of O_3 by controlling density of leaves in the plant canopy. Optimal plant nutrition is associated with denser canopies compared to low fertility. This may increase resistance to air pollutants by creating gradients in $[O_3]$ (resulting in protection of leaves deeper in the canopy) (Cowling and Koziol, 1982).

Early reviews on this topic concluded that the influence of air pollutants, including O_3, on nutrient relations as well as root growth and physiology have been largely neglected (Cowling and Koziol, 1982). A more recent review echoed this sentiment (Taylor and Ferris, 1996). In fact, we were unable to find a single article that attempted to examine the influence of O_3 pollution on nutrient uptake kinetics. Consequently, our current understanding of interactions of mineral nutrition with O_3 pollution comes exclusively from a handful of general studies in which differentially fertilized plants were grown in the presence of low or high O_3 levels.

Nutrition and O_3 Sensitivity

Interactions between plant nutrition and the effects of high $[O_3]$ have been investigated in tobacco. The majority of these studies found that the incidence and severity of O_3 damage increased with increasing soil fertility. In perhaps the most convincing study, Leone, Brennan, and Daines (1966), in 15 independent experiments, showed that leaf injury (necrosis and leaf fleck) was always more severe in tobacco fed an optimal supply of N compared with plants supplied with low N. Menser and Hodges (1967), who grew tobacco at four different N levels (15, 80, 160, 800 mg L^{-1}), reported generally similar results. They found leaf O_3 injury to be most severe in plants fed 160 mg L^{-1} N and least severe in the most N-deficient plants. In contrast, MacDowall (1965) found that leaf O_3 injury to tobacco was greatest with either the lowest (deficiency) or highest (excess) rates of N addition.

Several authors have attributed greater resistance to O_3 at low N levels to the higher carbohydrate levels that typically result from growth under N-limiting conditions (Leone, Brennan, and Daines, 1966; Leone and Brennan, 1970; Cowling and Koziol, 1982; Pell et al., 1990). Alternatively, plants growing with abundant N produce more rubisco, which has been implicated as the major target for O_3 effects (Pell et al., 1990). So, under high N conditions, there may simply be more target molecules for O_3 to react with, leading to greater damage. More work is needed in this area.

Even less work has been done with other nutrients. Ozone toxicity in rutgers tomatoes, spinach, and mangel increased with increasing P application (Brewer, Guillemet, and Creveling, 1961; Leone and Brennan, 1970); but in radish, no interactions of P level with damage severity was observed (Ormrod, Adedipe, and Hofstra, 1973). Bean and soybean ozone damage was greater at low compared to high K availability. Similarly, in bean plants exposed to O_3 for 2 hours, damage was inversely related to the rate of S application (Adedipe,

Hofstra, and Ormrod, 1972). These authors concluded that O_3 sensitivity may have been reduced as a result of a higher content of sulfhydryl groups (a key site of O_3 damage) in plants grown with optimal S. Finally, unfertilized soybean was always less sensitive to ozone compared with fertilized plants (Heagle, 1979).

Although experimental results can vary, O_3 damage is generally more severe in plants supplied with high fertility levels than those grown with lower fertility rates (oat, barley, spinach, and lettuce, Middleton, Darley, and Brewer, 1958; spinach and mangels, Brewer, Guillemet, and Creveling, 1961; spinach, Leone, Brennan, and Daines, 1966; radish, Ormrod, Adedipe, and Hofstra, 1973; radish, Pell et al., 1990) (but also see the 1992 review by Runeckles and Chevone, in which just the opposite conclusion was reached). Plant productivity under high ambient ozone levels might therefore be improved by selecting a lower fertilization rate than might be optimal under nonpolluted conditions (Grantz and Yang, 1996b).

O_3 Toxicity and Nutrient Uptake

In general, O_3 damage is more severe when plants are provided with optimal to supraoptimal quantities of fertilizer. Effects of O_3 on nutrient uptake are less known but are probably minor. For instance, $[O_3]$ of the air had only small effects on P, K, Ca, Mg, Mn, and Na concentrations in tissues of spring wheat (Vandermeiren et al., 1992); total plant P as well as P and K concentrations of flour were reduced by high O_3, but N concentrations of straw and flour were increased (total plant N content did not differ). Heagle, Miller, and Booker (1998) found that exposure of soybean to ozone increased foliar N, C, S, and B but did not effect P or K concentrations. In tomato, exposure to ozone was accompanied by increased leaf P, K, Ca, and N contents. Although this is a small number of studies, these data indicate that O_3-stressed plants generally contain more nutrients than non-O_3-stressed plants. Apparently, O_3 pollution reduces growth more than nutrient uptake from soil.

Warming

Atmospheric warming will affect metabolism of shoot systems, including their demand for nutrients. Higher air temperatures will also warm the soil, directly affecting root physiology. In some cases, soil temperatures determine where and when a given crop can be grown. In fodder rape, for example, inadequate P in shoots of plants grown in cold soil is the factor most limiting to growth (Cumbus and Nye, 1985). Low root temperatures often cause plants to display symptoms characteristic of nutrient deficiencies, such as low fresh weight:dry weight ratios, young leaves that turn dark green, and anthocyanin production in stems, petioles, and mature leaves (Cumbus and Nye, 1985).

Nutrient uptake generally increases as soils warm (Figure 7.4). In excised roots, active absorption of ions is coupled with temperature, a relationship characterized by a Q_{10} approaching two (Cumbus and Nye, 1982). That is, for a 10°C increase in temperature, nutrient uptake approximately doubles. Increases in uptake as soil temperatures rise has been reported in sesame (Ali et al., 2000), oilseed rape (MacDuff, Jarvis, and Cockburn, 1994; Ali et al., 1998), tomato (Ali et al., 1994) winter wheat (Miyasaka and Grunes, 1997), and barley (MacDuff and Jackson, 1991), but reductions in nutrient uptake have also been observed for several species, including tomato (Klock, Graves, and Taber, 1996) and sorghum (Murtadha et al., 1989). Obviously, results depend on the experimental temperatures used as well

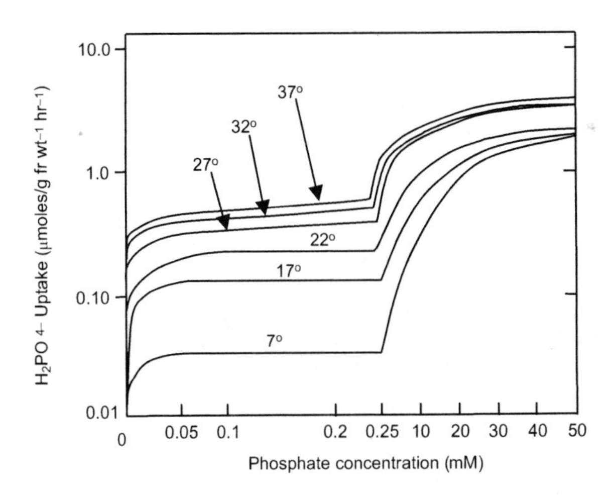

FIGURE 7.4. Example of temperature and concentration dependence of $H_2PO_4^-$ uptake in maize roots. Each point represents the mean of at least four determinations. Numbers indicate root temperature (in degrees Celsius). *Source:* Redrawn from Bravo-F and Uribe (1981).

as the soil thermal environment that a species is adapted for (Chapin, 1974).

Cool-season and warm-season plants have different soil temperature requirements and this will largely determine how nutrient uptake will be affected by soil warming. Soil warming will likely enhance nutrient uptake for crops growing in soils that are cooler than optimal (BassiriRad, 2000). For those already growing in soil of optimal or supraoptimal temperatures, higher soil temperatures may inhibit nutrient uptake; this is likely for cool-season grasses, for instance, because high soil temperatures already limit their growth (Huang and Xu, 2000). In any case, soil warming may prove more detrimental to plant success (including nutrient uptake) than air warming because roots have lower temperature optima than shoots.

Does Growth Control Uptake, or Does Uptake Control Growth?

The degree to which mineral nutrition is affected by warming may follow from direct effects of warming on sink activity (growth). For example, reduced nutrient uptake with decreasing soil temperatures could be linked to a decline in the rate of cell production by sinks (decreased nutrient demand) (Atkin, Edwards, and Loveys, 2000). Conversely, warming will likely stimulate cell division and expansion, creating greater demand for nutrients. If effects on sinks prove most important, then temperature-driven shifts in shoot and root nutrient demand must somehow be communicated to sites of nutrient uptake via chemical messengers that trigger up- or down-regulation of root transporters. Such a messenger has not been found.

Alternatively, temperature sensitivity of nutrient uptake might be explained by temperature sensitivity of root respiration, plasma membrane structure and function, function of nutrient transport proteins, and/or root hydraulic conductivity.

To What Extent Does Nutrient Uptake Depend on the Supply of Respiratory Products?

Root respiration is sensitive to temperature (Bravo-F and Uribe, 1981) (Chapter 5). The Q_{10} for respiration in maize roots is 3.6 from 2 to 13°C and 2.7 from 7 to 17°C. This may have implications for mineral nutrition, since ion uptake accounts for a significant percentage of respiratory energy generated in roots (Poorter et al., 1991). Temperature probably influences rates of active ion uptake into root cells indirectly by controlling respiration and ATP availability.

Predicting rates of nutrient uptake based solely on respiratory output may be unwise, however, because respiration and nutrient uptake rates often only roughly correspond (BassiriRad, 2000). Consider maize roots: P and K absorption reached a maximum at 25 and 35°C, while respiration rates did not reach a maximum until 40°C (Bravo-F and Uribe, 1981; BassiriRad, 2000). Furthermore, the Q_{10} values for P uptake were much higher than Q_{10} values for respiration. It is unlikely, therefore, that reduced ion uptake with decreasing temperatures (and increased uptake with increasing temperatures) is solely attributable to a drop (rise) in respiration.

Membranes

Root-zone temperatures may influence absorption of nutrients by changing the physical properties and function of root plasma membranes (Clarkson et al., 1988). As soil temperatures rise to optimal values, fluidity of membranes increases and ion transport by carrier proteins is improved. In several crop species, however, an abrupt increase in the activation energy of various membrane-bound reactions occurs at about 10°C and this apparently represents the upper temperature threshold for nutrient uptake. Above this temperature, membranes disorganize and roots quit functioning properly (Pike and Berry, 1980). Adaptation to high temperatures is associated with a higher phase separation temperature of membranes (Pike and Berry, 1980). It seems likely that warming could alter nutrient relations by affecting membranes and their associated protein transporters.

Discerning whether temperature-induced changes in nutrient uptake kinetics result from a direct effect on root transport properties or indirectly via a signal from the shoot has proven difficult (BassiriRad, Caldwell, and Bilbrough, 1993; Lambers, Chapin, and Pons, 1998). It is likely that short-term changes in soil temperatures will directly affect uptake by changing membrane properties and, in turn, the kinetics of enzyme function (irrespective of shoot demand) (Siddiqi, Memon, and Glass, 1984). In the long term, nutrient uptake by roots may be affected to a greater extent by chemical messengers from shoots resulting from growth demands (Engles, Munkle, and Marschner, 1992).

Hydraulic Conductivity

In general, root hydraulic conductivity increases as soil temperatures rise. In the case of intact plants and detached roots of barley, hydraulic conductance increased with temperatures up to 25°C (BassiriRad, Radin, and Matsuda, 1991). Similarly, increasing the root temperature from 20 to 30°C increased root hydraulic conductivity in wheat by 200 to 400 percent, which was attributed to greater conductance of protein water channels (Carvajal, Cooke, and Clarkson, 1996; see Chapter 4). Hydraulic conductance not only affects the rate of water passage through the plant but also may indirectly affect the transport of nutrients in soil solution to the membrane surface, where uptake occurs (see also Chapter 4).

Raising soil temperature alleviates the low hydraulic conductivity caused by insufficient nutrient availability as well as visible nutrient-deficiency symptoms (Carvajal, Cooke, and Clarkson, 1996). Hydraulic conductivity was reduced in N- and P-starved cotton, but was restored when root temperature was increased from 25 to 30°C for 4 h (Radin, 1990). Apparently, the fluidizing effect of higher temperature on membranes is responsible for this effect (Carvajal, Cooke, and Clarkson, 1996).

Nutrient Availability and Chemical Form Often Mediate Plant Response to High Root-Zone Temperatures

If soil temperature rises, plant preference for NO_3^- -N over NH_4^+ -N should increase (Clarkson and Warner, 1979). At low soil temperature, plants tend to favor NH_4^+ over NO_3^- (Clarkson, Hopper, and Jones, 1986), and at high temperatures NH_4^+ can become toxic. In carob, NH_4^+-fed plants displayed toxicity symptoms and were much smaller at a root temperature of 40°C compared to NO_3^--fed plants grown at the same temperature (Cruz, Lips, and Martins-Loucâo, 1993). NH_4^+ is often a better N source than NO_3^- at low, but not high, root-zone temperatures (Kafkafi, 1990).

SUMMARY

Crops must acquire enough nutrients from soil and air to meet growth and reproductive demands. Because most nutrients are needed at greater concentrations than are present in the soil solution, plants have evolved specific selective uptake mechanisms. High crop yield depends on an ability to assimilate and maintain nutrient levels between upper and lower thresholds.

Root nutrient-uptake kinetics, root morphology and architecture, as well as bulk flow of water through the SPAC all affect nutrient acquisition by roots. Environmental changes that influence these or other aspects of plant nutrition, such as nutrient transport and utilization within the plant, could shift crop-nutrient relations enough to influence yield quantity or quality.

Rising global atmospheric [CO_2] will improve growth and yield, but will decrease nutrient concentrations within most plant tissues. Reduced nutrient concentrations in a larger plant may be interpreted as reduced nutrient-uptake efficiency or, conversely, as greater nutrient-use efficiency. Regardless of how this phenomenon is viewed, crops may prove unable to acquire enough nutrients to keep pace with more C resulting from higher rates of photosynthesis.

Soil warming, an important component of global warming, will generally increase plant nutrient uptake capacity. It is unknown, however, whether the enhancement in nutrient uptake that occurs when soils are warmed is caused by changes in membrane-bound nutrient transporters or if higher uptake rates are the indirect result of changes in demand transduced from active sinks. It is likely that direct short-term affects of soil temperature on nutrient uptake may be negated by the long-term demand for nutrients from active sinks. It is uncertain if the temperature-driven increase in nutrient-uptake capacity will be of sufficient magnitude to meet the extra nutrient demand that will arise from greater photosynthetic rates and C availability in a higher CO_2 world.

Crops grown on saline lands suffer from the inability to adequately regulate ion uptake from soil. High soil Na^+:K^+ ratios favor uptake of Na^+ over K^+, resulting in electrophoretic flux of Na into root cells. Insufficient K^+ can lead to reduced turgor in young, expanding tissues and disturbed stomatal and enzyme function. Furthermore, as Na^+ floods into roots, plasma membranes depolarize, impairing function of various nutrient transport systems. Concentrations of most nutrients decrease in crops challenged with salinity. Addition of Ca^{2+}, however, can often restore function of nutrient-uptake machinery, balance plant-nutrient relations, and alleviate growth inhibitions associated with NaCl toxicity (to some extent). Ca^+ effects this change by stabilizing membranes while also triggering a cascade of events leading to more efficient Na^+ compartmentalization. It is very difficult to speculate on how nutrient problems associated with farming on salinized lands will be affected by rising temperatures and elevated CO_2. Rising temperatures have the potential to alleviate certain nutrient deficiencies caused by NaCl stress, but could exacerbate others. Rising atmospheric [CO_2] could reduce Na uptake, lessening some of its toxic effects, but it could also magnify deficiencies of essential nutrients that often occur in plants grown in salty soils. These interactions all require study.

Finally, tropospheric O_3 pollution is unlikely to result in any serious direct nutritional imbalances in crops. Because O_3 does not penetrate the soil and is not transported to roots from leaves, any effects of high O_3 on plant nutrition will be indirect, mediated mainly by changes in assimilate supply to roots. Decreases in assimilate supply might result in reduced nutrient uptake, but too few convincing data exist to conclude this with any confidence. Some evidence indicates that soil nutrient levels do play a role in determining sensitivity of plants to O_3 damage. In most experiments, plants receiving more fertilizer suffered greater O_3 damage compared to those receiving less. Where O_3 levels are high, concentrations of nutrients considered to be optimal might need to be adjusted down.

Chapter 8

Vegetative Growth and Development

The most important feature of crops is their growth. Photosynthesis assimilates the carbon used in growth, and respiration converts that assimilated carbon into usable energy and metabolic intermediates required for growth, but it is growth that matters most. We define growth as the increase in size of an organ or a tissue (Box 8.1). Various aspects of size are important to crops. For example, shoot height and leaf surface area are critical to solar radiation absorption and therefore photosynthesis; root length, depth, and surface area affect absorption of water and mineral nutrients; and volume can be meaningful to storage of water and other substances. But increase in dry mass is usually the most significant measure of crop growth, with the chemical composition and energy content (energy being dependent on composition) of that dry mass also of basic importance.

GROWTH PATTERNS

The balance of the effects of an environmental change on growth can be determined by measuring the products of growth in an experiment that alters environmental conditions. For example, biomass produced by rice raised in different [CO_2] environments can be harvested, dried, and weighed. This approach is typical and may provide a simple and important summary of the effects of the environment on a crop. Thus, it can be said that under the experimental conditions used by Allen, Baker, and Boote (1996), elevated [CO_2] caused an increase in rice growth (Figure. 8.1). While valuable, this is just an empirical description of an experimental result. Its generality is unknown, though when a large number of experiments exhibit broadly similar results, confidence emerges that the result is general in qualitative and/or quantitative terms.

BOX 8.1. Definitions

Determinate: Growth well defined by genetics. Typical for annual plants and leaves and flowers of perennials.
Development: Change in form and function resulting from cell divisions, cell enlargement, and cell differentiation beginning in a zygote and continuing for the life of a plant.
Differentiation: Process of cell specialization. Includes structural, functional, and/or biochemical transitions. Associated with loss of meristematic ability.
Embryo: Young plant contained within the seed before germination.
Growth: Increase in size primarily as the result of cell expansions. Dry mass of structure is usually the most appropriate gauge of "size" in crops.
Indeterminate: Growth not well defined by genetics. Typical of perennials and stems and roots of some annuals.
Intercalary meristem: Meristematic region inserted (intercalated) between regions of nondividing cells. Generally located at the base of a leaf blade and/or sheath of many monocots, such as wheat, rice, maize, and sorghum.
Meristem: Embryonic or dividing group of cells giving rise to similar cells or to cells that differentiate to produce specific tissues or organs. Meristems may be capable of producing new cells indefinitely.
Phenology: Periodic developmental phenomena that are related to (or correlated with) environmental conditions or time. Often used to relate biological phenomena to climate.
Primary meristem: Meristematic tissue derived from the apical meristem. Contributes to an increase in length (height).
Primordium: Localized collection of meristematic cells. Formation of a primordium is the first visible (microscopically) sign of organ initiation.
Storage: Organic components of a cell that exist only temporarily and are used for growth or respiration, or are translocated, at a later time. An example is starch in a chloroplast.
Structure: Organic components of a cell that are not normally broken down to supply energy and carbon skeleton intermediates used for growth. Examples are membrane lipids, structural proteins, cellulose, hemicelluloses, and lignins.

A more useful approach would be to relate the effect of an environmental change on growth to the underlying processes involved. In that way, an explanation of the effect of an environmental change on a crop can be provided. This is important because an explanation (compared to a mere description) carries with it an ability to extrapolate experimental results in space and time (i.e., to predict outside the en-

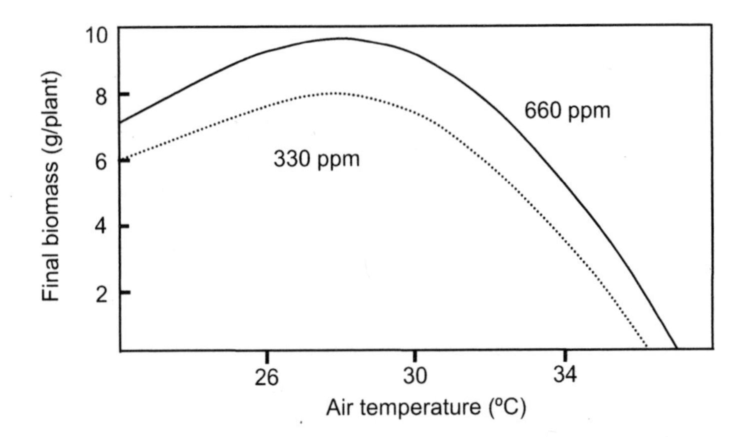

FIGURE 8.1. Relationship between temperature and final rice biomass at two concentrations of CO_2. Lines represent a second degree polynomial fit of rice biomass at maturity versus weighted mean day/night air temperature. Plants were grown at either 660 or 330 ppm CO_2 in five separate experiments. *Source:* Figure is adapted from Allen, Baker, and Boote (1996).

velope of past experimental conditions) (Amthor and Loomis, 1996). It is important, therefore, to not only understand how environmental changes will affect growth but also to understand what the underlying mechanisms causing those growth changes are.

The great variety of shapes, structures, and functions of cells and organs among different species results from flexibility in cell division, enlargement, and differentiation. Each of these three processes may be affected by environmental change, and each is important to yield. Each is also constrained by the genetic limitations of each crop species and cultivar.

MERISTEMS, ORGANS, AND DEVELOPMENT

Meristems

Plants continue to grow and develop throughout their lives (i.e., unlike animals, they are continuously embryonic). They are capable of this because meristems are always present, unless the plant is unusually and severely damaged by an outside agency (and even then, many "mature cells" can dedifferentiate and become meristematic). Both root and shoot tips, or apices, contain meristems (Box 8.1), broadly defined as groups of dividing cells. Apical or primary meristems are initiated during embryo development in seeds on the parent plant and later give rise to cells responsible for apical growth. Meristems are also found just above nodes in grass stems and at the base of grass leaves (Figure 8.2). Woody perennials (cotton is an important woody perennial crop, though it is typically grown for only one season) contain lateral meristems called vascular cambium that produce new phloem and xylem cells each year. That vascular cambium, and meristems between grass stem nodes and at the bases of their leaves, are called secondary meristems. All new cells are born in meristematic regions through repeated cell divisions (the cell cycle, Figure 8.3).

The Cell Cycle

The life of a cell can be described as a cycle beginning with the formation of a cell (after completion of the division of the mother cell)

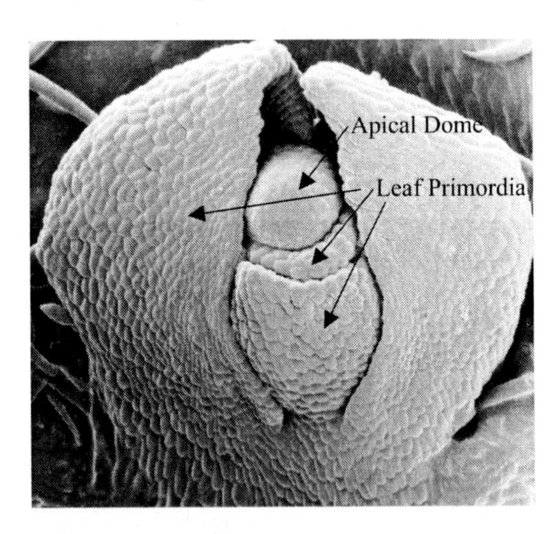

FIGURE 8.2. Scanning electron micrograph of a lateral shoot meristem of a monocot (maize). Shoot apical meristems are similar in appearance and function. Note that the leaf primordia are beginning to form sheaths around the apical dome proper (arrow). *Source:* Photo by Michael J. Scanlon, University of Georgia.

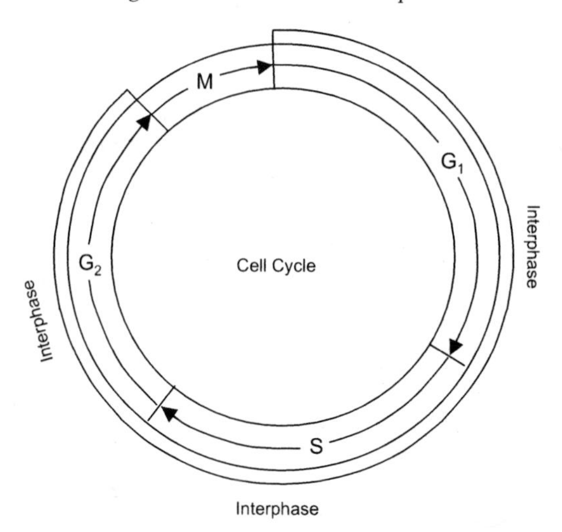

FIGURE 8.3. Typical representation of the cell cycle. The cell cycle is divided into two major phases: mitosis (M) and interphase. During mitosis, nuclear division (separation of daughter chromosomes into different nuclei) and cytokinesis (physical division of the new daughter cells) occurs. After new daughter cells are formed, cells enter interphase, in which they both recover from the division process and carry out normal metabolic function. The first stage of interphase is referred to as the Gap 1 (G_1) stage. During this stage, a significant increase in cytoplasmic mass occurs. Some cells undergo cell-cycle arrest in the G_1 stage, eventually differentiating and maturing into specialized cell types, but most cells transition into the DNA synthesis (S) stage. During the S stage, DNA as well as other components of the nucleus are replicated. Some of these cells (especially those destined to undergo rapid, intense metabolic function, such as root hairs or trichomes) undergo repeated DNA replication, sometimes forming many copies of DNA (a process called endoreduplication). Cells prepare to divide in the G_2 phase by synthesizing proteins required for the division process. Among other proteins, critical concentrations of cyclin proteins must be present to cause cells to enter mitosis. Those cells whose fate is to mature into specialized cell types undergo cell-cycle arrest in the G_1, S, or G_2 stage without entering mitosis.

and ending with completion of division of that cell (Figure 8.3). In indeterminate meristems, such as those found at shoot tips, it is often the case that of the two daughter cells produced by mitosis, one completes the cell cycle by eventually dividing into two new daughter cells, while the other is arrested at some stage of the cycle and subsequently differentiates into a mature specialized cell. In this fashion, meristems maintain a more or less constant number of meristematic cells and also produce other cells that mature and make up the plant body. In meristems that give rise to determinate structures such as leaves or flowers, cells are programmed to divide a finite number of times and then eventually differentiate into specialized cell types, such as mesophyll, epidermal, or vascular tissue cells.

At several critical points in the cell cycle, cells are signaled to either progress to the next stage of the cycle or differentiate into mature, nonmeristematic cells. The first occurs in very early interphase during G_1 (Fosket, 1994). At this point, cued by both endogenous and exogenous signals, cells either replicate their DNA and proceed through stage S, or they may leave the cell cycle to differentiate and mature into specific cell types. Cellular polyamine concentrations evidently play a significant role in triggering this transition (Galston and Sawhney, 1990; Poljakoff-Mayber and Lerner, 1999; Kakkar et al., 2000).

A second important control point in the cell cycle is the G_2/M transition, which is influenced by both hormones and cyclins. Cyclins, a class of regulatory subunits of a family of protein kinases, are required at threshold concentrations for cells to transition from the G_2 to the M phase of the cell cycle (Fosket, 1994; Soni et al., 1995; Jacobs, 1997; Mironov et al., 1999). Cellular cyclin concentration has been observed to increase throughout the life of a cell (i.e., through the G_1, S, and G_2 stages) until mitosis begins, at which point their concentrations drop precipitously. When cell division is complete, their levels begin to rise again. Evidently, the rate of buildup of cyclins is linked to the overall rate of new cell production and therefore growth. It has been suggested that crops and other plants match their growth patterns to the fluctuating environment by adjusting cyclin metabolism (Doerner et al., 1996). How environmental conditions influence regulation of polyamines and cyclins is unknown, but phytohormones are the likeliest signal molecules.

Cytokinins and auxins are phytohormones involved in growth regulation. Both, for example, affect expression of cyclin genes (Renaudin et al., 1994; Kouchi, Sekine, and Hata, 1995; Kende and Zeevaart, 1997), and both modulate cell division, cell expansion, and protein synthesis in meristematic regions. Auxins influence the events leading up to DNA replication (i.e., the $G_1 \rightarrow S$ transition) while cytokinins mediate events leading up to initiation of mitosis (i.e., the

$G_2 \rightarrow M$ transition; Fosket, 1994). Either directly or indirectly (e.g., as a result of changes in availability of substrates for growth), environmental signals can trigger a change in hormone metabolism, which in turn leads to reactions or cascades that affect cell divisions.

Availability of photosynthate, water, and mineral nutrients also influence the duration of the cell cycle (rate of cell division) through their effects on cell expansion. In favorable environments, cells grow and expand rapidly and cells quickly reach the size threshold beyond which the nucleus can no longer effectively govern cell function (Jacobs, 1997). Beyond this threshold, expanding cells are forced to duplicate their organelles and partition the growing space with new cell walls (i.e., they must complete mitosis). Taking this view, cell expansion rates may affect cell-cycle durations.

Cell-cycle lengths are highly plastic, and this plasticity is reflected in the broad range of growth rates and patterns exhibited by plants growing under different environmental conditions. The cell cycle may take hours (in rapidly growing plants or tissues) to months (in slowly growing plants or tissues) to complete. Furthermore, cell cycles also differ with respect to the duration of component phases (i.e., M, G_1, S, and G_2), both within and between tissues (Francis and Barlow, 1988).

Cell Expansion

Cell division creates two daughter cells, which combined occupy the same volume—each half the size—as the parent cell. So, in one sense, cell division makes no direct contribution to growth. Rather, it is the subsequent expansion of cells, involving an increase in the surface area of cell walls, coupled with the swelling of vacuoles that causes plants, or their component parts, to become larger (Pritchard, 1994). A plant cell may increase in size between a hundred- and a thousandfold as it expands and differentiates (McQueen and Rochange, 1999).

Cell expansion is controlled by cell-wall loosening, wall extensibility, and the uptake rates of water and solutes (Cosgrove 1993, 1997; Ferris and Taylor, 1994; Taylor et al., 1994). Greatly simplified, the growth rate of cells (r) can be predicted with the following equation (Lockhart, 1965; Poljakoff-Mayber and Lerner, 1999):

$$r = m(P-Y) \quad [8.1]$$

where:

r = growth rate
m = cell-wall extensibility
P = turgor pressure
Y = yield pressure (turgor pressure required to bring about expansion)

Mechanical properties that govern cell-wall extensibility (m) are more important for growth regulation than the other terms in Equation 8.1 (Cosgrove 1993, 1997; Taylor et al., 1994). Cell-wall extensibility is controlled by the pH of cell walls. Cell walls are capable of expanding at acidic pH but tend to be rigid at neutral pH (Rayle, Haughton, and Cleland, 1970). Environmental conditions effect changes in auxin metabolism, triggering the excretion of H^+ ions from the cytoplasm into cell walls (by activating a plasma-membrane proton pump). Resulting acidification of cell walls then causes breakage of hemicellulose cross-links between adjacent cellulose microfibrils within the cell-wall matrix, leading to cell-wall loosening and, ultimately, enhanced cell expansion. This is called the acid growth hypothesis (Cleland, 1980).

Two classes of very similar proteins have been identified as the primary cause of acid-induced growth, α-expansins and ß-expansins. Expansins may be involved both in cell-wall disassembly and cell-wall elongation (McQueen and Rochange, 1999). There is also evidence that other enzymes, most notably xyloglucan endotransglycosylase (Ranasinghe and Taylor, 1996), which is a cell-wall-loosening enzyme thought to function by cutting and rejoining xyloglucan molecules, and ß-glucanases (Nicol and Hofte, 1998), which also influence cellulose-hemicellulose interactions (McQueen and Rochange, 1999), may also be important players in cell expansion.

When environmental conditions influence expansive growth of cells, they do so by affecting the availability of resources needed for growth or plant hormone metabolism, which in turn influence molecular and biochemical events controlling growth, such as cell-wall loosening, as discussed here. So far, little progress has been made in linking the effects of the environment on whole-plant growth to the underlying growth mechanisms (Ackerly et al., 1992).

Determinate and Indeterminate Meristems

Whole plants, and their organs, may exhibit either determinate or indeterminate growth (Box 8.1). Determinate meristems, such as those from which flowers and leaves develop, are not able to function (i.e., produce new cells) indefinitely. Many annual plants are determinate because their development is under strict genetic control. Sunflower is a good example: early in its development, the shoot apical meristem of this crop already contains concentric rings of cells, each of which is predestined to develop into specific tissues and organs of the mature plant (Fosket, 1994). This implies that little variation in growth habits (i.e., node and leaf number) can be expected from this crop. In an even more extreme example of a determinate plant, the fern *Azolla* contains a single apical initial cell that is genetically programmed to divide exactly 55 times. In *Azolla,* and probably other determinate species, there is a genetic "clock" that keeps track of, and limits, the number of cell divisions, nodes, and/or leaves that the plant is capable of producing before flowering is initiated. However, it is important to remember that two genetically identical plants with exactly the same number of cells, leaves, nodes, branches, and so on could still differ in size by 50 percent (for example) if their average cell size differed by 50 percent. So even though *developmental* plasticity may be limited in determinate species or organs, significant variations in morphology and yield are possible. Environmental conditions during development simply determine to what extent a determinate plant or organ is able to reach its potential size (Hay, 1999).

Perennial plants and root axes of some annual plants are characterized by indeterminate growth. Indeterminate plants or structures have persistent (permanent) apical initials that maintain the ability to divide and produce new cells indefinitely. Many crop varieties have been bred to exhibit determinate or indeterminate growth patterns, as is the case for cotton and soybean cultivars, but the distinction between determinate and indeterminate crop varieties is not always clear-cut. As a general rule, vegetative growth of determinate varieties ceases when reproductive growth is initiated. Indeterminate varieties, however, continue to produce vegetative biomass throughout reproductive growth phases (e.g., flower initiation, anthesis, setting and filling of bolls in cotton). Indeterminate species or varieties typically exhibit greater phenotypic plasticity than determinate ones because the switch to flowering in determinate annuals such as cereals requires that the apical meristem change into a determinate flowering structure.

SHOOTS

Leaves

The surface area of leaves exposed to sunlight and capable of acquiring CO_2 is an important determinant of crop yield. Environmental factors impact the development and growth of photosynthetic leaf area by changing rates and durations of leaf initiation and expansion. Leaf area depends on leaf number and area as follows (Sadras and Trápani, 1999):

leaf number = rate of leaf primordia initiation × duration of leaf primordia production
leaf area = rate of expansion × duration of expansion

A leaf first appears as a protuberance, called a leaf primordium, on the shoot apex or along the apex of a lateral bud or branch (Figure 8.2). Leaf primordia form in spatial arrangements that are characteristic of each species. That spatial arrangement is called phyllotaxis. Control of phyllotaxis may involve diffusion of a substance (or substances) from the shoot apex and/or existing leaves that either promotes or inhibits cell division and primordia formation at relative distances from the source of the substance(s). The time interval between inception of initiation of two successive leaf primordia is called the plastochron.

Development and growth of leaves (based on cell divisions and expansions) occurs in myriad ways, giving rise to the abundance of leaf shapes produced by different species. Within grasses, meristematic activity occurs throughout the leaf when it is on the order of 1 mm long. As a grass leaf elongates, meristematic activity in the distal end ceases, even though cell elongation and enlargement occurs for a time there, but meristematic activity continues at the leaf base. That is, grass leaves grow from their base. The basal meristem (an intercalary meristem) remains potentially active even after leaf "maturation." The grass leaf is divided into a blade, at the end, and a sheath, at the base. The basal meristem is at the base of the sheath. Leaf width in grasses arises from cell divisions (followed by cell growth) along the leaf axis, but these divisions cease well before the leaf or leaf segment matures. In many grass crops (e.g., wheat, barley, oat, sorghum, maize, and rice) the leaf completely encircles the stem, which results from a primordium that existed around the entire shoot axis. Grass leaf primordia appear as ridges around the whole shoot axis (Figure 8.2). Typically, the basal leaf meristem lies just outside another intercalary meristem, giving rise to new stem cells. The combination of leaf and stem intercalary meristems, along with adjoining cells, creates the internodes found along the shoot central axis.

Cell divisions in dicot leaves typically end by the time the leaf is less than half its mature size (Dale, 1988). Dicot leaves have a palisade layer (or layers) of mesophyll cells, whereas monocots do not. Mesophyll cells in dicot leaves stop growing before epidermal cells stop expanding, so the continued expansion of the epidermis usually pulls the mesophyll cells apart from one another, giving rise to considerable intercellular spaces.

Stems

Stem cell division and growth (elongation and dry-mass increase) occurs some distance below the apical meristem, depending on species (Sachs, 1965). In grasses, stems are composed of repeating node-internode segments. During early growth of a node-internode segment, cell divisions occur throughout the segment. Later, cell division is limited to intercalary meristems at the base of the internodes. As with grass leaves, grass internodes contain their oldest cells near their top.

An advantage of "placing" the meristems near the base of leaves and stem segments is that leaf and stem elongation can continue even if the top of the leaf and/or stem segment is detached (e.g., if they are eaten by an animal). Presumably, this is an important evolutionary feature of grazer-plant interactions.

Tillers

The main stem of monocots as well as dicots is formed in the embryo. In monocots, buds may arise in the axils of lower leaves, forming secondary shoots, and these shoots can produce adventitious roots. These are called tillers. Tillering is an important developmental stage in many cereal crops such as wheat. Often, changes in tillering represent the most obvious growth response in environmental change experiments, especially those conducted with widely spaced plants. This is typically less important in "normally" spaced crops, but changes in tiller number can also be important then.

ROOTS

A Word About Roots

Roots acquire water and dissolved ions from the soil and anchor plants to the ground. Some roots even store food (e.g., sweet potato), while others function in vegetative reproduction. Worldwide, root and tuber crops provide more food energy (Cal d^{-1}) than milk and cheese, pork, fruits, vegetables, poultry, beef, eggs, or seafood (Table 1.2). In spite of their importance, belowground plant structures are very often neglected in studies of effects of environmental change on crops.

Articles on roots are underrepresented in the literature compared to those on shoots because roots are difficult to study. Most of the methods developed to study root growth and development represent a compromise between what is experimentally possible or practical versus what is likely to reflect root growth conditions existing in the farm field. For example, plants are easily grown hydroponically, and this technique allows ready access to their roots. Unfortunately, roots grown under such conditions may behave very differently from roots growing in soils, which represent considerable mechanical impedance to growth, and which teem with microorganisms—some beneficial and others pathogenic. Studies on plants rooted in soil likely approximate normal root function as occurs in the field, but these roots cannot easily be seen or measured. Because of these experimental hurdles, our understanding of roots is lagging compared to our knowledge of shoots.

Root Meristems

Within the seedling, there may be one or several main roots derived from the original seed. Some of these primary roots produce lateral roots, and these laterals may in turn produce other laterals, and so on until a highly branched root system has evolved. A root system is essentially a collection of meristems that forage through the soil "searching" for water and nutrients. Most monocots, including cereal crops, have fairly shallow, highly branched, fibrous root systems that are composed of many adventitious roots more or less equal in diameter and length. Dicot crops such as cotton and soybean have taproot systems that are dominated by a single, large, primary root from which many smaller lateral roots have branched off. Root systems of dicot crops sometimes grow several meters into the soil.

Root growth is driven by orderly cell proliferation in apices. Cell proliferation, through repeated transverse divisions, results in columns of cells along the length of the root, and each cell in the column is more or less developmentally advanced than the last (Schiefelbein, Masucci, and Wang, 1997). With increasing distance from the root tip, the frequency of cell division decreases, cells appear larger, and they are more highly differentiated.

The root tip can be divided into four spatially and functionally distinct zones: the root cap, apical meristem, region of cell elongation, and region of cell differentiation. The root cap protects the apical meristem from mechanical damage and also secretes substances that lubricate the soil as it is penetrated by the root. The region of elongation, located just behind the meristem, is where undifferentiated cells expand lengthwise. The region of differentiation, sometimes called the root-hair zone, is where cells differentiate into mature cells of epidermal, ground, or vascular tissue. Root hairs produced in the region of cell differentiation play important roles in resource uptake and anchorage (Hofer, 1996).

The root apical meristem is less structurally complex than the shoot apical meristem. In shoots, lateral organs such as branches, flowers, and leaves are initiated by the apical meristem, but in roots, lateral organs arise in the pericycle, a layer of cells located some distance behind the root apical meristem. It is important to note that root developmental plasticity is controlled mostly by activation and elongation of lateral roots (Smucker, 1993; Kerk, 1998; Pritchard and

Rogers, 2000). In fact, the single most important question to be resolved concerning root responses to environmental change may be: What sequence of events triggers the activation of pericycle initials, and why are some initials activated to eventually form new lateral roots while others remain inactive?

DEVELOPMENTAL STAGES

Crop development can be described in terms of discrete events related to meristematic activity and properties of organs arising from meristems. The events, and in particular the time between events, may be called phenostages (for phenological stage) (Box 8.1). The rate of progress between events is the developmental rate. Numerical (quantitative) phenological scales have been developed for most crops. The Feekes' scale for wheat and similar cereals (Large, 1954) was an important innovation that was developed further by Zadoks, Chang, and Konzak (1974). Cereal scales emphasize the transition from vegetative to reproductive growth and the development of grain. For other crop types (e.g., tuber crops such as potato), different developmental stages may be more important.

Phenological scales are generally based on visible events. For example, the emergence of a young leaf from the collection of older leaves enclosing it or the emergence (appearance) of an inflorescence. It is generally the underlying (and preceding) event that is more important, however, and it is the underlying event that can directly respond to environmental change. For example, meristematic initiation of an inflorescence occurs before the inflorescence is visible, and the time of that initiation is affected by environmental conditions prior to inflorescence visibility that modify the genetic program. It is, therefore, the environmental state near the time of initiation that is often more critical than the environmental state when the eventual result of a developmental event is visible. In practice, however, it is often difficult to relate developmental events to environmental conditions because meristematic activity is generally "invisible."

ENVIRONMENTAL CHANGES

Warming

Growth and Development Are Controlled by Meristem (Not Air) Temperatures

It is critical to realize that meristem temperature, not air temperature, is directly related to development. Because meristem temperature commonly differs from air temperature measured at the standard 1.5 m above the soil surface, growth and development are usually not strictly related to air temperature, although they are influenced by it.

Between germination and emergence, and through the development of several leaves, apical meristem temperature is controlled mainly by soil temperature (because the seed and meristems are in contact with the soil, not the air 1.5 m above it). Following canopy development, apical meristem temperature will be related to canopy temperature, which can differ significantly from ambient air temperature, although in many grasses meristems remain close to the ground throughout much of development (Bowen, 1991).

Canopy warming from reduced transpiration (stomatal closure) or greater radiation absorption (more leaf area) can affect shoot meristem temperature, and therefore developmental rate, without changing measured air temperature (Chapter 4). Most important for understanding the effects of environmental change on crops, however, is the realization that warming air and surface soil as a result of an enhanced global greenhouse effect will warm meristems (for the same plant geometry and size), and this will directly affect developmental rate.

Germination

Germination is highly temperature dependent. The minimum temperature capable of supporting seed germination is generally 0 to 5°C, the maximum is ~45 to 48°C, and the optimum is ~25 to 30°C (Hall, 2001). At suboptimal temperature ranges (between lowest and optimal temperatures for germination), germination rate often exhibits a positive linear relationship with temperature. At supraoptimal temperatures (between optimal and maximum temperatures for germination), germination rate exhibits a negative linear relationship. The range of temperatures capable of supporting germination varies from crop to crop. The range in chickpea is far wider than for the other grain legumes lentil, soybean, or cowpea, for instance (Covell et al., 1986).

Cool-season crops have lower temperature optima than warm-season crops. For example, the minimum temperature for most of the cool-season crops listed in Table 8.1 is from 3 to 8°C, whereas minimum temperatures for germination of the warm-season crops is substantially higher (e.g., 18, 16, and 14°C for cowpea, cotton, and maize, respectively; Hall, 2001).

Knowledge of the effects of temperature on the physiology of germination might become increasingly important if global temperatures rise as projected. Applying mathematical models to characterize relationships between germination and temperature may prove

TABLE 8.1. Examples of crop plants adapted for cool and warm seasons.

Cool-season crops	**Warm-season crops**
Barley	Amaranth
Brassicas	Cassava
Canola (rape)	Common bean
Chickpea	Cotton
Fava bean	Cowpea
Flax	Cucurbits
Lentil	Lima bean
Lupine	Maize
Oat	Mung bean
Pea	Pearl millet
Potato	Pepper
Rye	Pigeonpea
Sugarbeet	Rice
Triticale	Sesame
Turnip	Sorghum
Wheat	Soybean
	Sugarcane
	Sunflower
	Sweetpotato
	Tomato

particularly useful for selecting cultivars and varieties better suited for higher temperature (Garcia-Huidobro, Monteith, and Squire, 1982; Covell et al., 1986; Ellis et al., 1986; Ellis, Simon, and Covell, 1987; Roberts, 1988).

Development

Temperature is crucial for the advancement of crops through sequential phenological stages of development. So important, in fact, that developmental progress is often gauged by thermal units or degree days instead of calendar days. The first formal expression of the relationship between temperatures and the rate of crop development, the so called "heat-unit law," was put forth in 1735 by Réaumer (Grace, 1988). Réaumer observed that a requisite amount of heat had to accumulate throughout the growing season before grapes and wheat would mature. This law has since led to the widespread adoption of degree-days to gauge crop developmental progress. Degree days may be expressed as follows:

$$S = \sum_{day=1}^{day=j} (\bar{t} - t_0)$$

where:

S = sum of degree days (or thermal time),
t_0 = the threshold temperature for the onset of development (or base temperature), and
$\bar{t}$ = mean temperature of the day.

Greater accuracy can be obtained by accounting for the often nonlinear relationship between temperature and development by calculating thermal units several times during each day (i.e., hourly or more frequently) and summing these over the day (Amthor and Loomis, 1996).

As illustrated for pearl millet (Figure 8.4), thermal units are often a better predictor of development than calendar days. Similar relationships of degree days with shoot development have been established for many other crops, including wheat (Hay, 1999), oats (Peltonen-Sainio, 1999), maize (Tollenaar and Dwyer, 1999), and cotton (Heitholt, 1999). In maize, for instance, the rate of development increases more or less linearly from about 10 to 30°C.

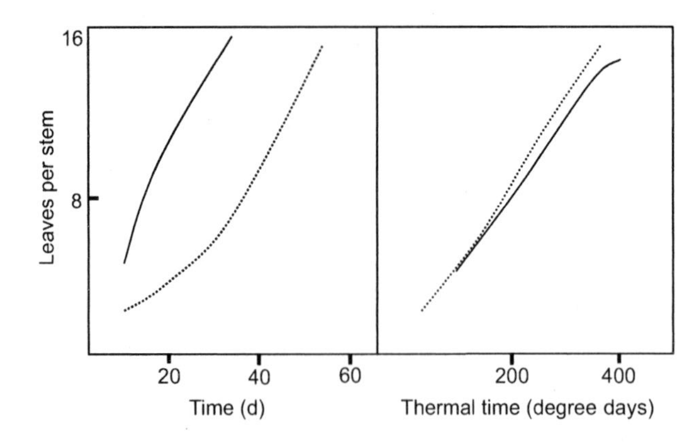

FIGURE 8.4. Stage of crop development is gauged more accurately by thermal time than calendar days. These data are from pearl millet planted late, i.e., at high temperature (solid lines), or early, i.e., at low temperature (dotted line). *Sources:* Data from Ong (1983a) adapted from Grace (1988).

Faster development means shorter growth durations. Shortening of development caused by high temperatures has been observed for single cells (discussed later), organs (such as cereal grains), phenostages, as well as the duration of the entire crop life cycle. In wheat, for example, a 1°C increase in temperature resulted in a 21-day (~8 percent) reduction in crop duration (Lawlor and Mitchell, 2000). In another example, the time to flowering was shortened by ~10 days in rice grown at 34/27/31°C (day/night/paddy temperature) compared to rice grown at 25/18/21°C (Allen, Baker, and Boote, 1996). This relationship has significant implications for crop growth in a warmer world; if the period of growth is reduced, the amount of growth may be reduced too.

Thermal time also appears to be applicable to root-system development (i.e., branching patterns). In pearl millet, the number of root axes and lateral roots were linearly related to the accumulated meristem temperature ($r^2 = 0.92$ and $r^2 = 0.85$, respectively, Gregory, 1986).

Photoperiod interacts with temperature to control development. Although tracking thermal time can help one to predict the stage of development for many crops, development of others is controlled by photoperiod (Grace, 1988). For these crops, degree days may either play no role in advancing plants from one phenostage to the next or it may simply fine-tune the flowering response (Evans, 1993).

In general, the developmental rate of most modern cultivars of cucumber, sweet corn, tomato, pea, cotton, maize, spring wheat, and some cultivars of rice, sorghum, and cowpea is temperature-driven (i.e., they are daylength neutral) (Hall, 2001). But even for day-neutral crops, exposure to high temperatures sometimes leads to a requirement for short days before a particular growth stage (e.g., flowering in soybean and tubering in potato) is initiated (Went, 1959; Dow el-madina and Hall, 1986).

Growth

Cell division. Plant growth rate increases with rising temperatures because sink metabolism is enhanced, particularly as a result of faster cell divisions (Farrar and Jones, 2000). In onion, for example, the cell cycle took 100 hours to complete at 5°C compared to just 8 hours at 35°C. The optimal temperatures for mitosis (i.e., the temperature at which the cell cycle is of the shortest duration) varies among species (e.g., 28°C in bean and 32°C in onion). For root meristems, the Russian scientist V. G. Grif (1981; cited in Francis and Barlow, 1988) advocated expressing the duration of cell cycles according to incremental 5°C deviations from optimal temperatures (Q_5) in a similar fashion as Q_{10} values are used to characterize enzyme function (Table 8.2). The further temperatures deviate from optimal (i.e., the higher the absolute value of Q), the greater the coefficient by which the cell cycle is lengthened. The absence of values greater than one increment above the optimal temperature ($>Q_{+1}$) in Table 8.2 is related to the observation that the optimal temperature for cell division is typically very close to the maximum temperature at which mitosis can occur (at least this appears to be the case for the majority of flowering plants surveyed thus far) (Francis and Barlow, 1988).

It is uncertain what phase of the cell cycle and, more specifically, what specific cellular process is most sensitive to temperature. At supraoptimal temperatures (at which cell cycling is inhibited), evidence suggests that the M-phase of the cell cycle may be the most sensitive and that microtubule disorganization in particular may be involved (Cherry, Heuss-LaRosa, and Mayer, 1989; Smertenko et al., 1997). In

TABLE 8.2. The coefficients (Q_5) by which the cell cycle is influenced by the temperature of meristems.

Coefficient[a]	Interval above or below optimal temperature (°C)	Species: onion	Species: sunflower	Species: faba bean
Q_{-5}	−20 to −25	2.4	—	—
Q_{-4}	−15 to −20	1.7	1.9	2.4
Q_{-3}	−10 to −15	1.5	1.8	1.6
Q_{-2}	−5 to −10	1.3	1.4	1.5
Q_{-1}	0 to −5	1.1	1.1	1.3
Q_{+1}	+5 to +10	1.2	—	1.6

Source: Table adapted from Grif (1981) which was translated from Russian and presented in Francis and Barlow (1988).

[a]The subscript after Q is the number of five-degree increments the coefficient is above (+) or below (−) the optimal temperature for cell division.

tobacco, disorganization of phragmoplast microtubules in heat-stressed cells was tightly correlated with inhibition of cell division during heat stress (Smertenko et al., 1997). In addition, organization of mitotic microtubule arrays was more heat sensitive than interphase microtubules.

Cell expansion. Cell expansion seems to be less sensitive to temperature than cell division. At high temperatures, cell division is enhanced and cell elongation rate increases, but the duration of cell expansion decreases. This sometimes results in smaller cells (e.g., in wheat roots; Burstrom, 1956). Similarly, just the opposite has been observed at low temperatures; cell division in wheat, for example, slows while the duration of cell expansion increases, resulting in fewer, but larger cells (Huang, Taylor, and McMichael, 1991).

Shoots. Temperature affects shoot morphology and architecture by influencing the initiation of lateral buds, branches, and leaves (development), as well as the rates of expansion of those organs (growth). Over a temperature range typical of northern temperate summers (6 to 18°C), relative plant growth rate (RGR) typically increases sharply with temperature (Grace, 1988). A nearly linear rise in RGR of rape, sunflower, and maize was observed as temperatures were increased from 10°C to 28°C. RGR then decreased at temperatures greater than 28°C.

Even though the *rate* of shoot growth increases with temperature, total biomass accumulation over the course of the growing season sometimes decreases (Gregory, 1986). Figure 8.1 shows the final biomass of rice grown over a range of temperatures at two atmospheric CO_2 levels. Growth temperatures above ~28 clearly result in a rapid decrease in biomass accumulation, and, in this case, higher atmospheric CO_2 levels did little to offset these growth reductions.

Because developmental stages are shortened by high temperatures, the temperature optima for accretion of biomass are typically lower than those which result in the maximum relative growth and maturation rates (Squire et al., 1984). Again, we see that although higher temperatures increase the rate of shoot growth, they often decrease the duration of time in which light, water, and nutrients are harvested, and, therefore, dry-matter production can suffer. For example, an average increase in temperature of 3.5°C throughout the growing season reduced final biomass of wheat by 16 percent (Mitchell et al., 1993). In corn, leaf length decreased ~2 percent for every degree °C above 20 (Ritchie and NeSmith, 1991). The implications of this phenomenon for reproductive growth and yield will be expanded upon in Chapter 9.

Roots. The temperature optima for root growth differs significantly for different crops (Table 8.3). Cool-season plants, or plants originating from cool climates, generally have lower temperature optima than warm-season plants and plants adapted for warm climates (McMichael and Burke, 1996). Potatoes, for example, which originated in the high Andes of South America (cool climate) have significantly lower soil-temperature optima than the root crops sweet potato, which originated from the hill area of Central America (warm climate), and cassava, which is adapted for the humid tropics of the Amazon basin (hot climate) (Figure 8.5) (Sattelmacher, Marschner, and Kühne, 1990). As seen in Figure 8.5, growth of belowground structures of potato, cassava, and sweetpotato increased with temperatures, then decreased as temperatures were increased beyond the growth optima (McMichael and Burke, 1996). Similar relationships of temperature with growth also hold for crops that lack large underground storage organs. In soybean, for example, rates of lateral root

TABLE 8.3. Optimum root-zone temperatures for root growth.

Crop	Root parameter	Optimum temperature (°C)	Source
Sunflower	Root elongation	20	A
Cotton	Root elongation	33	B
Maize	Root elongation	30	C
	Root mass	26	D
Soybean	Taproot extension	25	E
Oat	Root mass	5	F
Rice	Root growth	25-37	G
Tomato	Root elongation	30	H
Rape	Root extension	23	I
Oil palms	Root growth	30-35	J
Ryegrass	Dry weight	17	K

Sources: Table adapted from McMichael and Burke (1996) and Glinski and Lipiec (1990). (A) Galliger (1938); (B) Arndt (1945); (C) Anderson and Kemper (1964); (D) Walker (1969); (E) Stone and Taylor (1983); (F) Nielsen and Humphries (1966); (G) Kar et al. (1976); (H) White (1937); (I) Moorby and Nye (1984); (J) Agamathu and Broughton (1986); (K) Clarkson, Hopper, and Jones (1986).

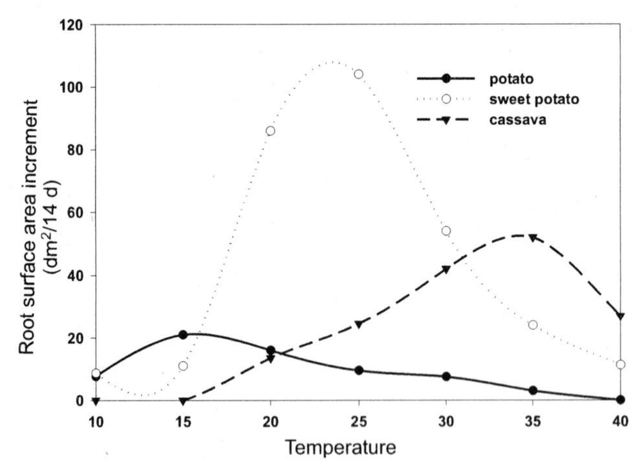

FIGURE 8.5. Influence of root-zone temperatures on root growth of three economically important root and tuber crops. All results are expressed as surface area increment over 14 days. *Source:* Data from Sattelmacher, Marschner, and Kühne (1990). Original data have been fitted with a spline curve.

elongation were 12, 30, 45, and 50 mm day^{-1} at temperatures of 17, 21, 25, and 29 °C (Stone and Taylor, 1983).

Normally, root diameters are inversely related to root-zone temperatures. High soil temperatures typically stimulate lateral root initiation (branching), resulting in root systems with higher length:weight ratios (because successive orders of roots become finer and finer) (Bowen, 1991; Kaspar and Bland, 1992; but see Pahlavanian and Silk, 1988). Decreasing root diameters with increasing temperatures have also been attributed to a decline in the proportion of cell divisions longitudinal to the files of cells at the quiescent centers of root apical meristems (Clowes and Wadekar, 1988). Whatever the cause, finer roots mean higher specific root lengths (SRL, length:dry wt. ratio), and roots of high SRL are often more efficient at acquiring water and ions (Eissenstat, 1992; Fitter, 1996; Eissenstat et al., 2000; see also Chapter 7 for a discussion of the influence of soil temperatures on mineral nutrient uptake). Root length is especially important for the acquisition of ions that diffuse slowly in soil, such as P (Barber et al., 1989; Bowen, 1991).

Roots apparently sense vertical gradients in soil temperatures and may be able to grow away from soil of unfavorable temperatures and/or toward soil of favorable temperatures. In general, cooler soil temperatures cause roots to grow more horizontally compared to warm soils, which favor vertical growth. This phenomenon has been characterized best for maize but has also been observed in other field crops such as soybean (Mosher and Miller, 1972; Onderdonk and Ketchenson, 1973; Kaspar, Woolley, and Taylor, 1981; Fortin and Poff, 1990; Tardieu and Pellerin, 1991).

Water Stress

In some parts of the world, warming might be accompanied by reduced soil moisture (evapotranspiration increasing more than precipitation; see Chapter 4). This will generally warm crops further, accelerating development even more. Other effects of water stress are important, too. Most notably, effects on photosynthesis (substrate supply) and growth. Growth can be affected by water stress directly, or indirectly through reduced substrate supply (including diversion of some substrates from growth to osmotic adjustment, though this is quantitatively small in most cases). Water limitations lead to inhibition of both cell expansion and cell division (Jones, 1985). Leaf growth is particularly sensitive to water stress (Boyer, 1968; Yegappan, Paton, and Gates, 1982; Taylor et al., 1994). Root growth, however, is sometimes stimulated by water stress, and this presumably represents a mechanism to compensate for limited water supply.

Elevated Atmospheric Carbon Dioxide Concentration

Elevated [CO_2] usually influences the number of organs formed, their growth rates, and/or their final sizes. These effects are caused mainly by an increased supply of carbohydrates available for growth (as a result of increased photosynthesis; Chapter 5). Qualitatively, vegetative growth is virtually always stimulated by CO_2-enrichment, but the magnitude of this response is complicated by many factors. For example, it is often the case that plant growth is stimulated by CO_2-enrichment only until other resources, such as nutrients (especially N or P), become depleted. However, increased root and shoot growth spurred on by extra photoassimilates might enable plants to capture additional resources, preempting such an imbalance between mineral nutrients and carbohydrates (see Box 7.1). Crop growth responses to elevated [CO_2] are also dependent upon water availability (Table 8.4) and temperature (Figure 8.1).

Because elevated CO_2 affects growth primarily by increasing carbon assimilation, species in which the photosynthetic rates are normally limited by [CO_2] exhibit greater growth responses than those in which leaf [CO_2]s are nonlimiting. For example, in soybean, a C_3 crop, photosynthesis is stimulated 40 percent or more in twice-ambient (720 μmol mol^{-1}) compared to ambient concentrations of CO_2. In contrast, the C_4 crop maize realizes only small, if any, stimulation of photosynthesis at similar [CO_2]s (Chapter 5). As a result, soybean growth is greatly increased by high CO_2 (~40 percent; Rogers and Dahlman, 1993), whereas maize growth is only marginally enhanced. (Although photosynthesis is not as sensitive to [CO_2] in C_4 crops, modest growth enhancements are sometimes observed, and these effects are attributed to increased WUE.)

Furthermore, the efficiency of carbohydrate transport differs between species (depending on phloem structure and function; Chapter 6), as do patterns of carbon allocation from sources to sinks (for example, potato compared to tomato). That is, long-term growth

TABLE 8.4. Effect of elevated atmospheric [CO_2] on growth (biomass accumulation of roots and shoots) of several important crops.

Crop	Daytime [CO_2] (high/low; ppm)	Biomass in high [CO_2] relative to low [CO_2] (unitless)	Notes	Source
Wheat	550/350 (shoots)	1.09[a]	FACE, ample irrigation	A
		1.14[b]	FACE, limited irrigation	
	550/350 (roots)	1.28	FACE, ample irrigation	B
		1.23	FACE, limited irrigation	
Rice	589/389 (shoots)	1.15[b]	FACE, very high N	C
		1.09[b]	FACE, ample N	
	589/389 (roots)	1.12[b]	FACE, very high N	
		1.19[b]	FACE, ample N	
		1.22	FACE, low N	
Sorghum	550/350 (shoots)	1.03[b]	FACE, ample irrigation	D
		1.16[b]	FACE, limited irrigation	
	720/360 (fine roots)	1.58[c]	OTC, conventional tillage	E
		1.00	OTC, no tillage	
Potato[e]	560/360 (shoots)	0.80[b]	FACE	F, G
	700/350 (shoots)	0.97[b]	OTC	H
Sweetpotato[e]	665/354 (shoots)	0.91[c]	OTC	I
Sugarbeet[e]	600/372 (shoots)	1.16	OTC, ample N	J

TABLE 8.4 *(continued)*

Crop	Daytime [CO_2] (high/low; ppm)	Biomass in high [CO_2] relative to low [CO_2] (unitless)	Notes	Source
	600/372 (shoots)	1.09	OTC, low N	I
	700/381 (shoots)	1.26	OTC, ample N	I
	700/381 (shoots)	1.13	OTC, low N	I
Cotton	550/350 (shoots)	1.34[a]	FACE, ample irrigation	K, L
		1.27[b]	FACE, limited irrigation	L
	550/350 (taproots)	1.78[d]	FACE	M, N
	550/350 (fine roots)	1.52[d]	FACE	M, N
Soybean	700/350 (roots)	1.82	Phytotron, ample N	O
		1.43	Phytotron, low N	
		1.76	Phytotron, low P	

Sources: (A) Kimball, Kobayashi, and Bindi (2002); (B) Wechsung et al. (1999); (C) Kim et al. (2001); (D) Ottman et al. (2001); (E) Pritchard (unpublished data); (F) Bindi et al. (1998); (G) Bindi et al. (1999); (H) Schapendonk et al. (2000); (I) Biswas et al. (1996); (J) Demmers-Derks et al. (1998); (K) Mauney et al. (1992); (L) Mauney et al. (1994); (M) Rogers et al. (1992); (N) Prior et al. (1994); (O) Israel, Rufty, and Cure (1990).

[a]Average of four years.

[b]Average of two years.

[c]Values indicate seasonal root-length production measured with minirhizotrons.

[d]Average of three years.

[e]Data for yield of root crops is shown in Table 8.5.

responses not only depend upon photosynthetic capacity but sink strength as well as location are equally important. Fast-growing crops with large sinks often realize a greater growth stimulation in CO_2-enriched environments than slower growing plants with smaller sinks. Similarly, indeterminate plants are expected to respond more strongly than determinate ones.

Development

Little, if any, firm evidence indicates that developmental rate is affected by elevated [CO_2] per se (for example, Pleijel et al., 2000). In most cases, the effects of elevated [CO_2] on development are small, and are probably attributable to indirect effects. Effects of [CO_2] on development, germination, time to flowering, as well as other aspects of plant development are highly variable (Reekie, 1996; Andalo et al., 1998). In general, development is governed primarily by temperature, photoperiod, or time (but see Reekie [1996] for a discussion on effects of [CO_2] and photoperiod, perhaps mediated by phytochrome, on development of short-day compared to long-day plants).

Indirect effects of elevated [CO_2] on development will likely overshadow and complicate the interpretation of any direct effects. For example, changes in plant water relations brought about by elevated CO_2-induced stomatal closure (Chapter 4) might affect growth and/or development. Decreased stomatal conductance may also cause leaf and canopy temperatures to rise (because transpiration has a cooling effect), and this should result in accelerated development (Pinter et al., 2000). Teasing out interactions of elevated [CO_2] with photoperiod, temperature, and mineral nutrition, as well as separating direct from secondary effects, has proven difficult (Lawlor and Mitchell, 2000).

Growth

Total plant growth (dry-matter production) of crops growing at twice-ambient CO_2 concentrations (i.e., 720 ppm) is increased an average of 35 percent (Kimball, 1983; Table 8.4). The most obvious explanation for increased growth follows from the increased supply of assimilates to meristems (Pritchard et al., 1999). Assimilates not only represent the building blocks required for growth (e.g., respiratory substrates and carbon skeletons for expanding cell walls), but increases in their concentrations could also enable expanding cells to maintain a higher turgor (i.e., because of a more negative solute potential) (Hsiao and Jackson, 1999). Growth processes might also benefit from improvement of plant water status accompanying growth in high [CO_2] (Chapter 4). This is probably the mechanism behind growth increase in C_4 crops such as maize. Higher atmospheric CO_2 concentrations should also (at least theoretically) favor a lower cell-wall pH, and, as dictated by the acid growth theory discussed earlier, this could lead to biochemical changes in cell walls favorable for expansive growth (Hsiao and Jackson, 1999). Increases in cell turgor in growing regions (Sasek and Strain, 1989), a greater proportion of dividing cells within meristems, shorter cell cycles (Kinsman et al., 1997), enhanced cell expansion (Madson, 1968; Taylor et al., 1994; Ranasinghe and Taylor, 1996), and shifts in root-to-shoot cytokinin flux (Yong et al., 2000) have all been observed in growing regions of plants exposed to high CO_2, and all likely play a role in stimulating growth.

A better understanding of the mechanisms of increased crop growth in high [CO_2] will likely require knowledge of how carbohydrates, especially sucrose, function both as substrates for growth and as regulatory molecules. Increased sucrose transported to active meristems might have both direct (functioning as a growth substrate) and indirect (functioning as a signal molecule) effects on cell division and expansion (Francis, 1992; Williams, Winters, and Farrar, 1992). For instance, sucrose concentrations might function as a chemical control point of the cell division cycle, perhaps by affecting expression, concentration, or activity of cyclin genes or their products (Kinsman et al., 1997). Soluble carbohydrate concentrations in leaves undoubtedly play a role in regulating the activity of enzymes important for carbon fixation, and therefore can have feedback effects on photosynthetic capacity (Chapter 6). Finally, there is some evidence that sucrose stimulates the production of the hormone ethylene (Foster, Reid, and Pharis, 1992). This is interesting because ethylene has been shown to increase tillering in cereals, and increased tillering is one of the most universal morphological consequences of cereal growth in CO_2-enriched environments (Conroy et al., 1994).

Shoots. Growth in high [CO_2] results in taller plants with longer internodes. For example, height was increased 15 percent in soybean and 17 percent in wheat, but the number of nodes was not changed in either crop. Node number, in fact, rarely changes (Pritchard et al., 1999). Branch number and length often increase, as do stem diameters. The growth stimulation by elevated [CO_2] in cereals often results from increased tillering with more panicles present on which grain can develop (Kendall, Turner, and Thomas, 1985; Baker et al., 1996). Increased tillering and branching may indicate that exposure to elevated [CO_2] may weaken apical dominance (Enoch and Zieslin, 1988; Pritchard et al., 1999). And indeed, high sugar concentrations often repress the effects of auxins, including apical dominance, which leads to a more spreading, procumbent growth habit (Koch, 1996).

Crops grown in high [CO_2] commonly have more leaves per plant, greater leaf area per plant, and larger, thicker leaves. Total leaf area per plant increases an average of 37 percent, while specific leaf area (SLA = leaf area/leaf dry wt.) decreases an average of 6 percent in crops grown at 720 μmol mol^{-1} compared to 365 μmol mol^{-1} CO_2 (presumably because of greater storage of nonstructural carbohydrates and/or thicker leaves; Pritchard et al., 1999). Other changes in leaf anatomy, including reduced stomatal densities (Beerling and Woodward, 1995), increased mesophyll cell size (Radoglou and Jarvis, 1990), and formation of an extra layer of palisade mesophyll cells (in dicots such as soybean) (Thomas and Harvey, 1983), are also common.

Roots. Often, growth in high CO_2 stimulates root growth more than shoot growth (Chapter 6; Figure 6.2). All aspects of root growth, including length, branching, diameter, and rate of growth, have been shown to increase with supraambient CO_2 (Rogers, Runion, and Krupa, 1994; Rogers et al., 1997). Increased growth of tubers (number, dry weight, and diameter) and nodulation (number, activity, and dry weight) are also typical belowground responses to elevated atmospheric CO_2 (Rogers, Runion, and Krupa, 1994). The stimulation of belowground structures is especially important for root crops such as potato, sweetpotato, and sugarbeet, considering their large contribution to human diets worldwide (Table 8.5).

Most studies have shown that the initiation and extension of lateral roots (as opposed to extension of primary roots) is the most common root response to high CO_2. This results in more highly branched roots and a disproportionate stimulation of roots at shallow depths (Pritchard and Rogers, 2000). This has been shown for spring wheat (Van Vuuren et al., 1997), winter wheat (Chaudhuri, Kirkham, and Kane-

TABLE 8.5. Effect of elevated atmospheric [CO_2] on yield of several root crops.

Crop	Daytime [CO_2] (high/low; ppm)	Yield in high [CO_2] relative to low [CO_2] (unitless)	Notes	Source
Potato	543/398	1.26	OTC, ambient [O_3]	A
	694/398	1.41		
	543/398	1.37	OTC, high [O_3]	
	694/398	1.30		
	700/350	1.38[a]	OTC	B
	660/360	1.33[b]	FACE	C
	680/360	1.32	OTC	D
Sweetpotato	665/354	1.61	OTC	E
Sugarbeet	650[c]/375	1.26	OTC, ample N	F
	650[c]/375	1.12	OTC, low N	

Sources: (A) Donnelly et al. (2001b); (B) Schapendonk et al. (2000); (C) Miglietta et al. (1998); (D) Finnan et al. (2002); (E) Biswas et al. (1996); (F) Demmers-Derk et al. (1998).

[a]Value is averaged for two years.

[b]Authors found a 10 percent increase in tuber yield per 100 ppm increase in [CO_2] above ambient (up to 660 ppm).

[c]This datum represents an average of a three-year study. The elevated [CO_2] in the first year was 600 ppm but was increased in years two and three to 700 ppm.

masu, 1990; Fitter et al., 1996), cotton (Rogers et al., 1992), and sorghum (Chaudhuri et al., 1986; Pritchard, unpublished data). More highly branched root systems may have the capacity to capture more total soil resources (per plant), but they are usually less efficient (per unit of root length), probably because of greater overlap between adjacent branches (Fitter, 1996). The implications of root structural changes for function is discussed in Chapter 7.

Salinity

Development

Crop development is sometimes hastened by high salt levels in the growth medium. In wheat, salt stress accelerated development of the shoot apex by 18 days and significantly reduced the time to flowering (Grieve, Francois, and Maas, 1994). Shorter time to flowering was also reported in sweet clover (Volkmar, Hu, and Steppuhn, 1998). Although somewhat variable, length of developmental stages usually either remain the same or become slightly shorter in plants grown in saline soils (Francois et al., 1989; Grieve et al., 1993).

Effects of salinity on growth and development begin very early in germination and persist through physiological maturity. Salinity reduces the percentage of seeds that germinate and slows the rate of those that do germinate (Wahid, Rasul, and Rao, 1999). During germination, imbibition of water is impeded, as is metabolism associated with respiration, transport, and cell expansion within embryonic tissues. A wide range of effects on protein (Petruzzelli et al., 1991), carbohydrate (Lin and Kao, 1996), nucleic acid (Gomes Filho and Sodek, 1988), and polyamine metabolism (Friedman, Altman, and Levin, 1989; Wink, 1997) have been documented in germinating seeds challenged with high salt. Germination of large seeds seems to be less sensitive to salinity than smaller seeds, perhaps because large seeds have greater energy reserves with which to activate stress-response mechanisms (Chapter 3) (Wahid, Rasul, and Rao, 1999). High temperature and water stress appear to magnify the negative effects of salinity on germination (Hampson and Simpson, 1990).

Although salinity impedes germination, crops are usually able to germinate at higher levels of salinity than are tolerable during later vegetative and reproductive growth stages (Francois and Maas, 1999). Therefore, many salt-sensitive plants that do germinate in salinized soil do not live through the seedling stage, which tends to be most sensitive. After the seedling stage, salt tolerance generally increases as plants advance through later vegetative, reproductive, and grain-filling stages (Francois and Maas, 1999). This has been shown for several major crops, including wheat (Maas and Poss, 1989), barley (Ayers, Brown, and Wadleigh, 1952), maize (Maas et al., 1983), rice (Pearson and Bernstein, 1959), sorghum, and cowpea (Hall, 2001). Perennial crops, such as asparagus, seem to be more salt tolerant the second year than the first (Francois, 1987). This indicates that the quality of irrigation water becomes less and less crucial as crops mature and become less sensitive to salinity.

Over weeks to months, however, significant amounts of Na^+ and Cl^- are transported via transpiration into shoots. These salts accumulate in older leaves, often leading to premature leaf senescence (reduced leaf-area duration). So although development is less sensitive at later phenostages, premature death of older source leaves often limits the amount of assimilates that can be transported to active sinks such as developing grain (Figure 8.6). If the rate of leaf-area mortality exceeds the rate of new leaf production, then the amount of carbohydrate required to sustain further growth may become inadequate. Whole-plant death may follow. Salinity also hastens aging of root tissues (i.e., they mature more quickly) (Reinhardt and Rost, 1995), but we still do not know how root longevity is affected.

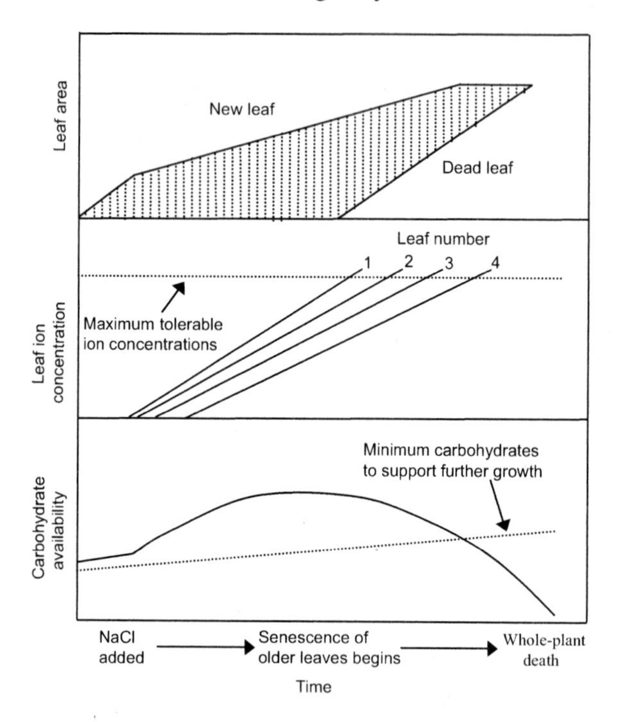

FIGURE 8.6. Long-term exposure to salinity often leads to the accumulation of ions in leaves (middle graph). Once maximum tolerable levels of ions are reached, leaf senescence begins and functional leaf area begins to decline (top graph). If leaf senescence exceeds production of new leaf area, carbohydrate availability often dips below the critical levels required to support growth (bottom graph). Time scales and NaCl concentrations required to elicit such a response vary from species to species (depending upon salt tolerance as well as growth habit). *Source:* Figure modified from Munns and Termaat (1986).

Growth

Salinity reduces whole-plant growth, tiller numbers, leaf numbers and sizes, and height. Four mechanisms have been hypothesized to account for salinity-induced growth reductions (Lazof and Bernstein, 1999): (1) salinity might depress photosynthesis (Chapter 5) and reduce supply of carbohydrate to growing tissues (Chapter 6), (2) salinity could inhibit expansive growth by reducing turgor pressure in meristematic regions (Chapter 4), (3) salinity might decrease the flexibility of cell walls in growing structures, and (4) mineral imbalances (toxicities or deficiencies) in meristems may constrain growth (Chapter 7). All four mechanisms are probably involved, but to differing extents, and over different time scales.

Immediately after the initiation of a salt (or any other osmotic) stress, plants become temporarily unable to acquire enough water to maintain turgidity in expanding cells. This seems to be the case in the short term (0 to 24 h), but not over periods of days to weeks (Munns et al., 2000; Passioura and Munns, 2000). Over several hours, osmotic adjustment in growing zones occurs and turgor generally recovers (presumably because the osmotic potential of the plant itself becomes lower).

Although turgor supplies the *energy* for cell expansion, it does not generally *control* it, and typically does not limit it either. Control is more likely exerted by the cell-wall characteristics that either allow cell walls to stretch and expand (to grow), or keep them rigid (prohibit them from growing). Hence, long-term reductions in cell growth following salinization probably involve cell- wall hardening brought about by biochemically regulated changes in the elastic properties of root (Neumann, Azaizeh, and Leon, 1994) and shoot (Cramer and Bowman, 1992) cell walls. Hardening of cell walls could involve a change in the properties or numbers of load-bearing bonds between adjacent cellulose microfibrils that comprise cell walls. This hypothesis has been supported by observations that the apparent yield threshold of cell walls is increased in salinized plants (that is, greater turgor pressures are required for cells to stretch and grow) (Cramer, Schmidt, and Bidart, 2001). A number of changes in cell walls have been observed in salinized crops (Zhong and Läuchli, 1993; Kafkafi and Bernstein, 1996), but reports are too sporadic and contradictory to allow any meaningful generalizations to be drawn.

Cell division also is reduced in salt-stressed crops (Zhu, 2001). Abscisic acid, a hormone typically up-regulated in plants subjected to salt or drought stress, has been shown to induce expression of ICK1 (Wang et al., 1998). ICK1 is a molecule that inhibits growth by reducing the activity of cyclin-dependent protein kinases that are required for cell division. Although mounting evidence points toward salinity-induced inhibition of both cell expansion and cell division, we are still a long way from a complete understanding of the cellular basis for reduced growth in salinized plants (Cramer, Schmidt, and Bidart, 2001).

Shoots. Salinity reduces shoot growth, leaf expansion, and retards leaf initiation (Bernstein, Silk, and Läuchli, 1993). The rates of leaf emergence are particularly sensitive to salinity. Growth very early in the development of embryonic leaves is more severely retarded than subsequent expansion. In sorghum, for example, the expansion of leaves is inhibited most when leaves are still enclosed in the sheath created by older leaves (Bernstein, Silk, and Läuchli, 1993). Similarly, expansion of younger sunflower leaves (numbers 14 to 19) is inhibited more than expansion of older leaves (8 to 14) (Rawson and Munns, 1984).

Roots. Root growth and development is less sensitive to salt stress than shoot growth, in spite of the fact that it is the roots, not the shoots, that are directly exposed to NaCl in the soil (Kafkafi and Bernstein, 1996). At low levels of salinity, root growth is often unaffected (and sometimes is enhanced), even while shoot growth is significantly reduced (Munns and Termaat, 1986). Nevertheless, moderate to high salinity generally does suppress root growth.

Salinity affects both the rate of root cell production (division) as well as expansion of newly formed cells. Often, the length of the elongation zone is shortened in roots of salinized crops (Zhong and Läuchli, 1993; Pritchard, 1994), final cell size is reduced (Kurth et al., 1986), roots are shorter (Kafkafi and Bernstein, 1996), and thicker (Zhong and Läuchli, 1993). Little information is available concerning the influence of salinization on root branching. Of the few existing studies, initiation of lateral roots has, on several occasions, been shown to increase in salinized plants. For example, pericycle cell divisions and lateral root initiations were stimulated and located much closer to the root tip in cotton (Reinhardt and Rost, 1995). This resulted in more lateral roots per unit length of primary root; these laterals failed to elongate normally, however, so the root system was still smaller than the control.

Shorter and thicker roots not only explore a smaller soil volume, but are also less efficient in the assimilation of water and mineral nutrients. When environmental stress impacts growth, the result is smaller or fewer leaves and fewer or shorter roots with which to acquire resources required for further growth. Thus, over time, limited substrate supply can further confound the direct toxic/osmotic effects that resulted in reduced growth in the first place.

Elevated Atmospheric Ozone Concentration

Atmospheric O_3 diffuses into leaves through stomates. In the leaf interior, ozone oxidizes (damages) proteins and lipids of membranes, increasing their permeability and lowering photosynthetic capacity (Chapter 3). Short- and/or long-distance transport of metabolites throughout the plant is also affected (Chapter 6). Furthermore, ozone often causes premature leaf senescence, which reduces whole-plant leaf area (Sanders et al., 1992). The effects of ozone on leaf duration and the manner in which this changes long-term plant function is very similar to the situation described earlier for salinity, which also leads to premature leaf senescence (but through the accumulation of toxic ions in leaves). The effects of ozone on leaf duration and how this affects leaf area, therefore, is very similar to the situation described in Figure 8.6. So, exposure to O_3 not only limits source strength by directly damaging the photosynthetic apparatus but also attenuates long-term, whole-plant source strength by reducing leaf area. Detoxification and repair processes such as up-regulation of antioxidant systems (Chapter 3) also require energy, further reducing the pool of respiratory substrates and photoassimilates available for growth.

In the previous section on the effects of elevated atmospheric [CO_2] on growth and development, we stressed that obtaining a mechanistic understanding of the events leading to increased growth would require a better understanding of how sucrose functions both as a substrate for growth and a regulatory molecule. This was because high [CO_2] stimulates growth mainly through increased photosynthesis and larger carbon pools. Similarly, understanding how O_3 affects growth and development must also involve an understanding of how sucrose, but in this case sucrose limitations, affects growth processes (cell division, expansion, and differentiation). We will not reiterate what is known, or what was discussed earlier, about how carbohydrate pools affect growth, but will simply suggest that substrate limitations are expected to detrimentally affect both cell division and cell expansion. Unfortunately, too little information is available concerning the effects of ozone on these processes to support a more detailed discussion (Bambridge, Macleod, and Harmer, 1995).

Development

The most consistent effect that elevated [O_3] has on development is to decrease leaf duration by causing premature senescence and to increase the length of the vegetative growth period (i.e., flowering is delayed; Amundson et al., 1986; Black et al., 2000). Premature leaf mortality is sometimes compensated for by accelerated development of new leaves. These shifts, for example, led to a delay in the maximum leaf area ratio (LAR) from 13 to 21 days after planting in radish grown with O_3 pollution (Walmsley, Ashmore, and Bell, 1980, cited by Runeckles and Chevone, 1992). On the other hand, O_3 pollution also may cause stomatal closure and reduced transpiration, and this can decrease plant evaporative cooling, leading to higher leaf and perhaps canopy temperatures and their associated effects on development.

It is important to recognize that O_3 effects on crop growth change as plants develop. This is probably due to source-sink shifts rather than specific effects on development per se (Runeckles and Chevone, 1992). But we still have only limited data from which to draw; most ozone studies have focused primarily on how yield is affected, with only passing attention paid to growth or development (Runeckles and Chevone, 1992). We will see in the next chapter that O_3 exposure, in addition to delaying flowering, has numerous other effects on reproductive development. We will also see in the following chapter that O_3 exposure can have very different effects on growth and development in determinate compared to indeterminate species and cultivars.

Growth

Shoots. Total shoot dry matter is typically decreased in crops exposed to O_3 pollution (Table 8.6), but the effects on rate of leaf area expansion and whole-plant leaf area are highly variable. In fact, plants sometimes produce more leaves to compensate for shorter leaf duration when exposed to O_3 pollution. In radish, for example, plants exposed to ozone grew leaves faster than the controls, but at the

TABLE 8.6. Effect of elevated atmospheric [O_3] on biomass accumulation (vegetative growth) of selected crops.

Crop	O_3 treatment	Treatment duration	Effect on growth (plant part)	Source
Soybean	35 versus 17 ppb (7-h daily means)	69 d	no change (shoots)	A
	60 versus 17 ppb (7-h daily means)		–18% (shoots)	
	84 versus 17 ppb (7-h daily means)		–18% (shoots)	
	122 versus 17 ppb (7-h daily means)		–43% (shoots)	
	66 versus 23 ppb (7-h daily means)	119 d	–16% (shoots) (350 ppm CO_2)	B
			–7% (shoots) (400 ppm CO_2)	
			–2% (shoots) (500 ppm CO_2)	
	90 versus 50 ppb (6.8-h daily means)	56 d	–31% (shoots) (well watered)	C
			–8% (shoots) (water stressed)	
			–6% (roots) (well watered)	
			–2% (roots) (water stressed)	
	130 versus 50 ppb (6.8-h daily means)	56 d	–40% (shoots) (well watered)	
			–20% (shoots) (water stressed)	
			–37% (roots) (well watered)	
			–17% (roots) (water stressed)	

TABLE 8.6 *(continued)*

Crop	O_3 treatment	Treatment duration	Effect on growth (plant part)	Source
Cotton	54 versus 10 ppb (12 h means)	102 d	+5%[a] (shoots) –3% (roots)	D
	64 versus 10 ppb (12 h means)		–2% (shoots) +22% (roots)	
	90 versus 10 ppb (12 h means)		–32% (shoots) –34% (roots)	
Wheat	1.5 × ambient (12 h means)	98 d	no change (shoots) (ambient +263 ppm CO_2)	E
			no change (shoots) (ambient CO_2)	
			–35% (roots) (ambient +263 ppm CO_2)	
			–5% (roots) (ambient CO_2)	
	60 versus 29 ppb (12 h means)	134 d	–14% (shoots) ([CO_2] = 379 ppm)	F
	61 versus 27 ppb (12 h means)		no change (shoots) ([CO_2] = 680 ppm)	
	2.5 × ambient (6 or 7 h d^{-1})	43 d	–6% (shoots)	G
Rice	50 versus 5 pph (7 h d^{-1})	89 d	–16% (whole plant)	H
	100 versus 5 pph (7 h d^{-1})		–32% (whole plant)	
	200 versus 5 pph (7 h d^{-1})		–52% (whole plant)	
	70 versus 10 ppb (hourly averages)	28 d	–37% (shoots) ([CO_2] = 400 ppm) –10% (shoots) ([CO_2] = 700 ppm) –58% (roots) ([CO_2] = 400 ppm) –8% (roots) ([CO_2] = 700 ppm)	I

Sources: (A) Kohut, Amundson, and Laurence (1986); (B) Mulchi et al. (1992); (C) Amundson et al. (1986); (D) Temple (1990); (E) Fangmeier et al. (1996); (F) Mulholland et al. (1997a); (G) Ollerenshaw and Lyons (1999); (H) Ming-hong, Zong-wei, and Fu-zhu (2001); (I) Olszyk and Wise (1997).

[a]Average of four cultivars.

expense of stem and root tissue (Held, Mooney, and Gorham, 1991). This reflects optimization of resources. That is, plants under stress tend to allocate resources preferentially to that plant organ through which the most limiting resource is acquired. In the case of O_3 pollution, carbon is growth limiting because of the direct toxic effect of oxidants on the photosynthetic apparatus. Other environmental conditions that limit carbon assimilation, shade, for example, also result in preferential allocation of resources to leaves over stem and root tissue. Further discussion regarding the effects of environmental change on partitioning of resources can be found in Chapter 6.

Roots. Total root biomass is virtually always decreased in plants exposed to high atmospheric [O_3]. In fact, most investigators feel that, generally, since shoots have priority over roots in the utilization of limited photoassimilates, root growth and development should decline in favor of leaf growth and maintenance in plants exposed to high O_3 (Rennenberg, Herschbach, and Polle, 1996; Taylor and Ferris, 1996; Andersen, 2003). Data summarized in Table 8.6, however, do not uniformly support this generalization. Clearly, further experiments on root responses to high atmospheric [O_3] in crops are needed to resolve this issue. Even less information is available concerning the specific shifts in root morphology and anatomy caused by atmospheric O_3 pollution.

SUMMARY

Growth and development of crops, even determinate ones, will be affected by environmental change. The great variety of shapes, structures, and functions of cells and organs among different species results from flexibility in cell division, expansion, and differentiation. Each of these processes may be altered, to different degrees depending on the crop, by environmental changes. Nearly all cells are derived in shoot or root meristems by cell division. Those cells then increase in size by a hundred- to a thousandfold as they expand and differentiate.

In day-neutral crops, degree-days (i.e., heat sums) control crop development. Therefore, rising global air and soil temperatures are expected to increase the plant developmental rate. Faster development will speed germination rate and will abbreviate the vegetative growth period and time to flowering. Hastening of development could result in smaller crops, other factors being equal.

Rising [CO_2] has the potential to counteract some of the effects of rising temperature. Although effects on development are minimal, C_3 crops grown at 720 ppm CO_2 usually grow 30 to 40 percent larger compared to those grown at today's [CO_2] (365 ppm). Growth stimulation of C_4 crops such as maize and sorghum, however, is expected to be substantially weaker. Crops exposed to high [CO_2] are typically taller, have more leaf area, and larger root systems. Root growth often increases more than shoot growth. The stimulation of growth by elevated [CO_2] is driven primarily by increased photosynthesis and larger pools of substrates to support construction of new tissues and respiration. Rising [CO_2] also reduces the amount of water transpired per molecule of CO_2 assimilated during photosynthesis; improved water relations likely contribute to faster growth.

Ozone damage inhibits growth, mainly through its effects on substrate supply. Elevated [O_3] reduces whole-plant carbon assimilation by inhibiting leaf-level photosynthesis (by damaging the photosynthetic apparatus) and reducing whole-plant leaf area (through premature leaf senescence). Reductions in photosynthesis, along with diversion of available assimilates away from growth for repair and detoxification, reduce substrates available for growth. And root growth often suffers more than shoot growth when crops are exposed to high [O_3]. Developmental responses to elevated [O_3] are complex and vary considerably among crops and with the timing, duration, and concentration of the O_3 exposure.

Mechanisms of salinity-induced growth reductions appear to be multifaceted. Nutrient supply to active meristems may be decreased, which could result in less growth. Reduced growth has also been correlated with inhibited cell expansion caused by cell-wall hardening. Many biochemical and mechanical changes in the cell walls of salinized crops have been reported. In the long term, accumulation of salts in older leaves results in premature leaf senescence, a reduction of whole-plant leaf area, and sometimes plant mortality.

Growth responses to environmental change will be complex and highly variable, not only between different crop species but also among cultivars or varieties of the same crop. Broad generalizations about how environmental change will affect vegetative growth and development are therefore difficult to make. Responses to environmental changes can sometimes be predicted based on the functional and structural attributes of a particular crop; characteristics such as sink strength (e.g., fast versus slow growing, size of storage organs), source capacity (e.g., C_3 versus C_4), assimilate uptake and transport capacity (root morphology and plasticity), and growth habit (e.g., determinate versus indeterminate) all play a role in controlling crop response to environment. The greatest hurdle for understanding how crop vegetative growth will be affected by salinity, rising temperatures, ozone pollution, and rising atmospheric CO_2 levels rests in the ability to unravel mechanisms that cause growth or development to change. It is easy to measure growth changes during the course of an experiment, but it is quite another matter to understand *how* those changes came about.

Chapter 9

Sexual Reproduction, Grain Yield, and Grain Quality

> There are times, aren't there, when plants shoot but do not flower, and when they flower but do not produce fruit?
>
> Master Kong (Confucius),
> *Analects* 9:22

Humans depend on seeds (grain) and fruits produced by plants, with about 75 percent of world food supply coming directly or indirectly from the seeds of just three crops: wheat, maize, and rice (Chapter 1; Egli, 1998). Crop seeds and fruits also provide beverages such as coffee and beer, fibers such as cotton, medicines, biofuels, and oils for food and industrial purposes (e.g., rape, cotton, sunflower, and maize oils). Successful sexual reproductive processes by crops are therefore critical to humanity.

Biological grain yield is the total mass of all grain produced by a crop per unit of ground area. Harvested yield may be considerably less than biological yield due to various losses before and during harvesting. Both biological and harvested yields could be affected, positively or negatively, by environmental change. For example, increased atmospheric [CO_2] might enhance crop growth and stimulate biological yield, whereas more frequent or intense periods of high wind could reduce harvested yield because of greater lodging.

Grain quality (Box 1.1) includes the physical and chemical properties (e.g., types and amounts of proteins, carbohydrates, and lipids) of harvested seeds. Seed composition varies considerably among crop species (Table 9.1). Grass seeds (i.e., cereal grains such as wheat, maize, rice, and barley) are generally high in carbohydrates and low in proteins and lipids. In contrast, seeds of most pulse crops (e.g.,

TABLE 9.1. Dry-matter composition of harvested seeds, fruits, and tubers.

Crop	Carbo-hydrate (%)	Protein (%)	Lipid (%)	Digestible energy (MJ kg^{-1})	Source
Cereals					
Barley (dehusked)	78	14	2.3		A
Buckwheat	70.8	13.1	2.7	15.7	B
Maize (field)	71.6[a]	8.7	4.0	16.4	C
Maize (popcorn)	77.6	13.2	5.2	16.8	B
Maize (sweet)	80	13	4		A
	78.4	12.8	3.7	14.7	B
	75	14	4.6		D
Millet (pearl)	78	13	5.7		A
Oat	75	13	8.2		A
Rice (brown)	86.9	8.5	2.2	17.1	B
Rice (white, unenriched)	91.0	7.6	0.5	17.3	B
Rye (whole seed)	80.2	13.6	1.9	15.7	B
Sorghum *(S. vulgare)*	82.0	12.2	3.9		D
Wheat (hard red spring)	76.8	16.1	2.5	15.9	B
Wheat (hard red winter)	79.3	14.1	2.1	15.8	B
Wheat (soft red winter)	81.2	11.9	2.3	15.9	B
Wheat (white)	83.1	10.6	2.3	15.8	B
Wheat (durum)	78.5	14.6	2.9	16.0	B
Oilseed Crops					
Coconut (copra)	19.8	7.5	67.3	28.7	B
Cotton (seed)	46.6	24.9	24.7		D
Flax	32.3	25.6	38.3		D
Peanut	18	29	48		A
	17.2	27.5	50.3	25.0	B
Rapeseed *(Brassica napus)*	24.6	22.5	48.2		D
Rapeseed *(B. campestris)*	22	24	41.0[b]		E
Safflower (whole seed)		13.5	32.8		D
Safflower (kernel)	13.1	20.1	62.6	27.1	B

Crop	Carbo-hydrate (%)	Protein (%)	Lipid (%)	Digestible energy (MJ kg^{-1})	Source
Sesame	17	21	53		A
Soybean	31.8	37.9	19.7	18.7	B
		40.7	21.3		G
Sunflower (kernel)	25	29	39		A
	17.8	25.2	49.7	24.6	B
Winged bean	37	38	19		A
Tubers, Beets, and Tap Roots					
Carrot	70	10	0	14	A
Cassava (tapioca)	93	1.8	0.5		A
	87	3	1		F
Potato	85	10	0		A
	82.2	10.4	0.5	15.7	B
	78	9	0		F
Red beet	77	14	0		A
	71.7	12.6	0.8	14.2	B
Sugar beet	82	5	0		F
Sweetpotato (*Ipomoea* sp.)	87.1	5.8	1.4	16.2	B
	84	5	2		F
Yam (*Dioscorea* sp.)	89	7.4	0.7		A
	80	6	1		F
Fruits and Vegetables					
Apple	86.5	1.3	3.8	15.6	B
Apricot	83.0	6.8	1.4	14.5	B
Avocado	18.1	8.1	63.1	26.9	B
Banana	89.3	4.5	0.8	14.6	B
Cranberry	77.7	3.3	5.8	15.9	B
Cucumber	57.1	18.4	2.0	12.8	B
Date	91.1	2.8	0.6	14.8	B
Eggplant	61.8	15.8	2.6	13.8	B
Fig	84.9	5.3	1.3	14.9	B
Grape (American type)	82.1	7.1	5.4	15.7	B
Guava (whole, common)	55.3	4.7	3.5	15.3	B

TABLE 9.1 *(continued)*

Crop	Carbo-hydrate (%)	Protein (%)	Lipid (%)	Digestible energy (MJ kg^{-1})	Source
Kumquat	71.7	4.8	0.5	14.5	B
Loquat	88.1	3.0	1.5	14.9	B
Mango	86.9	3.8	2.2	15.1	B
Okra	59.5	21.6	2.7	13.6	B
Orange (peeled)	83.6	7.1	1.4	14.6	B
Papaya	80.5	5.3	0.9	14.4	B
Peach	83.5	5.5	0.9	14.6	B
Pear	82.7	4.2	2.4	15.2	B
Pepper (hot, pod, dry)	38.4	14.8	10.4	15.4	B
Pepper (sweet, red)	58.1	15.1	3.2	13.9	B
Plantain	94	3	0.6		A
Plum (hybrid)	87.3	3.7	1.5	15.0	B
Prune (dehydrated)	91.4	3.4	0.5	14.8	B
Pumpkin	64.3	11.9	1.2	13.0	B
Raisin	93.3	3.0	0.2	14.7	B
Squash (summer)	60.0	18.3	1.7	13.2	B
Tomato	54	17	4		F
Pulses					
Bean (common, white)	64.0	25.0	1.8	16.0	B
Bean (lima)	70.5	23.7	1.5		D
Bean (lima, immature)	62.5	25.8	1.5	15.8	B
Bean (mung)	62.6	27.1	1.5	15.9	B
Bean (snap, green)	61.6	19.2	2.0	13.5	B
Broad (faba) bean	58.5	28.5	1.9	16.1	B
Chickpea	62.7	23.0	5.4	16.9	B
Cowpea	64.0	25.5	1.7	16.0	B
Lentil	63.2	27.8	1.2	16.0	B
Pea	62.7	27.3	1.5	16.1	B
Pigeon pea	63.6	22.9	1.6	16.0	B

Crop	Carbo-hydrate (%)	Protein (%)	Lipid (%)	Digestible energy (MJ kg^{-1})	Source
Nuts					
Almond	12.6	21	62		A
	17.7	19.5	56.9	26.3	B
Brazilnut	8.2	15.0	70.1	28.7	B
Butternut	8.7	24.6	63.6	27.4	B
Cashew	27	21	47		A
Hazelnut	14.5	13.4	66.2	28.2	B
Pecan	12.7	9.5	73.7	29.8	B
Pilinut	6.1	12.2	75.9	29.9	B
Pistachio	18.1	20.4	56.7	26.2	B
Walnut (black)	13.5	21.2	61.2	27.1	B
Fiber Crop					
Cotton (seed + lint)	40	21	23		F

Sources: (A) Platt (1962); (B) Watt and Merrill (1963); (C) Butzen and Cummings (1999); (D) Spector (1956); (E) Ohlson and Sepp (1975); (F) Penning de Vries, van Laar, and Chardon (1983); (G) Hurburgh (2001).

Notes: Composition of edible portion. In some cases, "carbohydrate" includes crude fiber (e.g., cellulosic material and lignins), but even when crude fiber is excluded, values may overstate actual carbohydrate fractions (Watt and Merrill, 1963). Fractions sum to less than 100% because ash (and usually crude fiber) is excluded. Digestible energy values are based on 4.184 J per calorie.

[a] Carbohydrate fraction is starch only. Total carbohydrate is greater.

[b] Indicates the midpoint of the range reported in the source.

bean, pea, and lentil) are high in proteins and have relatively low levels of carbohydrates and lipids, whereas oilseed crops (e.g., rape, sunflower, sesame, and safflower) have particularly high concentrations of oil (i.e., lipids) in their seeds.

The economic value of many grains is a strong function of seed quality. For example, wheat grain value is improved by higher protein concentration, while the value of barley for malting is decreased by high protein concentration. For rice, in which cooking and textural properties are valued most (as opposed to wheat, which is valued for breadmaking), carbohydrate concentration and composition are the most important components of quality. In sunflower, the quantity and quality of oil in the seeds is critical for many uses of the crop.

Subtle differences in seed quality can be important for marketing purposes because different end uses can demand fairly specific seed characteristics. Wheat used for making bread, for instance, requires a different protein concentration than wheat used for pasta, cookies, or crackers, because small changes in protein concentration can have large effects on dough strength, a breadmaking property. In England, grain protein concentration differs by only 5.6 percent between high-quality breadmaking wheats and the lower-quality varieties used as animal feed and in industrial applications (Thompson and Woodward, 1994).

Environmental changes that affect any crop physiological process have the potential to ripple through subsequent growth and development, ultimately affecting both grain quantity (i.e., biological and harvested yields) and quality. This is so because each point along the developmental continuum, from seed germination to physiological maturity and harvesting by humans, and then forward to planting of the next crop, plays some role in yield formation (Figure 9.1). If, for example, early seedling development or growth is altered by environmental changes, the following phenostage may be affected, and that in turn can alter the next developmental stage, and so on, even spanning multiple crop generations.

Any change in yield brought about by an environmental change could be accompanied by a different change in quality. For example, wheat grain yield and grain protein concentration are often negatively correlated (e.g., Kibite and Evans, 1984), so an environmental change that enhances yield may reduce grain protein concentration, whereas a change that limits yield may increase protein concentration.

SEXUALLY REPRODUCTIVE DEVELOPMENT

The switch from vegetative to sexually reproductive development is a defining moment for grain crops. During that transition, shoot (apical) meristems begin to build flowers (i.e., the meristems form flower primordia), which eventually produce grain. Flowers are determinate organs in which male and female gametes (sperm and egg, respectively) are produced. Mechanisms facilitating the union of

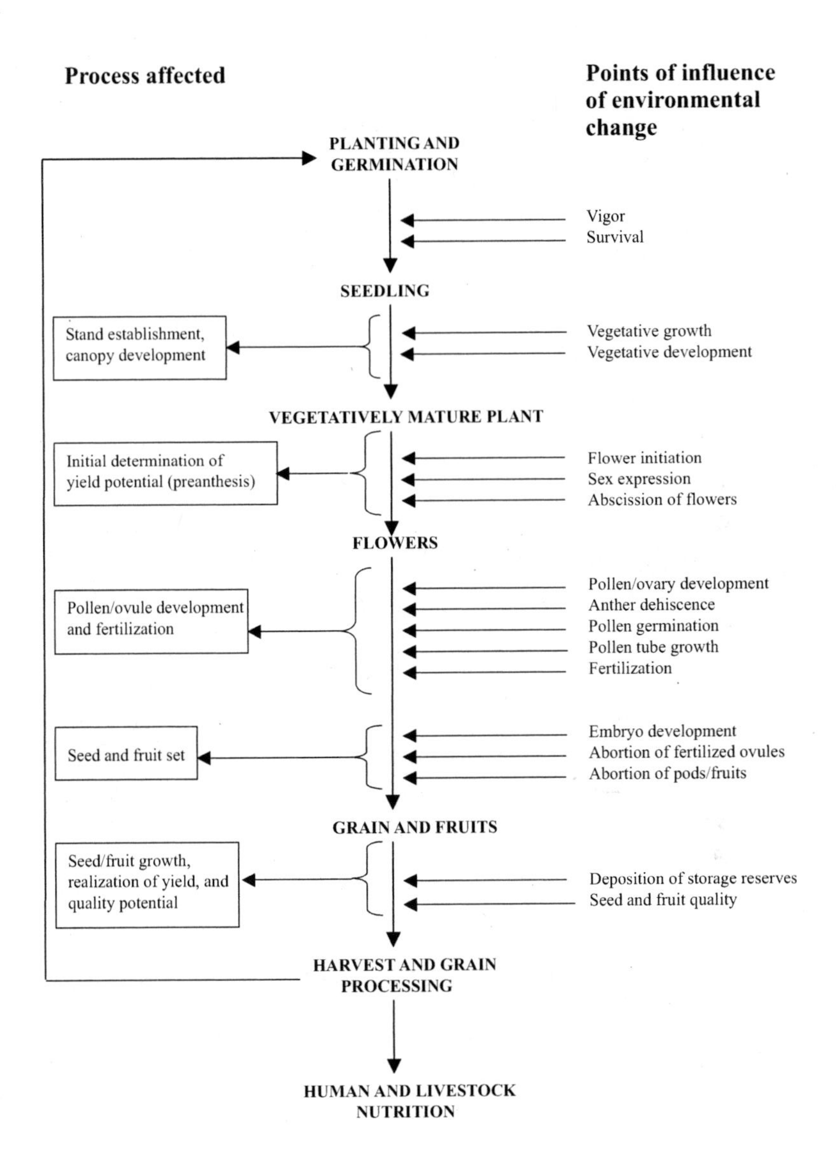

FIGURE 9.1. Potential points of influence of environmental change on crop reproduction, yield, and food quality. *Source:* Figure modified from Wolters and Martens (1987) and Black et al. (2000).

sperm with egg (i.e., pollination and fertilization) are complex. They are also easily disrupted by unfavorable environmental conditions.

Initiation of the production of flower primordia is triggered when certain conditions are met. In most modern crop varieties, the accumulation of a genetically related quantity of "heat sums" or "degree days" (Chapter 8) provides the stimulus to initiate sexually reproductive development (this can also correspond with the previous production of a given number of leaf primordia). In fewer crops, the onset of reproductive development is triggered by photoperiod (actually, night length), a phenomenon called photoperiodism. In other crops, a combination of accumulated heat units and night length serves as the cue to initiate flowers.

For crops sensitive to heat sums, global warming could alter the timing of flowering. For a given location, night length will not change in the future, but the combination of night length and heat sums may. Moreover, if the latitude at which a crop is grown changes because of human adaptation to environmental changes such as warming, that crop will experience different night lengths on specific calendar dates.

Male and Female Gametophytes

Male flower parts are called stamens, which are composed of anthers connected to the flower by a filament. The anther, or pollen sac, contains a column of many microspore cells (spores in heterosporous plants that give rise to male gametophytes and are usually smaller than megaspores) surrounded by a tissue region known as tapetum. As stamens mature, microspores undergo meiosis, producing four haploid cells. Developing microspores are supplied with nutrients and energy by tapetal cells. The tapetum is evidently vital for viable pollen formation, as abnormal tapetal development or functioning has been linked to male sterility in a number of crops. Each microspore eventually undergoes mitosis, forming a two-celled pollen grain, consisting of a vegetative and a generative cell. The generative cell eventually divides into two sperms, each of which is a nucleus with a small amount of associated cytoplasm. The resulting haploid structure is a partially developed male gametophyte (haploid type of plant).

Female flower parts are called pistils, which are made up of a stigma, a style, and an ovary. The stigma is the structure on which pollen grains land and germinate. The style, through which pollen tubes must grow to reach the egg, supports the stigma and attaches it to the ovary, the basal structure of the pistil. Within each ovary, one or more ovules (progenitors of seeds) are present. Ovules are connected to the ovary wall (placenta) by a funiculus.

A single megaspore cell (a spore in a heterosporous plant that gives rise to a female gametophyte and is usually larger than a microspore) differentiates from, and is surrounded by, the nucellar tissue within each ovule. As the pistil develops, the megaspore undergoes meiosis, producing four haploid cells. Three of these cells disintegrate, and the fourth undergoes three mitotic divisions to form the embryo sac, containing eight nuclei, which is the female gametophyte. When mature, the embryo sac contains seven cells and eight nuclei. One of those cells becomes the egg.

Pollination and Fertilization

For fertilization to occur, pollen must travel from the anther (where it is produced) to a receptive stigma associated with a viable egg. Passage of pollen from anther to stigma is most commonly facilitated by insects or wind. Once a pollen grain lands (or is deposited) on a stigma, it germinates and develops a pollen tube, which, if successful, penetrates the stigma, grows through the style into the ovary, and then grows into an ovule (usually through a micropyle). The elongating pollen tube carries the two sperm nuclei and discharges them into the embryo sac. The germinated pollen grain with a pollen tube is a fully developed male gametophyte.

Once inside the embryo sac, one sperm nucleus fuses with the egg nucleus (i.e., fertilizes the egg), yielding a zygote that later develops into an embryo, and the other sperm fuses with two other nuclei of the embryo sac, forming a triploid nucleus. This process of zygote and triploid nucleus production is called double fertilization, and occurs only in higher plants. The triploid nucleus divides repeatedly to form a nutritive tissue called endosperm, which envelopes the embryo. Typically, the outer layers of the ovule eventually form a toughened seed coat. What is usually called a seed includes the embryo, endosperm, and seed coat. Herein, the term seed will include caryopses (i.e., grass "seeds") and cypselae (i.e., one-seeded fruits of Asteraceae) in addition to true seeds.

Embryogenesis and Seed Development and Growth

Five main stages of seed development can be distinguished (see Figure 9.2 for a representative dicot field crop, soybean). After an egg is fertilized by a sperm, the zygote that is formed begins to divide. This marks the onset of embryogenesis. During embryo develop-

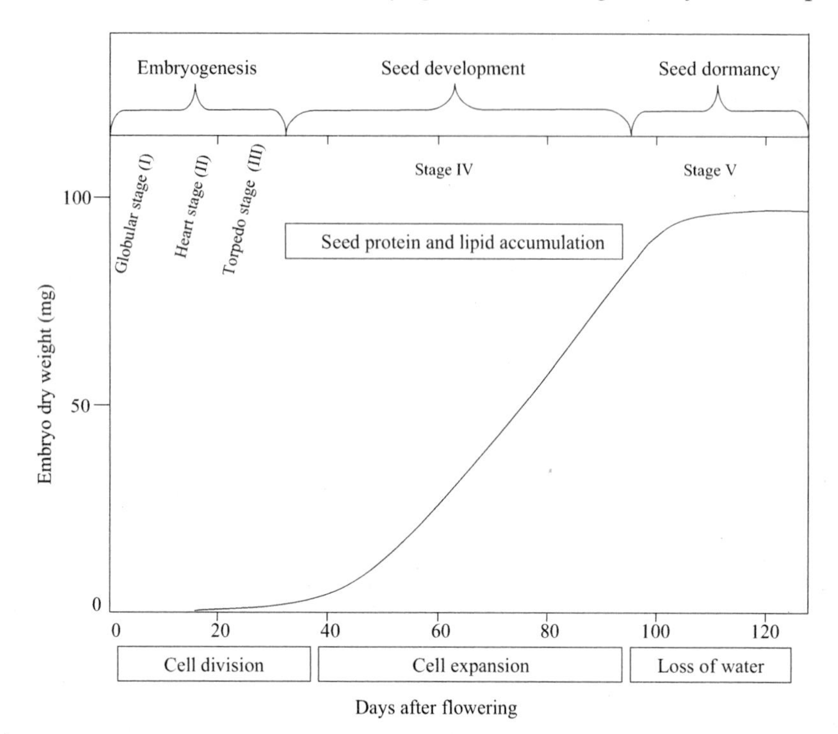

FIGURE 9.2. Stages and durations of soybean embryogenesis and seed formation. The three stages of embryogenesis, in which differentiation of the embryo and cotyledons occurs, are called the globular, heart, and torpedo stages. These names reflect the morphology of the developing embryo. At the end of Stage III, at which time all cells of the mature embryo have been formed, the developing embryo shifts from a period of controlled cell divisions to a time of rapid cell enlargement (Stage IV). Seed proteins and lipids accumulate and are deposited in the cotyledons during Stage IV. During the dormancy phase (Stage V), metabolism greatly slows, the embryo loses most of its water, and the seed coat assumes its final form. Endosperm development accompanies the development of the embryo. In soybean, the endosperm plays an important role in supporting the development of the embryo. In contrast, in most monocots, the endosperm tissue represents the main storage tissue of the seed. *Source:* Figure simplified from Goldberg, Barker, and Perez-Grau (1989).

ment, cells divide repeatedly and basic tissue patterns take shape. By the end of stage III (Figure 9.2), all cells and structures of the mature seed are present, including root and shoot apical meristems and several leaf primordia. This period of cell division is critical because it sets an upper limit on seed sink strength; the number of endosperm and embryo cells is generally positively related to seed sink strength, or potential growth.

As the period of seed cell division nears completion, substantial influx of raw materials from the mother plant begins. This marks the beginning of stage IV, which is characterized by expansion of the developing seed and the accumulation of substantial carbohydrate, protein, and/or lipid reserves. There is no vascular connection between the mother plant and the developing seeds, so all raw materials (mainly sucrose and amino acids) entering the seed must do so apoplastically, down concentration and/or pressure gradients. During stage V, metabolic functions slow and the seed enters dormancy.

There is substantial qualitative and quantitative variability in patterns of storage of reserve compounds in seeds (humans generally harvest grain crops for those reserve compounds). Some crop species deposit the majority of seed storage reserves in endosperm tissue, while others accumulate them in the cotyledons (see Box 4.3) of the embryo. In most monocots, the majority of material is stored in endosperm tissue and the cotyledon remains small. In many dicots, however, the cotyledons make up the bulk of the mature seed because their cotyledons, not endosperm, supply most of the raw materials required for early seedling growth.

Seed growth is a complex process influenced by many factors. Photosynthesis, nutrient and water acquisition, translocation of raw materials into developing seeds, and conversion of precursors into storage compounds by the seed itself all contribute to seed size and quality. Seeds of most crops are able to synthesize a diverse array of complex organic molecules from a few simple precursors. For example, soybean cotyledons can grow normally in a nutrient media containing only sucrose, one amino acid, and some minerals (Thompson, Madison, and Muenster, 1977).

Ovule position within an ovary can be important to both seed development and growth. Although ovules at the stylar end of the locule are first in line for fertilization, ovules at the opposite end may be first in line to receive substrates of growth from the mother plant. This effect can be striking in maize, where the kernels at the distal end of the cob often do not receive sufficient substrates to develop fully. So, depending on whether an environmental change affects pollen success or supply of substrates, the fertilization and growth of different ovules may be affected differently.

Seed growth is not only of immediate importance to human food supply, but also to future crops. Even small differences in seed size can translate into significant differences in seedling establishment. And differences in seedling success resulting from seed size often persist through to maturity of the plant; if other factors are equal, plants grown from large seeds can out-yield plants grown from small seeds (Fenner, 1992).

YIELD COMPONENTS AND VARIATION

Variation in yield of grain crops is caused by differences in seed number per unit ground area and/or differences in mean seed mass. Seed number and mean mass of individual seeds are often referred to as yield components. Final biological grain yield is the product of final seed number per unit ground area, the mean seed-filling duration (SFD), and the mean growth rate during seed filling (i.e., mean seed-filling rate, SFR). By definition, mean seed mass equals SFD × SFR. To understand effects of environmental change on yield, effects on each component of yield must be known.

Seed Number

Final seed number per unit of ground area is the number of seeds set (i.e., number of ovules fertilized) minus the number that are later aborted. Variation in yield is most often attributed to changes in number of seeds harvested per unit of ground area. The number of ovules capable of becoming mature seeds is particularly important, and is probably determined by canopy photosynthesis (amount of carbohydrate produced) before and during flower growth. Conditions associated with high rates of photosynthesis, such as high soil nitrogen levels, abundant light, and high atmospheric [CO_2], almost always result in a greater number of ovules. Conditions that limit canopy photosynthesis, whether through effects on leaf area or biochemical aspects of

CO_2 assimilation, are generally associated with fewer ovules. Seed set is especially sensitive to environmental conditions around the time flowers develop because that is when pollen is produced, young ovules are formed, and pollination and fertilization occur.

Crops typically retain only the number of seeds that can successfully mature under a given set of environmental conditions. Producing abundant ovules and then aborting some that are fertilized provides an efficient (and inexpensive) mechanism for balancing sink strength with source activity and ensures that growth by individual seeds is not limited by raw materials (significant environmental variation or extremes during flower development or seed growth can upset these balances, however). This probably explains why SFR is usually not limited by source activity and why variation in seed size is normally small in most crops.

Seed-Filling Duration

Seed-filling duration (Stage IV of seed development) is the time between the initiation of seed filling and the initiation of seed dormancy. Seed filling can overlap with vegetative growth in indeterminate crops, but in determinate crops, initiation of seed filling generally corresponds to cessation of vegetative growth. Changes in SFD can cause substantial changes in final seed size and yield.

A critical regulator of SFD is temperature. In particular, SFD in many crops is inversely related to temperature (Evans, 1975). This implies that a temperature-driven biological clock exerts influence on SFD such that an increase in temperature reduces SFD (the biological clock runs faster than the "wall clock" when temperature increases, so maturation is reached earlier with respect to the wall clock) and a temperature decrease increases SFD. Control of SFD may also be exerted through leaf longevity, water relations of the growing seed, or mechanical limitations to expansive seed growth (discussed in detail by Egli, 1998). Supply of growth substrates in particular, as it is influenced by leaf-area duration, may play a role in controlling SFD. A lack of response of SFD to experimental manipulation of source/sink relationships indicates that the duration of photosynthesis (i.e., leaf-area duration) is probably more important than instantaneous rates of photosynthesis. This indicates that leaf senescence may play a role in triggering cessation of seed growth and initiation of seed dormancy.

Alternatively, as seeds mature, their water potential declines to levels that seem to be relatively constant across different species (Egli and TeKrony, 1997), and it has been suggested that this attainment of a particular water potential triggers seed dormancy (Adams and Rinne, 1980; Ahmadi and Baker, 2001). Finally, physical restrictions on seed growth may be exerted by the seed coat, carpels, cotyledon or endosperm cells, or other structures (Egli, 1998). Under favorable environmental conditions, genetically defined physical limitations probably limit seed size, and for a given SFR, this might define SFD. Perhaps leaf-area duration, declining seed water content, and mechanical factors share control of SFD.

Whatever the controlling mechanisms are for a particular crop, SFD is sensitive to many environmental factors (especially temperature), and changes in SFD often translate into larger or smaller seeds, with corresponding effects on yield. Water stress, nutrient limitations, and high temperature, for example, all lead to abbreviated SFDs, smaller seeds, and limited yield in most grain crops. Indeed, effects of water stress and nutrient limitations on SFD may be mediated through changes in canopy temperature resulting from changes in canopy energy balance (e.g., Seligman et al., 1983).

Seed-Filling Rate

Seed-filling rate is the average rate at which seeds accumulate biomass during the seed-filling period. Seed-filling rate can be strongly affected by temperature. The SFR-temperature relationship can be positive or negative, depending on the temperature range and crop considered.

Under some circumstances, SFR is limited by supply of carbohydrates, amides, minerals, and other materials from the mother plant. This was proven by manipulating the balance of sources and sinks. For example, removing some growing seeds increases the supply of raw materials available for the growth of remaining seeds (Wardlaw and Wrigley, 1994; Egli, 1998). Conversely, shading or removing leaves decreases carbohydrate supply and reduces SFR. Rarely, however, is the source-sink balance perturbed enough in nonexperimental fields to significantly affect SFR (Egli, 1998). Barring substantial loss of leaf area to herbivores, SFRs are usually controlled by genetics more than by substrate availability. Control over seed number (i.e., through abortion) largely eliminates significant variation in SFR.

While SFR is usually unaffected by the supply of raw materials from the mother plant, it is sensitive to the number of endosperm or cotyledon cells available for expansion. Because cell number is set by the environmental conditions that prevail during the period of embryo and endosperm cell division (Stages I-III), any environmental factor that influences cell division in the embryo (e.g., high temperature) also often affects the potential for filling of individual seeds. Several reports for wheat have indicated that large seeds are mainly associated with a large number of endosperm cells (Wardlaw, 1990).

ENVIRONMENTAL CHANGES

Elevated Atmospheric Carbon Dioxide Concentration

It has been well-known for many decades that elevated atmospheric [CO_2] increases photosynthesis, at least in C_3 species. It has also been clear for a long time that elevated [CO_2] reduces stomatal aperture and increases WUE in both C_3 and C_4 plants. These facts imply that grain yield should increase with an increase in [CO_2] in many instances because both source activity (photosynthesis) and WUE can be important to seed growth and yield (Evans, 1993; Polley, 2002). In fact, a voluminous scientific literature on the subject indicates that yield is significantly enhanced by experimental increases in atmospheric [CO_2] (e.g., Tables 9.2 and 9.3; Figure 9.3; Lawlor and Mitchell, 1991). Yield increases brought about by experimentally increasing [CO_2] are especially strong in C_3 crops, but can also occur in C_4 crops. Indeed, elevated [CO_2] is so beneficial to reproductive growth that CO_2 has been pumped into glasshouses for many years to improve flowering characteristics (including quality and number) and fruit yields of several horticultural crops (Enoch and Zieslin, 1988; Murray, 1995).

Effects of elevated [CO_2] on grain yield can vary greatly among varieties of the same crop species. For 17 rice varieties, effects of elevated [CO_2] (i.e., 664 compared to 373 ppm) on yield ranged from a 10 percent decrease to a 343 percent increase (Ziska, Manalo, and Ordonez, 1996). Yield enhancements of nine soybean cultivars re-

TABLE 9.2. Effect of elevated atmospheric [CO_2] on yield of C_4 grain crops.

Crop	Daytime [CO_2] (elevated/ ambient; ppm)	Yield in elevated [CO_2] relative to ambient [CO_2] (unitless)	Notes	Source
Maize	520/340	1.40	OTC[a]	A
	718/340	1.50		
	910/340	1.38		
	600/350	1.09	Laboratory chamber[b]	B
	850/350	1.00		
	500/350	1.01	OTC[c]	C
Sorghum	485/330	1.19	Field chamber[d]	D
	660/330	1.31		
	795/330	1.59		
	713/359	1.16	OTC[e]	E, F
	561/368	0.96	FACE[f], ample irrigation	G
		1.20	FACE, limited irrigation	

Sources: (A) Rogers et al. (1983); (B) King and Greer (1986); (C) Rudorff et al. (1996); (D) Chaudhuri et al. (1986); (E) Amthor et al. (1994); (F) Reeves et al. (1994); (G) Ottman et al. (2001).

[a]Open-top chambers with plants grown in pots.

[b]Results averaged over three irrigation levels.

[c]Open-top chambers with charcoal-filtered air (i.e., subambient [O_3]).

[d]Plants grown in a rhizotron covered with closed-top chambers.

[e]Open-top chambers with plants grown in the ground. Mean of two experiments.

[f]Free-air CO_2 enrichment.

sulting from CO_2-enrichment ranged from 20 to 90 percent (Figure 9.4). For all crops, the positive effects on yield of a given ppm increase in [CO_2] are greater at subambient [CO_2]s compared to superambient [CO_2]s (e.g., Figure 9.3).

TABLE 9.3. Effect of elevated atmospheric [CO_2] on the yield of C_3 crops.

Crop	Daytime [CO_2] (elevated/ ambient; ppm)	Yield in elevated [CO_2] relative to ambient [CO_2] (unitless)	Notes	Source
Rice	664/373	1.48	GH[a], 29/21°C	A
		1.00	GH, 37/29°C	
	589/389	1.14	FACE, high N	B
		1.10	FACE, medium N	
		1.07	FACE, low N	
		1.08	FACE, high N	C
		1.10	FACE, medium N	
		1.03	FACE, low N	
Cotton (bolls)	550/350	1.22	FACE	D
	550/350	1.47	FACE[b], ample irrigation	E
		1.43	FACE[b], limited irrigation	
(lint)	550/350	1.60	FACE	F
Soybean		SEE FIGURE 9.4		
	710/400	1.40	GH[c]	G
	630/315	1.32	Literature[d]	H
	various[f]	1.24	Literature	I
Bean	700/350	1.30	GH[e], 28/18°C	J
Bean (pods)	459/372	0.87	OTC	K
Barley	459/372	1.17	OTC[g], cv. Alexis	K
	539/372	1.38		
	459/372	1.39	OTC[g], cv. Arena	
	539/372	1.80		
Wheat		SEE FIGURE 9.3		

Sources: (A) Ziska, Manalo, and Ordonez (1996); (B) Kobayashi, Lieffering, and Kim (2001); (C) Kim et al. (2001); (D) Mauney et al. (1992); (E) Mauney et al. (1994); (F) Pinter et al. (1996); (G) Ziska, Bunce, and Caulfield (2001); (H) Allen (1990); (I) Ainsworth et al. (2002); (J) Prasad et al. (2002); (K) Manderscheid and Weigel (1995).

[a]Glasshouse with plants grown in pots; values are averaged over 17 cultivars (range of response at 29/21°C was -10 to +343%); temperatures are day/night.

TABLE 9.3 *(continued)*

[b]Averaged over two growing seasons (1998 and 1999).

[c]Glasshouse with plants grown in pots: values are averaged over 9 cultivars (yield increases ranged from 20-90%).

[d]Averaged over three cultivars, four locations, 3 years, and two CO_2-enrichment systems.

[e]Glasshouse with plants grown in underlying soil.

[f]Mean of 57 studies with average "high" value of 689 ppm CO_2 and with an unspecified average ambient [CO_2], though presumably it was about 350 ppm CO_2.

[g]Open-top chambers with plants grown in pots

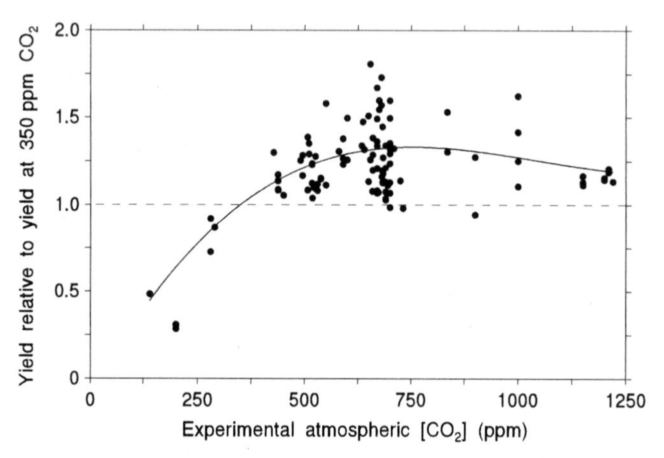

FIGURE 9.3. Effect of [CO_2] in the range 140-1220 ppm on wheat yield relative to yield at 350 ppm CO_2. Results shown are for wheat supplied with ample water and nutrients and with the most favorable temperature and [O_3] used in each of 48 studies. Data are from laboratory chamber, glasshouse, OTC, closed-top field chamber, and FACE experiments (yield enhancement by elevated [CO_2] may be greater in laboratory and glasshouse experiments than in field experiments [Table 2.1]). The line shows an empirical cubic polynomial fitted to the data. It passes through unity at [CO_2] = 350 ppm. Results of FACE experiments all fell below the polynomial. According to the polynomial, the positive effects of [CO_2] on yield are greatest at a [CO_2] of about 750 ppm, with declining yield above that [CO_2]. The relative effects of a change in [CO_2] were greater at subambient than superambient [CO_2]. Two data were not shown because their relative yield was greater than 2.0 (2.05 at 735 ppm CO_2 and 2.11 at 900 ppm CO_2), but they were included in the calculation of the polynomial coefficients. *Sources:* Studies and assumptions are listed in Amthor (2001), with the addition of data from Moot et al. (1996) and deletion of data from Heagle, Miller, and Pirsley (2000) for reasons given in Amthor (2001).

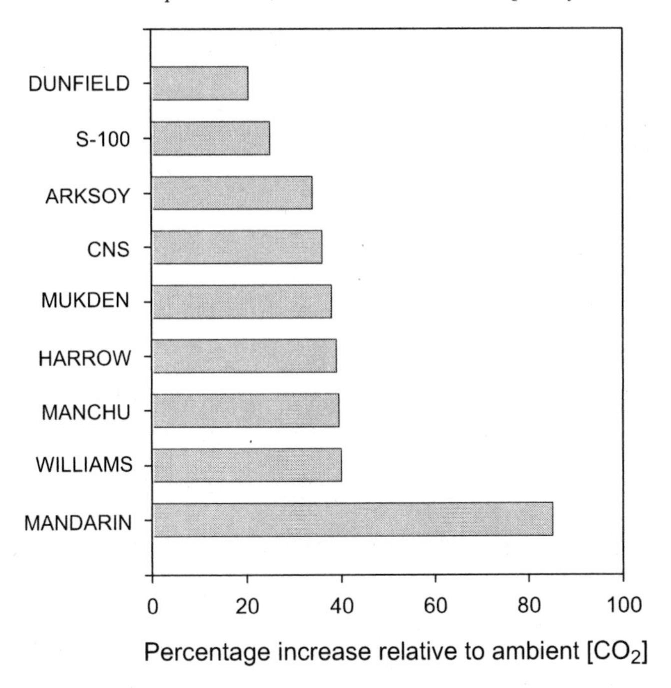

FIGURE 9.4. Percentage yield increase for nine soybean varieties grown in CO_2-enriched air (710 ppm) compared to yield from the same varieties grown in ambient air (400 ppm CO_2). The experiment was conducted in a glasshouse with pot-grown plants. The experiment was repeated three times. These data illustrate the large intraspecific differences in the effect of elevated atmospheric [CO_2] on yield. *Source:* Figure redrawn from Ziska, Bunce, and Caulfield (2001).

Most experiments have involved [CO_2]s from present ambient values up to about 700 ppm, and the yield of many crops is stimulated by [CO_2] increases up to about 1000 to 1500 ppm (e.g., Figure 9.3; effects of elevated [CO_2] on yield of most C_3 grain crops is similar to that shown for wheat). But [CO_2]s greater than 1000 to 2000 ppm can be deleterious to yield (see data for wheat and literature reviews for other crops in Grotenhuis and Bugbee [1997] and Reuveni and Bugbee [1997]). Effects of such high [CO_2]s (i.e., >2000 ppm) on yield are probably irrelevant to the foreseeable future because atmospheric [CO_2] is unlikely to be allowed to significantly exceed 1000 ppm (i.e., governmental actions will eventually be implemented to prevent such a high global atmospheric [CO_2] in order to avoid large and undesirable climatic changes). In short, increasing atmospheric [CO_2], acting alone, would significantly benefit the yield of most grain crops. This is the good news about increasing [CO_2].

Components of Increased Yield

Many factors can contribute to enhanced yield with CO_2-enrichment. Through its stimulatory effects on both photosynthesis per unit leaf area and leaf-area growth (or LAI), elevated [CO_2] results in increased canopy photosynthesis, at least in C_3 crops. This enables crops to grow larger, producing more flowers and fruits compared to plants grown at today's atmospheric [CO_2] because ovule and seed numbers are positively related to the amount of canopy photosynthesis.

In the cereal crops, elevated [CO_2] usually stimulates growth of seeds on tillers (Pleijel et al., 2000) more than on the main stem; e.g., +200 percent and +20 percent increases were reported for tiller and main-stem grain production in wheat (Schütz and Fangmeier, 2001). Similarly, an increase in soybean yield caused by elevated [CO_2] was explained by greater seed production on axillary branches (Ziska, Bunce, and Caulfield, 2001), and yield increases in bean were the result of more pods per plant. In rice, a 47 percent increase in yield brought about by elevated [CO_2] was due almost entirely to more panicles per plant (Lawlor and Mitchell, 1991). In general, elevated [CO_2] increases yield by causing crops to produce more seeds or fruits per plant rather than larger individual seeds (McKee, Bullimore, and Long, 1997). Jitla and colleagues (1997) argued that rice tiller initiation, and hence number of potential seeds per plant, was significantly stimulated by elevated [CO_2] during the two weeks following germination.

Seed size is sometimes affected by elevated [CO_2], but in an inconsistent fashion. A great deal of the variability in the effects of elevated [CO_2] on individual seed size is probably due to variable effects on seed-filling processes. For example, SFR is sometimes increased by elevated [CO_2] (Wheeler et al., 1996; Li et al., 2000; Li, Hou, and Trent, 2001), while SFD sometimes decreases. These effects may cancel each other, eliminating a significant net effect on final seed size. Effects of elevated [CO_2] on SFR and SFD are probably regulated in part by other factors, such as nutrient availability, temperature, and soil moisture. Moreover, effects of elevated [CO_2] on reproductive development (i.e., time to flowering and length of the reproductive phase) is difficult to sort out because of opposing claims made by different investigators. Although a number of researchers have noted that flowering is advanced in CO_2-enriched crops anywhere from days to weeks, indicating faster development (Sionit, Strain, and Flint, 1987; Enoch and Honour, 1993; Horie et al., 2000; Wagner et al., 2001), others reported that CO_2-enrichment does not consistently affect time to flowering (Wheeler et al., 1996). Once flowering has occurred, the grain-filling period is usually either shorter in CO_2-enriched crops (indicative of faster development) or remains about the same (Mulholland et al., 1998b). An increase in plant (i.e., canopy) temperature caused by stomatal closure and reduced transpirational cooling caused by elevated [CO_2] (Chapter 4) would contribute to accelerated reproductive development. For this reason, when CO_2-enrichment hastens reproductive development, it might be most easily explained by changes in canopy energy balance resulting from changes in stomatal physiology. Other factors, however, might also be involved. For example, shortened SFDs caused by elevated [CO_2] might be related to earlier plant senescence. If elevated [CO_2] causes an increase in seed number, nitrogen demand for seed growth could increase. Because as much as 70 percent of seed nitrogen is derived from disassembly of leaf photosynthetic proteins during leaf senescence (Fangmeier et al., 2000), increased seed growth might hasten canopy senescence and ultimately limit SFD. Such a chain of events is most likely in determinate crops unable to assimilate large amounts of nitrogen during the grain-filling period. In fact, elevated [CO_2] can extend the canopy lifetime in soybean (a crop with indeterminate varieties and able to assimilate atmospheric N_2 during grain growth) in spite of increased canopy temperature resulting from stomatal closure (S. P. Long, personal communication, 2003).

Grain Quality

A general response of plants to elevated [CO_2] is a decrease in nitrogen concentration of vegetative organs. Since nitrogen transport to developing grains depends largely on amino acids exported from leaves, grain nitrogen (protein) concentration might decrease as well, affecting the quality of grains used in some food products (Conroy et al., 1994).

Changes in grain nitrogen (protein) concentration caused by elevated [CO_2] may be species specific. Grain nitrogen concentration was unaffected by elevated [CO_2] in rice and soybean, reduced slightly in cotton, and reduced significantly in barley and wheat (Jablonski, Wang, and Curtis, 2002). In wheat, and presumably other crops, reductions in grain protein concentration caused by elevated [CO_2] can be minimized through the application of extra nitrogen fertilizer (Kimball et al., 2001). Effects of elevated [CO_2] on C_4-grain nitrogen concentration are relatively unstudied.

It is important to recognize that although CO_2 enrichment sometimes decreases grain nitrogen concentration, the total grain nitrogen content per unit of land area can increase because grain growth can be stimulated to a larger degree than grain nitrogen concentration is reduced.

Grains harvested from CO_2-enriched crops have higher carbohydrate concentrations, which, in many cereal crops, make up about 80 percent of seed mass (Thompson and Woodward, 1994; Table 9.1). This indicates that starch-based industrial applications may benefit significantly from increased atmospheric [CO_2]s while industries relying on proteins might not. Similarly, an increase in carbohydrate content relative to protein will increase the quality of barley for malting as well as the cooking and eating qualities of rice. Breadmaking quality of wheat may decline, however.

Elevated [CO_2] had only small effects on wheat-grain lipid concentration and composition (Williams et al., 1995).

Warming

The focus herein is on the effects of warming on crops where they are now grown. For example, a comprehensive analysis of long-term yields in the U.S. corn belt demonstrated that a summer temperature above "normal" reduced yields of both maize (Figure 9.5) and soybean (Figure 9.6). It is noteworthy that soybean was adapted to the prevailing temperature, so warming *or* cooling reduced yield (any change was detrimental), whereas relatively cool Julys or Augusts stimulated maize yield.

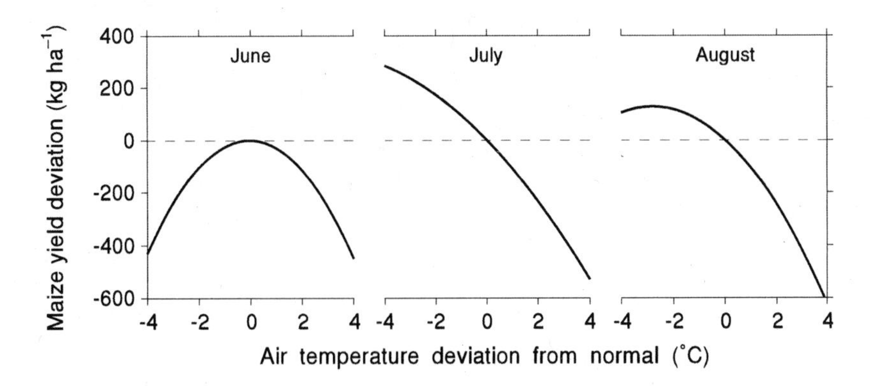

FIGURE 9.5. Quadratic regression relationships between air-temperature deviation from "normal" and maize-yield deviation from "normal" in the five U.S. corn-belt states, Illinois, Indiana, Iowa, Missouri, and Ohio, during the period 1930-1983. Normal temperature was the mean from 1891 to 1983. Effects of technology on yield increase during the study period were accounted for. Separate analyses were conducted for temperature during June, July, and August. *Source:* Equations are from Thompson (1986).

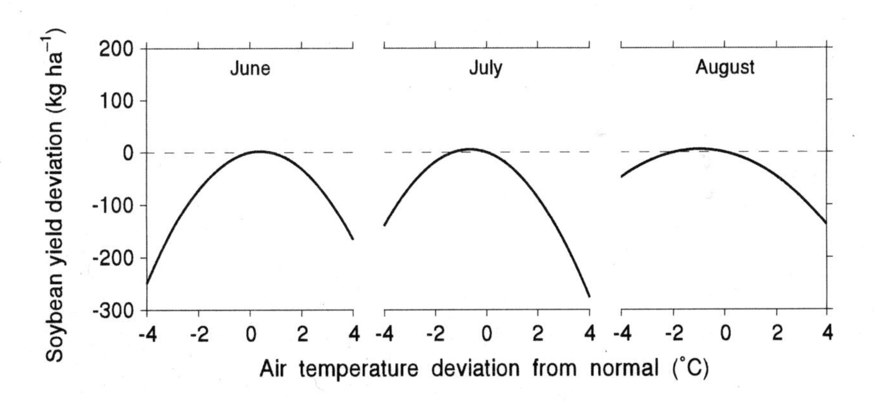

FIGURE 9.6. Quadratic regression relationships between air-temperature deviation from "normal" and soybean-yield deviation from "normal" in the five U.S. corn-belt states, Illinois, Indiana, Iowa, Missouri, and Ohio during the period 1930 to 1968. Normal temperature was the mean from 1930 to 1968. Effects of technology on yield increase during the study period were accounted for by assuming a constant rate of background yield increase over time. Separate analyses were conducted for temperature during June, July, and August. *Source:* Equations are from Thompson (1970).

This focus on the effects of a change in temperature on crops where they are now grown ignores several options available to farmers that might (partially) compensate for any negative effects of long-term warming, such as (1) a poleward shift in cultivation of a given species or genotype, (2) earlier planting of a given species or genotype, or (3) replacement of a current species or genotype with one better adapted to higher temperature. These options may be mainly applicable to temperate latitudes; warming in the tropics may be harder to address through management and species/genotype selection. In addition, although it appears that soybean in the U.S. corn belt is well adapted to prevailing temperature, and future warming might be met with further successful adaptation, this is not obvious for maize (Figure 9.5). It may also be significant that any "rapid" warming in the near future could challenge crop breeders and farmers with a "moving target" in terms of adaptation to temperature.

In spite of management practices and cultivar selection as potential means of alleviating the negative effects of unfavorable temperature, moderately high temperature already limits yield in several crops; 10 to 15 percent globally for wheat (Wardlaw and Wrigley, 1994) and perhaps as much for rice and maize (Stone, 2001). Thus, it is likely that global warming would have negative consequences for yield of some grain crops. On the contrary, warming may be beneficial to crops whose present yield is limited by low temperature, including limitations set by freezing and chilling injury in spring and/or autumn. So there may be bad and good news for grain yield in a warmer future, depending in part on the current geographic location of a given crop.

High Temperature During Vegetative Growth

As already discussed (Chapter 8), warming generally increases vegetative developmental rate and sometimes growth. A shorter vegetative growth duration can decrease the length of time canopies are able to intercept solar radiation, which may in turn reduce the total amount of absorbed energy available to drive photosynthesis (if increased leaf area does not compensate for reduced duration with warming). Because of this, integrated preanthesis canopy CO_2 assimilation may be reduced by warming. A result could be fewer and/or smaller organs (i.e., leaves, tillers, and flowering sites) to support sexual reproduction. Thus, preanthesis warming has the potential to limit yield by limiting time-integrated canopy leaf area and photosynthesis, independently of direct effects of temperature on sexual reproduction.

The preanthesis temperature threshold, above which development is significantly hastened and subsequent sexual reproductive potential becomes compromised, is about 15°C for wheat (Stone and Nicolas, 1995). For warm-season crops such as maize, reduced yields resulting from high preanthesis temperature can occur when temperature exceeds about 30°C (Stone, 2001).

High Temperature During Reproductive Growth: Seed Set and Number

Yield can be especially sensitive to high temperature around the time of anthesis (Savin, Stone, and Nicolas, 1996; Wheeler et al., 2000), often seen through effects on final seed number. Flower and ovule numbers, pollination, pollen tube growth, and fertilization can all be reduced by high temperature, with consequent negative effects on yield.

Pollen may be the most heat sensitive of all reproductive structures (Saini and Aspinalli, 1982; Sato, Peet, and Thomas, 2000; Aloni et al., 2001). The temperature threshold above which damage to pollen occurs is relatively low, and pollen viability is rapidly diminished at just a few °C above this critical temperature (about 30 and 34°C for wheat and rice, respectively [Stone, 2001]). High nighttime temperature may be particularly detrimental. In cowpea, night temperature above 30°C leads to low pollen viability and failure of anthers to dehisce (Ahmed, Hall, and DeMason, 1992). Loss of pollen viability is generally most severe when high temperature occurs five to ten days before anthesis (Warrag and Hall, 1984; Ahmed, Hall, and Demason, 1992).

High temperature during pollen development can lead to shriveled, fewer, and smaller pollen grains. The reason is unclear, but may involve premature degeneration of tapetal cells (Ahmed, Hall, and DeMason, 1992); recall that the tapetum mediates transfer of material from the mother plant to developing microspores. High temperature may also affect several pollen enzymes that are important in carbohydrate metabolism (Aloni et al., 2001). Whatever the mechanism, it appears that most damage occurs during meiosis, of microspore cells or after meiosis around the time when pollen microspores are released from tetrads (Hall, 2001).

Sensitivity of pollen to high temperature may be related to a lack of HSPs (see Chapter 3) in pollen (Dupuis and Dumas, 1990). Apparently, the ability of pollen to synthesize HSPs declines as it develops, becoming completely absent by maturity (Gagliardi et al., 1995).

High temperature can also interfere with ovule function. Excessive proliferation of the nucellus or integument tissues, resulting in crowding of the embryo sac, has been observed at high temperature (Stone, 2001). These changes may interfere with hormonal signaling from the embryo sac, which controls the direction of pollen tube growth. High temperature during very early seed development can lead to abortion of ovules (Cheikh and Jones, 1994). Grain set in wheat, for example, is especially sensitive to high temperature during the first three days after anthesis (Stone and Nicolas, 1995), when exposure to temperature above 30°C can reduce grain number by 70 percent (Stone, 2001).

The stigma and style are less temperature sensitive than anthers, pollen, and ovules (Warrag and Hall, 1984). In rice, for example, the upper temperature threshold for proper functioning of stigma and style is at least 6°C higher than for pollen (Stone, 2001). Thus, effects of warming on pollen, and to a lesser extent ovules, are probably responsible for (some of) the negative effects of high temperature on seed number.

Seed-Filling Duration, Seed-Filling Rate, and Yield

High temperature during the grain-filling period can reduce individual seed size at maturity, which may result from combined effects on SFD and SFR (Figure 9.7; and see Ong, 1983b). Effects of temperature on SFD may be most important to yield. In wheat, SFD decreased 3.1 days for each 1°C temperature increase (Weigland and Cuellar, 1981). The SFD of crops decreases over most of the temperature range, perhaps reaching a minimum (stable?) value at modestly high temperature (Figure 9.7). Shorter SFDs are unavoidably accompanied by a reduction in total solar radiation absorbed by a crop, and this is important to yield because of the strong positive relationship between yield and the time-integrated amount of solar radiation

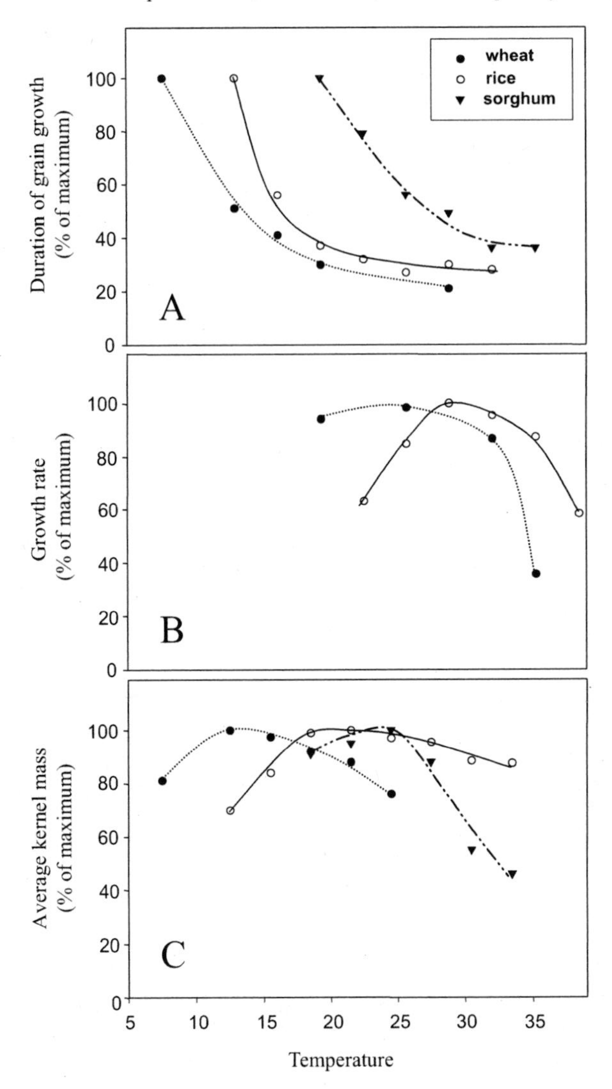

FIGURE 9.7. Relationships between temperature and (A) duration of the grain-filling period, (B) seed growth rates of wheat, rice, and sorghum, and (C) average mass per kernel. Values for rice are averaged for japonica and indica varieties (only seed-filling duration data). Temperatures are averages of day and night values in all cases. *Sources:* Data for (A) and (C) were calculated from Chowdhury and Wardlaw (1978); data for (B) are from Tashiro and Wardlaw (1989).

intercepted by a canopy. Less energy absorbed typically translates into less total (time-integrated) photosynthesis.

As temperature increases up to an optimum for the SFR, SFR increases. Above the optimum, SFR declines with further warming (Figure 9.7). Even if warming does not push crops past the optimum temperature for SFR, it is key that negative effects of warming on SFD are generally larger than positive effects on SFR. The net effect of warming is then usually an increase in the final size of individual seeds with temperature increase up to the optimum for seed size, which may be species specific (Figure 9.7), and a decrease in seed size with warming above the optimum. The main question with respect to global warming is whether warming will push crops beyond their optimum temperature and reduce yield (seed number × seed size). Even now, warm years may result in reduced yield in many crops (Fischer, 1983; Eastin, 1983; Figures 9.5 and 9.6). This is further demonstrated by warming experiments that show reduced yield with modest warming (e.g., Thorne and Wood, 1987; Table 9.4; Figure 9.8).

Decreases in SFRs at high temperature are often attributed to specific effects on the enzymes involved in converting assimilates provided by the mother plant into seed mass. The activity of soluble starch synthase (SSS), a key enzyme in amylopectin production, is particularly sensitive to temperature above 30°C (Wardlaw and Wrigley, 1994; Wilhelm et al., 1999). Since amylopectin accounts for 50 to 60 percent of the seed mass in maize, rice, and wheat, it is easy to see how changes in SSS could significantly affect yield. Other enzymes functioning in amylopectin synthesis appear less sensitive to high temperature. Implications for genetically engineering plants with a more heat-stable isoform of SSS are obvious.

In addition to the effects on SSS, high temperature during early kernel development (i.e., the first 10 to 15 days after pollination) can inhibit cell division and formation of endosperm tissue (Cheikh and Jones, 1994; Wardlaw and Wrigley, 1994; Ahmadi and Baker, 2001). This is important because the number of endosperm cells is one control on the ability of seeds to accumulate dry mass (i.e., kernel or seed sink strength).

Although different species can be differentially affected by a specific temperature, modestly high temperature generally reduces the yield of grain crops. For example, grain mass of wheat can decline

TABLE 9.4. Effect of experimental warming on wheat yield.

Temperature treatment[a]	Effect of warming on yield (%)	Notes	Source
1.1-1.6°C above ambient	−7	cv. Hartog	A
1.1-1.6°C above ambient	−0.5	cv. Late Hartog	A
1.5-2.2°C above ambient	−43	cv. Hartog	A
1.5-2.2°C above ambient	−30	cv. Late Hartog	A
2.0°C above ambient (flag leaf emergence to anthesis)	−14	1974 experiment	B
2.1-2.3°C above ambient	−16	cv. Hartog	A
2.1-2.3°C above ambient	+3	cv. Late Hartog	A
4°C above ambient	−18	448 kg N ha^{-1}	C
4°C above ambient	−22	83 kg N ha^{-1}	C
5.8°C above ambient (anthesis to maturity)	−18	1974 experiment	B
+1.6°C	−22	1996 experiment	D
+2°C (whole season)	−22	cv. Galahad	E
+2°C (whole season)	−6	cv. Hereward	E
+2°C (whole season)	−31	cv. Merica	E
+2°C (whole season)	−36	cv. Soissons	E
+2°C (whole season)	−48	cv. Hereward, 1995	E, F
+2.4°C	−23	1995 experiment	D
+3°C	−27		G
+4°C (anthesis to maturity)	−31	1992 experiment	H
+4°C (anthesis to maturity)	−46	1993 experiment	H

Sources: (A) Rawson (1995); (B) Fischer and Maurer O (1975); (C) Mitchell et al. (1993); (D) Van Oijen et al. (1999); (E) Batts et al. (1998); (F) Batts et al. (1997); (G) Moot et al. (1996); (H) Wheeler et al. (1996).

[a]For treatments listed as above ambient, the experiment involved an ambient (outdoor) temperature and a warming treatment tracking the ambient temperature. For other treatments, two mean temperatures were compared, with neither exactly matching a given ambient situation (see Amthor (2001) for details of some of the temperature comparisons).

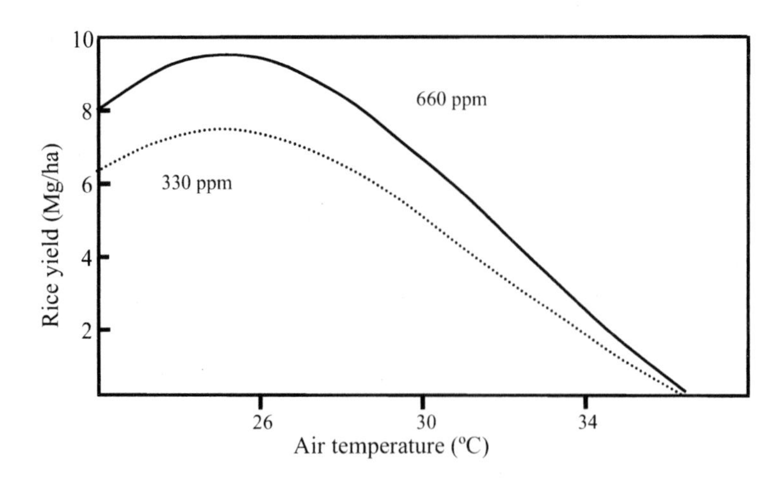

FIGURE 9.8. Relationship between temperature and yield of rice grown in two concentrations of CO_2 (i.e., 330 and 660 ppm). Lines represent a second-degree polynomial fit of yield versus weighted mean day/night air temperature. Data were obtained from five separate experiments. *Source:* Figure is adapted from Allen, Baker, and Boote (1996).

5 percent for each 1°C rise in air temperature from 17.7 to 32.7°C, and in rice grain mass can be reduced 4.4 percent for every 1°C temperature increase from 26.7 to 35.7°C (Tashiro and Wardlaw, 1989). For maize, it was estimated that a 6°C warming during grain filling resulted in a 10 percent yield loss (Thompson, 1986; Wilhelm et al., 1999), and for sorghum, increased nighttime temperature (above an optimum) can have deleterious effects on yield (Eastin, 1983). Bean yield declined about 11 percent for every 1°C warming above a daytime/nighttime temperature combination of 28/18°C in controlled-temperature experiments (Prasad et al., 2002), whereas field observations summarized by Evans (1993, p. 18) indicate that bean yield declines with increasing temperature above a mean growing-season temperature of about 20°C.

Seed Quality

It is unlikely that preanthesis temperature has much effect on seed quality, although that possibility has not been explored. Seed quality is, however, sensitive to temperature during the seed-filling period. Because high temperature can differentially affect the various processes involved in seed filling (for example, protein compared to carbohydrate synthesis, SFD compared to SFR), warming can affect seed composition. In addition, because seeds are often smaller at higher temperature, their milling quality can suffer (Keigley and Mullen, 1986; Wardlaw and Wrigley, 1994).

Grain protein concentration increases in most crops (e.g., soybean, maize, rice, wheat, and barley) with warming (Dornbos and Mullen, 1992). For wheat, higher protein concentration is particularly important because it correlates strongly with dough strength, which influences a host of important breadmaking traits. Dough strength improves with warming up to a threshold and then deteriorates. This threshold ranges from about 25°C in sensitive wheat cultivars to about 35°C in more heat-tolerant varieties (Finney and Fryer, 1958; Wheeler et al., 2000). In barley, even small increases in protein concentration decreases malting quality. In fact, effects of present temperature variability on barley quality already represent a significant difficulty for many maltsters.

In addition to increased protein concentration, warming often causes shifts in protein composition, and these shifts may be important to quality. Ironically, activation of HSPs in seeds exposed to high temperature, although providing protection to the developing seed against heat damage, also reduces the breadmaking quality of wheat. Evidently, temperatures high enough to elicit a heat-shock response tend to increase the amounts of proteins that weaken dough strength (e.g., gliadin), while decreasing other proteins that contribute to dough strength (e.g., glutenin). Furthermore, HSPs themselves may bind to other seed proteins in ways that interfere with grain processing (Wardlaw and Wrigley, 1994; but see Blumenthal et al., 1998).

Often, increases in grain protein concentration with warming are due to decreased starch content rather than increased protein content per se (i.e., protein amount per seed usually does not increase with warming). Apparently, high temperature interferes with the conversion of small molecular mass carbohydrates into starch (e.g., by reducing the activity of SSS, as mentioned previously). High-temperature disruptions in growing-seed carbohydrate metabolism were observed for a number of crops, including barley, rice, wheat, and maize. Wheat grown at high temperature, for example, often produces seeds exhibiting a chalky appearance attributed to small starch granules surrounded by numerous air spaces (Tashiro and Wardlaw, 1991; Tester et al., 1995). A similar chalky appearance has been observed for rice grown at high temperature, as a result of changes in both size and shape of starch grains. In rice, it was estimated that starch amylose content decreased 0.4 percent for every 1°C rise in temperature during the period five to 15 days after anthesis (Stone, 2001). Although probably of minor biological importance, even these small changes in starch properties are significant for the cooking of rice because the amylose content is positively correlated with expansion during cooking and is inversely related to stickiness and tenderness. For maize, similar effects on seed carbohydrates were predicted to reduce the quality of expanded foods such as some breakfast cereals and snack foods (Stone, 2001).

Effects of high temperature on oil content and quality have been documented for several oil crops, including flax, sunflower, cocoa, and jojoba (Fenner, 1992). In sunflower, the ratio linoleic acid:oleic acid is about 6:1 at 12°C and approaches 1:1 at 28°C. Since sunflower oil is valued more for linoleic acid than oleic acids, high temperature can significantly decrease the quality of sunflower oil. High temperature also reduces lipid accumulation in wheat kernels to the extent that flour properties can be affected (Williams et al., 1995).

Although changes in grain quality caused by warming might be important from a marketing (economic) standpoint, they are unlikely to significantly affect human nutrition. In fact, because many people lack sufficient dietary protein, warming-induced increases in grain protein concentration might at first appear to be beneficial. Unfortunately, increases in protein concentration in individual seeds caused by warming may be unable to compensate for decreases in total grain protein amount per unit of ground area, also caused by warming.

Elevated Atmospheric Ozone Concentration

Present tropospheric [O_3]s limit the yield of many crops (e.g., Heagle, 1989), and elevated [O_3]s can reduce yield further (e.g., Black et al., 2000; Table 9.5). In fact, a negative linear relationship between yield and [O_3], as illustrated for soybean in Figure 9.9, is typical for many crops. (Present ambient [O_3]s are more temporally and spatially variable than [CO_2]s, and it is impossible to state a "nor-

TABLE 9.5. Effect of elevated atmospheric [O_3] on yield of selected crops.

Crop	O_3 treatment	Treatment duration	Effect on yield (%)	Source
Barley	32 versus 6 ppb	108 d	–14	A
	44 versus 11 ppb	99 d	–8	
Bean (dry)	15 versus 116 ppb (12-h daily means)	54 d	–72	B
Cotton	90 versus 10 ppb	102 d	–4 to –70 (lint)	C
	58 versus 26 ppb	119 d	–43 (bolls)	D
	72 versus 26 ppb	119 d	–65 (bolls)	
	104 versus 26 ppb	119 d	–80 (bolls)	
Rape (winter)	77 versus 30 ppb ($6\ h\ d^{-1}$)	49 d	–14	E
Rice	0-200 ppb, ($5\ h\ d^{-1}$)	5 d $week^{-1}$ for 15 weeks	–12 to –21 at 200 ppb	F
Soybean	35 versus 17 ppb (7-h daily means)	69 d	–8	G
	122 versus 17 ppb (7-h daily means)		–41	
	66 versus 23 ppb (7-h daily means)	84 d	–29	H
Wheat	2.5 × ambient	43 d	–13	I
	62 versus 20 ppb	31 d	–46	J
	1.4 × ambient (daytime)		–12 (mean of 2 years)	K
	2.0 × ambient	72 d	–29 (well watered) –26 (water stressed)	L
	60 versus 26 ppb (7-h daily means)	emergence to maturity	–2.3	M
	62 versus 22 ppb daily (1h maximum [O_3])		–30 (at 375 ppm CO_2)	N

Sources: (A) Adaros, Weigel, and Jäger (1991); (B) Temple (1991); (C) Temple (1990); (D) Heagle et al. (1986); (E) Ollerenshaw, Lyons, and Barnes (1999);

TABLE 9.5 *(continued)*

(F) Kats et al. (1985); (G) Kohut, Amundson, and Laurence (1986); (H) Mulchi et al. (1992); (I) Ollerenshaw and Lyons (1999); (J) Gelang et al. (2000); (K) Fuhrer et al. (1992); (L) Fangmeier et al. (1994); (M) Mulholland et al. (1997a); (N) Tiedemann and Firsching (2000).

Notes: Table modified from Krupa and Jäger (1996). Results of several other studies were reviewed by Black et al. (2000); for example, they cited 22 reports of reduced wheat yield resulting from elevated atmospheric [O_3].

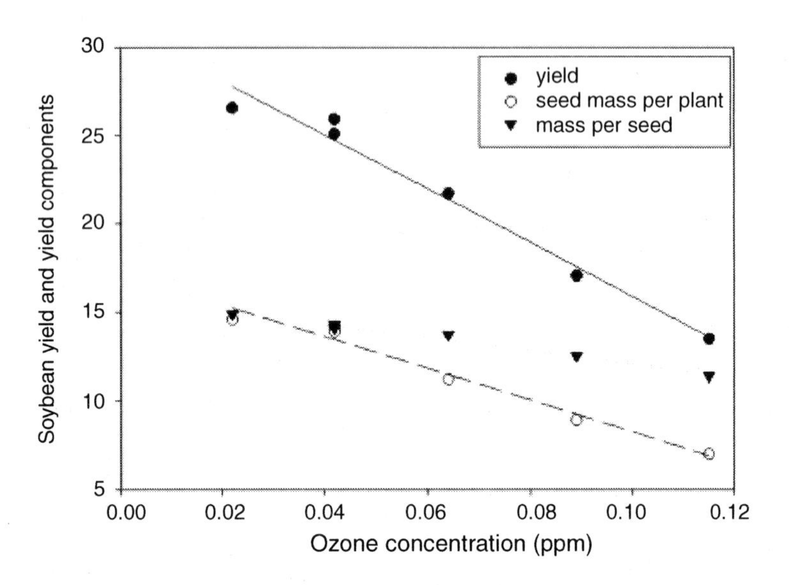

FIGURE 9.9. Dose-response curve for soybean exposed to 7 hours per day O_3 (cf. Figure 1.8). Yield = kg ha^{-1} ÷ 100; the unit for seed mass per plant is gram; mass per seed = gram × 100. Soybeans were planted in soil 65 km southwest of Chicago, Illinois, in open-top chambers. Ozone treatments were initiated five weeks after emergence. *Source:* Data from Kress and Miller (1983).

mal" or "global" tropospheric [O_3] applicable to crops in general; see Figure 1.8.) The fraction of yield lost per unit of [O_3] increase, however, varies substantially across crop species, crop cultivar, temperature, and management practices (i.e., water availability, soil fertility levels, and so on). For cotton, for example, yield lost to high [O_3] ranged from to 0 to 40 percent depending on cultivar (Temple, 1990). In general, both present ambient [O_3] and elevated [O_3] often have significant negative effects on yield of C_3 grain crops, but the C_4 crops maize and sorghum are relatively tolerant of present tropospheric [O_3]s (Heagle, 1989; Black et al., 2000). High-quality data about the effects of elevated [O_3] on yield of C_4 grain crops are limited.

All components of yield, including ear or pod numbers, seeds per ear/pod, total seeds per plant, and individual seed mass, can be affected by elevated [O_3]. The components of yield changed most depend on the timing (i.e., before, during, or after anthesis), duration, and intensity of exposure to elevated [O_3] (Ollerenshaw and Lyons, 1999; Meyer et al., 2000).

In general, exposure of crops to elevated [O_3] before anthesis reduces yield by reducing grain or fruit number. This effect is attributed to inhibited photosynthesis (Chapter 5), premature leaf senescence (Chapter 3), and diminished plant and canopy leaf area (Chapter 8), all of which are common in plants exposed to high [O_3]. Exposure to elevated [O_3] also interferes with the distribution of assimilates to active sinks (e.g., growing seeds; Chapter 6) (Bergweiler and Manning, 1999). All of these effects can lower yield by limiting assimilate supplies needed to establish fruiting sites or to grow seeds or fruits once they are initiated (Kohut, Amundson, and Laurence, 1986; Meyer et al., 2000). For these reasons, effects of elevated [O_3] on vegetative growth and development can alter yield potential in the absence of any direct effects of O_3 on reproduction proper.

Sexual Reproduction

Sensitivity of grain crops to elevated atmospheric [O_3] is often greatest between the induction of flowering and physiological maturity (Mulchi, Sammons, and Baenziger, 1986; Lee, Tingey, and Hogsett, 1988; Younglove et al., 1994; Pleijel et al., 1998; Tingey et al., 2002). Special sensitivity to elevated [O_3] around the time of anthesis indicates that sexual reproductive structures and processes might be particularly sensitive to O_3 (Mulholland et al., 1998b). Pollen and ovule initiation and development, timing of flowering and number of flowers produced, and seed development can all be sensitive to elevated [O_3] (Black et al., 2000).

Pollen germination and pollen tube growth are particularly sensitive to high [O_3] (Feder, 1968; Feder and Shrier, 1968; Stewart et al.,

1996; Black et al., 2000). Mechanisms to account for the effects of elevated [O_3] on pollen function are unclear, although some possible explanations have been offered. Perhaps O_3 directly deactivates or indirectly leads to deactivation of a specific regulator of pollen tube growth (Feder and Sullivan, 1969). Migration of pollen organelles away from the plasma membrane has been observed, and this could reflect some, as yet unknown, specific cellular response to O_3 or its products (Harrison and Feder, 1974). The influence of O_3 on pollen germination and pollen tube growth may also be attributable to changes occurring in the stigma or stylar tissue instead of the pollen itself. Little is known about the effects of elevated [O_3] on the structure and functioning of either of these tissues. Changes in the topography or chemistry of the stigma and style might change the behavior and success of pollen grains that land there.

Since pollen germination occurs on the stigma (directly exposed to atmospheric O_3), while pollen tube growth occurs within the style (protected from direct O_3 exposure by stylar tissue), it is likely that these processes have different responses to a given atmospheric [O_3] (Black et al., 2000).

During reproductive growth, current photosynthate is often rapidly translocated out of leaves and into growing seeds. This might reduce the amount of metabolic substrates available for repair of O_3 damage to leaves. Based on this reasoning, Tingey and colleagues (2002) proposed that the increased susceptibility of bean to elevated [O_3] during pod growth was due to loss of the leaf carbohydrates needed for repair of cellular injury in leaves, leading to reductions in whole-plant photosynthesis.

Progress toward teasing apart the effects of O_3 on vegetative growth from specific effects on sexual reproduction has been slow for three main reasons: (1) in most manipulative experiments, whole plants or canopies are exposed to elevated [O_3]s, as opposed to flowers or fruits alone; (2) experimental O_3 episodes are not usually planned to correspond with critical reproductive periods (for example, anthesis) (Stewart et al., 1996); and (3) most studies are done with small plants growing vegetatively because results can be obtained more quickly and easily than from studies with larger flowering plants. While research has provided substantial data about the influence of elevated [O_3] on yield, it has done little to shed light on the mechanisms involved.

Seed-Filling Duration and Seed-Filling Rate

Elevated atmospheric [O_3] affects the growth and development of individual seeds (Amundson et al., 1987), but smaller seeds usually contribute less to yield reductions than do reductions in fruit or seed numbers. For example, seed mass per plant declined to a greater extent than mass per individual seed in soybean as [O_3] increased (Figure 9.9).

Several studies have shown that smaller seeds on wheat plants exposed to elevated [O_3] were attributable to reduced SFR rather than shorter SFD (Black et al., 2000). Yet other investigators have attributed smaller fruits to shorter SFD caused by shorter leaf-area durations (i.e., premature leaf and canopy senescence) (Slaughter, Mulchi, and Lee, 1993; Gelang et al., 2000). In fact, elevated [O_3] probably reduces seed mass by affecting both SFD and SFR. Regardless of whether effects of elevated [O_3] are mainly on SFD or SFR, seed development and metabolism are unlikely to be influenced directly by O_3 because seeds are well protected from ambient air by several layers of cells. Intercellular transport of free radicals (produced from O_3) to developing seeds is also unlikely because the radicals would react with structures outside (enclosing) seeds (Grunwald and Endress, 1988).

Decreased SFRs are often attributed to a reduced supply of raw materials (mainly nitrogen and carbohydrates) from the mother plant, either because of lower canopy photosynthesis or inefficient assimilate translocation (Chapters 5 and 6). At this time, there are insufficient data to support a more detailed or concrete explanation.

Effects of Elevated Atmospheric Ozone Concentration on Determinate versus Indeterminate Crops

The influence of elevated atmospheric [O_3] on yield may depend on whether a crop or cultivar is determinate or indeterminate (see Chapter 8 for an overview of these developmental strategies). Indeterminate crops have the ability to adapt to environmental conditions by aborting unneeded or unsupportable reproductive sites or by producing additional reproductive sites to compensate for those lost or damaged by stress. This presumably is why indeterminate plants exposed to elevated [O_3] during early reproductive development often compensate for floral abortions or reproductive failures by producing new flowering branches or retaining a higher proportion of fertilized ovules later on.

In contrast, determinate crops such as the cereals wheat, rice, and barley simultaneously form all floral initials during early plant development, well before flowering (Gelang et al., 2000). In those species, an elevated [O_3] episode during a critical reproductive period that results in loss of reproductive sites can drastically reduce yield because compensatory adjustments (i.e., production of additional reproductive organs) are unlikely.

The importance of growth habit was perhaps illustrated best for cotton, a crop for which both determinate and indeterminate cultivars exist. When comparing the O_3 sensitivity of cotton cultivars with contrasting developmental strategies, the longer, and more flexible, flowering period of the indeterminate cultivars enabled them to recover from O_3 damage early in the growing season by producing more branches and bolls. Conversely, more determinate cultivars were unable to compensate for O_3 damage in that way, and they therefore suffered significant (40 percent) yield losses (Temple, 1990). Such differences may prove to be important when selecting crops or cultivars in areas that experience O_3 pollution.

Seed Quality

Exposing crops to elevated [O_3] commonly results in increased grain protein and mineral concentrations (Mulchi et al., 1988; Pleijel et al., 1991). Oil content, however, has been shown to decrease in oilseed rape and soybean exposed to elevated [O_3] (Kress and Miller, 1983; Ollerenshaw, Lyons, and Barnes, 1999). It is unclear whether these small changes significantly affect grain processing or human nutrition.

Although changes in the composition of seeds caused by elevated [O_3] are often minor, when combined with changes in yield they become more important. For example, the combination of decreased seed mass and decreased oil concentration combined to reduce total oil yield by 55 percent in oilseed rape exposed to elevated [O_3] (Bosac et al., 1998). In addition, even though grain protein concentrations usually are higher in crops exposed to elevated [O_3], probably because of the inverse relationship between grain yield and grain protein concentration (e.g., Kibite and Evans, 1984; Pleijel et al., 1999), less total grain protein is generally produced per unit of ground area because of fewer and/or smaller seeds.

Salinity

For most crops, yield is unaffected by soil salinity until some threshold is reached, and beyond that yield decreases about linearly with increasing salt concentration. Thus, two important parameters determine the effects of soil salinization on yield: (1) the damage threshold (i.e., level of salinity below which no yield limitation due to salinity occurs), and (2) the percentage decrease in yield per unit increase in salinity *above the threshold* (i.e., the slope of yield versus soil salinity). The salinity damage threshold and the slope vary appreciably between species and cultivars (Figure 9.10). Barley, oat, and cotton are relatively tolerant of soil salinity in terms of yield. Sorghum, soybean, and wheat exhibit intermediate tolerance (or intermediate sensitivity), while bean, maize, and rice are among the more sensitive crops. In general, yield of grasses is less affected by salinity than yield of legumes (Francois and Maas, 1999).

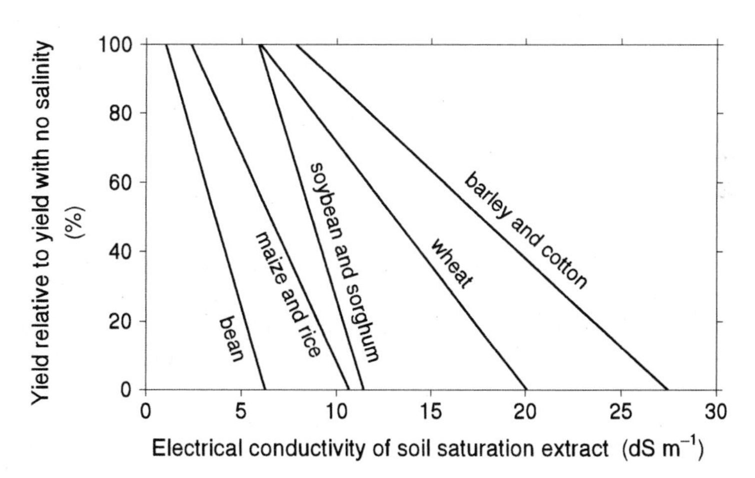

FIGURE 9.10. Sensitivity of yield of some crop species to soil salinity based on experimental results. The important values with respect to crop salt tolerance are the salinity thresholds beyond which yields drop below 100 percent and the slope of the yield decline as salinity increases above that threshold. *Source:* Derived from Table 3 in Francois and Maas (1999).

Most crops exhibit differential sensitivities to soil salinity at different times during their life cycles (Zeng, Shannon, and Lesch, 2001). Usually, sensitivity to salinity decreases as plants mature; the seedling stage is often most sensitive and the grain-filling period is least sensitive (rice is a notable exception, in which sensitivity during the reproductive stage is quite high). Because of this, yield is normally inhibited more when exposure to high soil salinity occurs prior to the developmental switch to reproductive growth.

Crops grown in salinized soil sometimes fail to set seeds. Indirect effects arising from reduced vegetative growth are mainly to blame. However, observations that seed set and development are vulnerable to soil salinity when an experimental salinity treatment is initiated during the reproductive phase implies that sexual reproduction may also be directly affected.

Reduced seed set is most likely caused by loss of pollen viability, a change in stigmatic receptivity, or a combination of both. Salinity might limit pollen function in several ways. It could induce pollen sterility, interfere with germination, or alter pollen tube growth (Abdullah, Ahmad, and Ahmed, 1978; Tyagi and Rangaswamy, 1993). Salinity has also been shown to decrease the number of pollen grains produced in maize, mung bean, field pea, and chickpea (Dhingra and Varghese, 1997).

Pollen from plants grown in saline media accumulates Na^+ and Cl^-, while concentrations of other essential ions (such as K^+) may be reduced. In addition, pollen grains exposed to Na^+ and Cl^- do not germinate properly and their pollen tubes do not grow well in vitro (Yokota, 1986). The strong correlation between Na^+ and Cl^- contents and pollen viability indicates that pollen dysfunction on saline soils is caused by ion toxicities (Reddy and Goss, 1971; Khatun and Flowers, 1995). Other changes in pollen composition have been observed to result from salinity, such as less starch and protein and more soluble sugars and free amino acids (Dhingra and Varghese, 1986, 1997). These changes could play a role in reducing pollen viability.

Although not well studied, the effects of salinity on female reproductive components (i.e., carpel function) may be more significant than the effects on pollen. The relative contributions of pollen and stigmatal dysfunction to reduced seed set caused by salinity were examined by Khatun and Flowers (1995) in a series of experiments with rice. When pollen from plants growing on nonsaline soil was used to pollinate female (emasculated) plants grown in 10 mol·m^{-3} Na^+ and 25 mol·m^{-3} Na^+ (equivalent to approximately 1.0 and 2.5 dS·m^{-1}), seed set was reduced by 38 and 72 percent, respectively. When pollen from plants treated with high salinity levels was used to pollinate females grown on nonsaline soil, seed set was reduced only 10 to 20 percent. When both parents were grown on saline soil, loss of seed set resulting from effects on pollen and stigmatic function appeared to be additive. From that series of experiments, Khatun and Flowers concluded that yield reductions in rice caused by salinity were linked to the effects on panicle development, stigmas, and grain filling, not to the effects on pollen function. That experiment has yet to be repeated with other crop species, however.

Salinity has inconsistent effects on seed size. In salinized barley, for example, seeds were smaller because of a slower SFR (Gill, 1979). Smaller seeds were also reported in rye (Francois et al., 1989), rice (Khatun, Rizzo, and Flowers, 1995), and wheat (Francois et al., 1994). Conversely, other reports on wheat (Labanauskas, Stolzy, and Handy, 1981) and rice (Zeng and Shannon, 2000) indicated that seed mass was unaffected by salinity. Maas and Grieve (1990) found that individual seed mass increased in salinized spring wheat. The effects of salinity on seed size are yet to be fully evaluated and may well vary considerably among cultivars and species, and with the severity and timing of exposure to salt.

Similar inconsistencies are found with regard to the effects of salinity on seed quality. For example, wheat flour quality was improved by saline irrigation water one year, but was adversely affected the next (Kelman and Qualset, 1993). In terms of protein and carbohydrate concentrations, salinity improves quality in some crops and adversely affects it in others (Dhingra and Varghese, 1997).

INTERACTIVE EFFECTS OF ENVIRONMENTAL CHANGES ON YIELD

Theoretical and experimental studies of the effects of multiple environmental changes on yield are less common than studies of the effects of changes in single environmental factors. As a result, generalizing about the effects on yield of simultaneous multiple environmental changes is more difficult than generalizing about effects of changes in single factors. This is the case in spite of the fact that simultaneous changes are occurring in several environmental factors that strongly affect yield, and it is the net effect of multiple changes rather than the effects of single factors that are of most importance to yield.

Elevated Atmospheric Carbon Dioxide Concentration in Combination with Elevated Ozone Concentration

Elevated atmospheric [CO_2] may remove some limits placed on yield by elevated atmospheric [O_3]. Because elevated [CO_2] commonly reduces stomatal conductance (i.e., it reduces stomatal aperture), it should reduce the amount of O_3 diffusing into leaves (for a given ambient [O_3]). Furthermore, higher [CO_2] should, in C_3 crops at least, stimulate photosynthesis, and this can counter (some of) the reduction in photosynthesis caused by elevated [O_3]. Nonetheless, crop losses to O_3 pollution may not be completely offset by rising [CO_2]. This is true in part because elevated [CO_2] is not expected to protect crops from any direct deleterious effects of elevated [O_3] on reproductive processes such as pollen germination and pollen tube growth (McKee, Bullimore, and Long, 1997; Mulholland et al., 1998a). Looking at this combination of changes in atmospheric composition another way, crops growing in areas where O_3 pollution is a problem may reap a smaller yield benefit from increasing atmospheric [CO_2] compared to areas with lower [O_3]s.

The effects of elevated [CO_2] and elevated [O_3] on grain quality differ; elevated [O_3] generally causes an increase in grain protein concentration whereas elevated [CO_2] may cause a reduction in grain protein concentration (e.g., Pleijel et al., 1999). As a result, the combination of elevated [CO_2] and [O_3] may result in only modest changes (if any) in grain protein concentration.

Warming in Combination with Elevated Atmospheric Carbon Dioxide Concentration

Warming often modifies the effects of elevated atmospheric [CO_2] on grain yield. Because the *relative* effect of elevated [CO_2] on C_3 photosynthesis is greatest at high temperature (in part because photorespiration is rapid at high temperature), the relative effect of elevated [CO_2] on the yield of C_3 crops should, perhaps, be maximal at high temperature (Ferris et al., 1999), but it is the absolute effect of changes in the environment on yield, relative to the unaltered environment, that is of practical importance. In that light, the combination of warming and elevated [CO_2] may be detrimental to yield. For example, in a review of experiments with wheat, elevated [CO_2], in the absence of a warming treatment, stimulated yield in 15 of 17 experiments (Amthor, 2001). Conversely, warming, in the absence of an elevated-[CO_2] treatment, inhibited yield in 16 of 17 experiments. The combination of elevated [CO_2] and warming, compared to ambient [CO_2] and normal temperature, reduced yield in 11 studies and increased it in six. This indicates that yield stimulation by elevated [CO_2] may be offset by yield reductions caused by concomitant warming (Batts et al., 1998; Amthor, 2001). For example, in wheat, an increase in the mean temperature of just 0.7 to 2.0°C during the growing season was enough to fully negate the yield enhancement caused by twice-ambient [CO_2] (Batts et al., 1998). Stated another way, rising [CO_2] may not offset yield reductions caused by warming (Ahmed, Hall, and Madore, 1993; Conroy et al., 1994). This may be the case for crops other than wheat also. In rice, warming above about 27.5°C eliminated the positive effect of doubled [CO_2] on panicle dry mass at crop maturity (Horie et al., 1995) (and see Figure 9.8).

Although global warming is related to rising atmospheric [CO_2] through the effects on regional and global atmospheric energy balances, crop warming can be even more directly (i.e., locally) affected by elevated [CO_2]. Because elevated [CO_2] causes stomatal closure, canopy temperature should increase in a CO_2-enriched environment because latent energy exchange (evaporative cooling) is reduced. This means that aside from the effects of atmospheric [CO_2] on global warming, elevated [CO_2] is expected to increase crop canopy temperature directly. This relationship between [CO_2] and canopy temperature may explain the observation that elevated [CO_2] increased the degree of sterility caused by high air temperature in an experiment with rice (Matsui et al., 1997).

Increased Soil Salinity in Combination with Other Environmental Changes

Negative effects of elevated [O_3] on yields are sometimes less severe under saline conditions, indicating that salinity increases tolerance to O_3. Even so, decreases owed to salinity offset any benefits associated with greater tolerance of O_3, so the combination of elevated $[O_3]$ with soil salinity limits yield to a greater extent than elevated $[O_3]$ or salinity acting alone.

Elevated atmospheric $[CO_2]$ may alleviate salt stress to some degree. The increase in tissue carbohydrate concentrations caused by elevated $[CO_2]$ might function as osmoticants, thereby enabling plants to better take up water in salty soils (Chapter 4). Furthermore, compartmentation of toxic ions into vacuoles and cell walls, away from cytoplasm, is an energy-requiring process that might be promoted by the higher carbohydrate levels generated in a CO_2-enriched environment. More important, though, elevated $[CO_2]$ reduces the amount of water transpired per unit of carbon assimilated (i.e., it enhances WUE). This reduces the buildup of salts in leaves (for a given amount of plant growth), which alleviates symptoms of salt stress to some extent (Munns and Termaat, 1986).

Warming might enhance damage caused by soil salinity. Warming may increase water use by crops and therefore increase salt uptake for a given amount of growth. Warming may also increase amounts of irrigation and therefore increase the amount of salt added to soils with irrigation water.

SUMMARY

Reproductive development and growth by crops is especially important for human welfare because we depend on crop fruits and seeds, directly and indirectly, for most of our food. Seed production by crops depends on vegetative (i.e., root, leaf, and stem) development and growth, development of pollen and egg, pollination, and fertilization. The final size of individual seeds generally hinges on cell division within the embryo (which is one limit on potential seed sink strength), followed by seed filling and maturation processes. Environmental conditions prior to the shift to reproductive development usually affect yield by influencing photosynthesis per unit of leaf area, canopy development and interception of solar radiation per unit of ground area, and initiation of potential fruiting sites; a strong positive correlation between canopy photosynthesis per unit of ground area and seed number exists for most crops. In contrast, post-flowering conditions affect yield mainly by influencing ovary or seed abortion, or by changing SFR or SFD. In general, SFD is more plastic than SFR. In spite of the importance of individual seed size, yield variation is more often due to changes in seed number per unit of ground area than to individual seed mass at maturity.

Grain yield is generally enhanced, especially in C_3 crops, by elevated atmospheric [CO_2], other factors being equal. Plants grown with CO_2 enrichment are larger and produce (or retain) more flowers and fruits compared to plants growing in ambient air. Stimulation of yield by CO_2 enrichment is most often due to more seeds (or fruits) per plant (or per unit of ground area) rather than larger individual seeds. In fact, elevated [CO_2] leads to smaller seeds about as often as it leads to larger seeds. Soil nitrogen availability may be the key to this variability.

Warming hastens crop development and therefore shortens the SFD. Seed number can be reduced via the direct effects of high temperature on reproduction, particularly pollen formation and function. Because warming speeds reproductive development, the seeds that do develop are often small. Although SFR is sometimes stimulated by warming, this effect often does not fully compensate for shortened SFD. For many crops where they are now grown, an increase of just a few °C significantly reduces yield.

Elevated tropospheric [O_3] also limits yield. The magnitude of yield reduction depends in part on the timing, duration, and concentration of O_3. Chronic elevated O_3 exposure before the switch to reproductive development reduces vegetative growth, which in turn limits the number of fruits or seeds that are able to form. Preflowering exposure to elevated [O_3] often reduces yields more than postflowering exposure. Exposure to high [O_3] around the time of anthesis interferes with pollen development, pollen tube growth, and therefore seed set. Effects of elevated [O_3] on SFR, SFD, and final seed size are, compared to the effect of elevated [O_3] on vegetative growth and development, relatively minor. When seed size is limited by elevated [O_3], this small effect is normally attributed to slower SFR rather than a shorter SFD.

High soil salinity limits yield, mainly through effects on vegetative growth. In fact, resistance to salinity usually increases as a plant ages, with the seedling stage considered most sensitive and the reproductive stage least sensitive. Nevertheless, irrigation with salinized water during reproductive development can limit yield through effects on fertilization, SFR, and SFD. Both pollen and stigmatic functioning appear to be somewhat sensitive to salt.

Because end uses of grain can require quite specific grain quality attributes, shifts in seed quality caused by environmental change could sometimes be as important as changes in yield per se. In general, grain protein concentration is inversely related to yield, and elevated atmospheric [CO_2] is expected to decrease grain protein concentration, while O_3 pollution and warming can increase protein concentration. The effects of salinity on grain quality are poorly understood. Changes in grain carbohydrate concentration are often opposite to changes in protein concentration. The effects of environmental changes on gain lipid concentration and quality must be studied further before generalizations can be made.

While considerable knowledge of the effects of changes in single environmental factors on yield exists, less is known about the effects of simultaneous changes in multiple factors. This is the case in spite of the fact that several important environmental changes are now occurring simultaneously, and most scientifically based predictions about the future indicate that this will continue for decades to come. Hence, the effects of changes in single factors on yield can in some cases become irrelevant in the context of the multiple changes occurring in our environment.

One key generalization is that the often large increases in yield caused by elevated [CO_2] will be moderated, or perhaps in some cases eliminated, by simultaneous warming, increases in tropospheric [O_3], and/or increased soil salinity. In the end, however, it remains unclear how combinations of environmental changes will affect the yield and grain quality of major crops in most regions. In some locations (i.e., regions or climates), one combination of environmental changes could enhance yield, while in another location the same combination of changes could limit yield. In short, considerable uncertainty surrounds the effects of environmental change on the yields of major grain crops, with sound cases to be made for enhancements of yield in some situations and equally sound cases to be made for new limitations placed on yield in other situations.

Chapter 10

The Biotic Environment

When the weather is good for crops it is also good for weeds.

Theodore Roosevelt Papers
August 23, 1902

In addition to constraints imposed on crops by their physical and chemical environments and their own physiologic and genetic limitations, crops face competition with weeds for physical and chemical resources, crops are eaten by insects and infected by diseases, and crops interact with a multitude of soil organisms—some beneficial, others not. Environmental changes are likely to alter many of the relationships between crops and other organisms.

WEEDS

By competing with crops for sunlight, water, and soil nutrients, weeds are responsible for significant limitations to crop yield throughout the world. If environmental changes affect crops, they might also affect weeds and crop-weed interactions, with either net positive or negative implications for the crop.

It has been stated that "many weeds are C_4 plants, while most major crops are C_3 plants" (Patterson, 1993, p. 273), with the implications that C_4 species (many weeds) would benefit most from warming and C_3 species (many crops) would benefit most from rising atmospheric [CO_2]. These may, however, be overly simplistic assumptions to use as a basis for understanding or forecasting effects of environmental change on crop-weed interactions, because maize, sugarcane, sorghum, and millet are C_4 crops and many important weeds depend on C_3 photosynthesis (Bunce and Ziska, 2000). Moreover, while many environmental-change experiments within laboratory chambers have involved both crop and weed species (e.g., Patterson, Highsmith, and Flint, 1988; Tremmel and Patterson, 1993, 1994), only a few experiments included crops and weeds growing in competition. Even fewer experiments involved crop-weed combinations in field or fieldlike settings. Worse, in some field experiments care is taken to eradicate weeds from experimental plots by strategies uncharacteristic for normal farms, such as hand cultivation or extra herbicide applications, thus eliminating opportunities to understand the effects of environmental changes on crop-weed interactions. A result of all these factors is that although there has been considerable speculation about the effects of environmental change on crop-weed interactions (e.g., Patterson, 1993), little is known from actual observation of crop-weed interactions in appropriate experiments.

Warming

Regardless of their photosynthetic pathway (C_3 or C_4), many important weeds originated in, or are now confined to, tropical and warm temperate regions (Patterson, 1993; Bunce and Ziska, 2000). The success of these weeds relative to crops, as well as their poleward (or upward, with respect to elevation) expansion, may be enhanced by warming.

Only a few studies of the effects of warming on crop-weed competition have been published. In one, pigweed growth was stimulated more than soybean growth by warming (comparing 26/17, 29/20, and 32/23°C day/night temperatures) (Flint and Patterson, 1983). In another study, a 3/2°C day/night warming increased the competitive advantage of lambsquarters grown with sugarbeet in laboratory chambers (Houghton and Thomas, 1996). In another, an 8°C increase in air temperature from 26/22 to 34/30°C day/night (albeit a rather extreme treatment) stimulated biomass production in the weeds sicklepod (150 to 200 percent) and Palmer amaranth (150 to 1600 percent), whereas soybean growth declined (10 to 20 percent) (Wright et al., 1999). In these three cases, the competitive ability of weeds improved with warming.

In other instances, the success of specific weed species might decline with warming because of enhanced competition with crops or other weed species (Patterson, 1993; Patterson et al., 1986). Cotton, a

C_3 crop adapted to the tropics, gained competitive advantage over the weed velvetleaf at 32/23 compared to 26/17°C day/night (Flint, Patterson, and Beyers, 1983). Clearly, the effects of warming on crop-weed interactions are likely to depend on the geographic (i.e., climatic) origins of the crop and the weed, in addition to their physiologic attributes (e.g., C_3 versus C_4 photosynthetic metabolism). The lack of a significant database of experimental studies of the effects of warming on crop-weed interactions limits the usefulness of predictions of the effects of global warming on the relative success of weeds, and the effect of that success on crop yield.

Elevated Atmospheric Carbon Dioxide Concentration

There is a notable lack of data for many important crops growing in competition with even a single weed species in elevated atmospheric [CO_2] (indicated by "no data" in Table 10.1). A priori, it is expected that rising [CO_2] per se will favor C_3 plants growing in competition with C_4 plants. This implies that the competitive advantage of C_3 crops might be improved relative to C_4 weeds, and vice versa, and this is supported by the available data (Table 10.1). But because the available database is small, tests of general notions about the effects of elevated atmospheric [CO_2] on C_3-C_4 (or C_4-C_3) crop-weed competition are limited. Experimental studies of the effects of atmospheric [CO_2] on crop-weed interactions when the crop and the weed use the same photosynthetic pathway are also limited (Table 10.1).

Although increasing atmospheric [CO_2] is likely to enhance crop growth and yield, it might be even more beneficial to weeds. Because the number of (potential) weed species in a location greatly exceeds the number of crop species, and because many weeds are "exploitive" species that take special advantage of enhanced resource (including CO_2) availability, the benefits of rising [CO_2] on *some* weeds will probably exceed the benefits to any particular crop in most cases. In addition, because increasing atmospheric [CO_2] enhances the growth of many weeds, it might increase both the geographic spread of weeds into new regions and the vigor of weeds relative to crops. For example, Ziska (2003) found that the growth of six C_3 weed species important in North America was stimulated to a greater degree than the growth of C_3 crops in controlled-[CO_2] experiments. In particular, the growth of weeds was especially stimulated by 380 ppm CO_2

TABLE 10.1. Competitive advantage imparted by elevated atmospheric [CO_2] on crop-weed combinations.

Crop	Weed	Competitive benefit imparted to	Method of controlling [CO_2]	Source
C_3 crop / C_3 weed				
Alfalfa	Dandelion	Weed	Field chamber	A
Barley		no data		
Bean		no data		
Cassava		no data		
Cotton		no data		
Peanut		no data		
Potato		no data		
Rice		no data		
Soybean	Lambsquarters	Weed	OTC	B
Sugarbeet	Lambsquarters	Crop	Laboratory chamber	C
Sunflower		no data		
Wheat		no data		
C_3 crop / C_4 weed				
Alfalfa	C_4 grasses	Crop[a]	Field chamber	D
Barley		no data		
Bean		no data		
Cassava		no data		
Cotton		no data		
Peanut		no data		
Potato		no data		
Rice	Barnyard grass	Crop	Glasshouse	E
Soybean	Johnsongrass	Crop	Laboratory chamber	F
	Redroot pigweed	Crop	OTC	B
Sugarbeet		no data		
Sunflower		no data		
Wheat		no data		
C_4 crop / C_3 weed				
Maize		no data		

Crop	Weed	Competitive benefit imparted to	Method of controlling [CO_2]	Source
Sorghum	Cocklebur	Weed	Glasshouse	G
Sugarcane		no data		
C_4 crop / C_4 weed				
Maize		no data		
Sorghum		no data		
Sugarcane		no data		

Sources: (A) Bunce (1995b); (B) Ziska (2000); (C) Houghton and Thomas (1996); (D) Bunce (1993); (E) Alberto et al. (1996); (F) Patterson, Flint, and Beyers (1984); (G) Ziska (2001).

[a]In two of three experiments during the first year of alfalfa establishment.

(i.e., about present atmospheric [CO_2]) compared to 284 ppm CO_2 (i.e., preindustrial atmospheric [CO_2]). Based on such results, Ziska (2003) speculated that increases in atmospheric [CO_2] during the 1900s may have contributed to the spread and success of invasive weeds.

Weeds apparently can affect the relative gains in crop yield caused by elevated atmospheric [CO_2]. When [CO_2] was about doubled relative to the present atmospheric level, yield of wheat grown in monoculture in pots in a glasshouse was enhanced 30 percent, but when the wheat was grown in competition with various weed species, the relative (and absolute) yield gains caused by elevated [CO_2] were reduced (Thompson and Woodward, 1994). Similarly, for soybean grown in OTCs without competing weeds, a [CO_2] increase of 250 ppm increased yield 23 percent, but when either lambsquarters (C_3) or redroot pigweed (C_4) were grown with the soybean in elevated [CO_2], yield was the same or less than yield in weed-free OTCs at ambient [CO_2] (Ziska, 2000). These results have implications for interpretation of results of elevated-[CO_2] experiments that involve abnormally low densities of weeds (e.g., through special actions taken to maintain weed-free experimental plots).

Many weeds propagate from (and overwinter as) rhizomes or other belowground organs. When elevated atmospheric [CO_2] enhances belowground growth of weeds (irrespective of any changes in root/shoot ratio), weed survival and propagation might be enhanced. This could increase weed-control requirements, either by mechanical means or with herbicides (Patterson, 1995; Bunce and Ziska, 2000).

To the extent that elevated atmospheric [CO_2] affects physiology and growth of weeds, it might also affect how well herbicides control weeds. For example, growth of the C_3 weed common lambsquarters was significantly stimulated by 720 compared to 360 ppm CO_2 in glasshouses, and elevated [CO_2] simultaneously increased tolerance of the weed to glyphosate (Ziska, Teasdale, and Bunce, 1999). (Glyphosate, marketed under the trade name Roundup, is globally the most-used herbicide; it is applied postemergence and is mobile in phloem.) Conversely, growth of the C_4 weed redroot pigweed was unaffected by elevated [CO_2], and increased tolerance of redroot pigweed to glyphosate caused by elevated [CO_2] occurred in only a few instances (Ziska, Teasdale, and Bunce, 1999). Elevated [CO_2] enhanced tolerance of Canada thistle, a C_3 perennial weed, to the herbicides glyphosate in OTCs, with greater tolerance enhancement in roots compared to shoots (Ziska, Faulkner, and Lydon, 2004). In glasshouse experiments, elevated [CO_2] sometimes increased the tolerance of quackgrass (a C_3 perennial weed in some forage and pasture crops) to glyphosate, and enhanced regrowth from root segments of glyphosate-treated plants (Ziska and Teasdale, 2000).

Little is known aside from these few examples about effects of rising atmospheric [CO_2] on the tolerance of weeds to herbicides, and it is unclear what mechanisms underlie increased tolerance to herbicides when it is observed. Possibly, reduced stomatal aperture in elevated [CO_2] reduces uptake of herbicides applied to leaves, or enhanced growth by C_3 weeds may itself enhance tolerance to herbicides. Because postemergence herbicides are usually most effective on small plants (e.g., seedlings), and because elevated [CO_2] may accelerate early-season growth of C_3 weeds (and warming could enhance it in C_3 and C_4 weeds through more rapid development), rising atmospheric [CO_2] (and warming) could enhance tolerance to herbicides through accelerated seedling growth and development (Bunce and Ziska, 2000). Also, reduced transpiration (hence water uptake from the soil) per unit of weed growth as a result of elevated [CO_2] might reduce effectiveness of soil-applied herbicides (Bunce and Ziska, 2000). Moreover, when elevated [CO_2] stimulates rhizome and tuber growth by weeds, it may enhance belowground carbohydrate reserves used by weeds for propagation and make it more difficult (or expensive) to control later weed growth because of enhanced weed regeneration and vigor (Bunce and Ziska, 2000). All these observed and potential interactions between elevated [CO_2], weeds, and herbicides indicate that increasing [CO_2] could cause important changes in herbicide-weed interactions.

Elevated Atmospheric Ozone Concentration

If there is a general influence of atmospheric [O_3] on weed-crop competition, it is unknown. Aside from a few studies that examined the effects of [O_3] on plant mixtures (mostly pasture crops), this topic is unstudied. The majority of studies on forage crops indicate that exposure of grass-legume mixtures to elevated [O_3] tends to favor the grass species over the legume (Runeckles and Chevone, 1992; Fuhrer et al., 1994), but exceptions to this pattern have been noted (Johnson, Hale, and Ormrod, 1996).

Crops especially sensitive to O_3, such as soybean, may be at a competitive disadvantage compared to more O_3-tolerant weeds in elevated-[O_3] environments. For crops less sensitive to O_3, competitiveness with respect to weeds could increase, decrease, or remain the same in elevated [O_3], even as crop yield declines. An important consideration is that if growth or vigor of one weed declines as a result of increasing O_3 pollution, other more O_3-resistant weeds could take its place. Thus, crops may be competing with a moving target with respect to weeds that might better tolerate rising atmospheric [O_3].

Salinity

Little attention has been paid to the influence of soil salinization on competitive interactions between field crops and weeds. It might be expected that the competitive advantage of relatively salt-tolerant weeds such as purslane increases relative to more salt-sensitive crops (Grieve and Suarez, 1997). Conversely, a salt-tolerant crop such as barley might be expected to outcompete most salt-sensitive weeds in salty soil.

INSECT PESTS AND DISEASES

Insect pests and microbial pathogens such as viruses, bacteria, and fungi can cause extensive damage to crops and may significantly limit yield through a limitation on the biosynthesis of desired plant parts such as grain and tubers, through the consumption or destruction of those parts after they are synthesized, or through both mechanisms. A significant percentage of total production costs in many crops is incurred controlling the growth and reproduction of such pests. Pest control can involve actions by farmers across multiple fronts: timely application of appropriate chemicals that are toxic or objectionable to pests (i.e., pesticides), use of crop varieties selected or genetically engineered to resist or repel pests, and other aspects of crop management such as cultivation, species rotations, and timing of sowing.

The battle against insect pests and diseases is made difficult by the great diversity of life histories and feeding habits of insects and pathogens. For example, some insects preferentially feed on leaves while others prefer roots, fruits, or stems, and different insects may feed by sucking out the contents of plant cells, tapping into phloem conduits, or consuming whole tissues and even organs. Some insects complete one life cycle per year (univoltine) while others may produce from two to 12 generations in a single season (multivoltine).

Environmental changes might affect the interactions between crops and pests by (1) modifying the chemistry of plant tissues, thereby changing their palatability or nutritional properties; (2) speeding or slowing the development of the crop, thereby affecting the timing of crop susceptibility to pests; or (3) altering the developmental rate or demography of pests themselves, thereby changing the timing or location of pest attacks. With the exception of warming, most environmental changes are likely to influence crop-pest interactions mainly by altering the physiology of the crop and the nutritional quality of the crop tissues.

The main determinants of plant tissue quality from an insect's or pathogen's perspective might involve the presence or absence of toxic secondary plant compounds or pathogenesis-related proteins (PR proteins), nitrogen concentration, toughness (e.g., degree of lignification), and/or the topography of the relevant plant organ's surface (e.g., the nature or extent of epicuticular waxes and trichomes). Changes in any of these characteristics could make plants more or less susceptible to diseases, and more or less nutritious/palatable to insect pests.

The effects of the environment on biochemical defenses of crops could be particularly important (Chiang and Norris, 1983; Hartmann, 1996). The three main groups of protective chemicals are the alkaloids, terpenoids, and phenolics. Alkaloids are nitrogenous compounds synthesized from amino acids, terpenoids are carbon-based molecules (lipids) synthesized via the mevalonate pathway, and phenolics are aromatic, carbon-based molecules formed via the shikimate pathway. Phenolics and terpenoids, often synthesized in fairly large quantities, are generally regarded as quantitative defensive compounds (their function in defense depends on their concentration). Quantitative defense compounds are expensive to synthesize and their synthesis is closely linked to carbohydrate availability (i.e., precursor amounts). Alkaloids are qualitative defensive compounds. Their function in defense is not concentration dependent, they require much smaller metabolic expenditures, and their synthesis is less sensitive to environmental fluctuations compared to the phenolics and terpenoids (Hashimoto and Yamada, 1994).

Most crops contain fewer types of secondary compounds, and less of each, compared to their wild progenitors. The same properties that make these compounds efficacious in deterring pests tend also to make them undesirable for crops. Furthermore, construction of some secondary compounds, especially phenolics and terpenoids, is costly in terms of the carbon and energy required for biosynthesis; carbon and energy that could instead be invested in grain or tubers. Nonetheless, secondary compounds play roles in plant protection in a number of important crops. Maize, for example, produces a chemical called DIMBOA (2,4-dihydroxy-7-methoxy-1,4-benzoxazin-3-one), a natural pesticide effective against the European corn borer as well as some bacteria and fungi (Yan, Liang, and Zhu, 1999). Cabbage and cassava synthesize the noxious chemicals glucosinilates and cyanogenic glycosides, respectively (Hashimoto and Yamada, 1994). Cotton, sorghum, and potato contain substantial amounts of phenolics which have antiherbivore properties (Friedmann, 1997). The solanaceous crops potato, tomato, and pepper contain alkaloids that are toxic to most insects. How the levels or effectiveness of these compounds might be affected by environmental change is largely unstudied but could be important.

Most of the chemical defenses synthesized by crops mentioned so far are constitutive; that is, they are always present in the crop. But plants also have protection mechanisms that are activated only by wounding, pathogen infection, or insect feeding. The hypersensitive response (HR) and systemic acquired resistance (SAR) are important examples (Sandermann, 2000). The HR involves programmed death of the cells in direct contact with a pathogen, resulting in necrotic lesions. By sacrificing the cells surrounding the site of infection, pathogens can be isolated and their ability to spread to other parts of the organ or plant is reduced.

Systemic acquired resistance involves activation of a number of plant defense mechanisms, both at the site of infection as well as in more distant parts of the plant (Schneider et al., 1996). It is a relatively disease-resistant state resulting from prior exposure to a pathogen. This state is associated with production of pathogenesis-related (PR) proteins, initiation of physical barriers in plant tissues, and production and translocation of signal molecules (Schneider et al., 1996). Presumably, when a plant is invaded by a pathogen, the infected cells produce signal molecules such as salicylic acid or ethylene, which are then translocated to other parts of the plant. These signal molecules may then "turn on" numerous defense mechanisms, including at least 20 genes encoding PR proteins. A role of PR proteins in limiting the multiplication and spread of invading pathogens has been shown for almost all important crop plants (Ohashi and Ohshima, 1992). As mentioned later, air pollutants such as O_3 cause biochemical changes that are nearly identical to the HR described here, while the effects of elevated CO_2, warming, and salinity stress are less clear.

Secondary compounds and inducible defense reactions are undoubtedly important to crop-pest interactions, but more fundamental things such as crop-tissue water content, nitrogen concentration, amino acid concentration, and C/N ratio may be even more important. For example, over 115 studies reported a significant positive correlation between plant nitrogen concentration and insect growth, fecundity, and population levels (Salim and Akbar, 1995). Because several environmental changes are likely to affect crop water content, nitrogen and amino acid concentrations, and C/N ratio, those same environmental changes may in turn alter crop-pest interactions.

Climatic Change

Warming of just a few degrees Celsius could influence crop-pest interactions in many ways. Perhaps most important, warming will directly alter the developmental rate, timing of developmental events, physiological rates, overwintering success, and geographic distribution of many insects (Patterson, 1993; Cannon, 1998; Bale et al., 2002). Warming could further affect crop-pest relationships by hastening crop development and changing crop tissue quality. Experts have concluded that warming is likely to intensify pest problems encountered by crops (Cannon, 1998).

A chief concern is that insect pests will develop more quickly with warming because the development of insects is so tightly linked with ambient temperature. Warming may even enable multivoltine insect species to complete more life cycles during a given year than is possible in current climates (Landsberg and Smith, 1992). Effects on univoltine species may be less dramatic, but phenology of those insects could change as well. For example, a 3°C increase in temperature has been projected to cause the cabbage root fly to become active a month earlier in the United Kingdom (Collier et al., 1991), and a 2°C warming might cause spittlebugs to complete their life cycles two to three weeks earlier (Whittaker and Tribe, 1996). Warming could advance flight phenology of aphid species by more than a month (Zhou et al., 1995). Similar phenologic shifts are possible for many other insect pests (Whittaker, 2001).

The overall effect that faster insect development will have on crops is unknown. The synchrony between insect and plant development will probably be the key issue. Because insects are likely to emerge earlier and develop more quickly, and because crops may also develop more quickly, it is difficult to predict whether insect and crop life cycles will overlap to a greater or lesser extent. If warming has a greater effect on the developmental rate of the crop than on the development of an insect pest, the insect may suffer. However, if warming has a greater effect on pest developmental rate, the crop may suffer greater herbivory (Bale et al., 2002).

Warming may enable many insects to increase their geographic distributions. Northerly spread of insects (including Odonata, Orthoptera, and Lepidoptera) was observed during warm summers in the early 1990s in the United Kingdom (Cannon, 1998; Morecroft et al., 2002), and there are other examples of increased geographic ranges of insects during particularly warm periods (Bradshaw and Holzapfel, 2001; Whittaker, 2001). Attempts to devise models for predicting how insect distributions and development might be affected by climatic change are being made (Landsberg and Smith, 1992; Yamamura and Kiritani, 1998).

Natural enemies of crop pests, such as beneficial insects and birds, could also benefit from warming—an effect that could offset additional herbivory pressure on crops caused by warming. Some aphid predators (Coccinellids), for example, are likely to become more efficient in controlling aphids in a warmer world (Skirvin, Perry, and Harrington, 1997). Range expansion of birds that feed on insects could also reduce insect damage to crops. (Expansion of ranges of birds that eat grain could reduce yield.)

It is more difficult to predict how microbial pathogens will respond to climatic change because those organisms are more sensitive to rainfall and humidity than are insects, and it is unclear how the hydrologic cycle will change in many cropped regions. In general, wetter and warmer conditions favor microbial pathogens compared to dryer and cooler conditions.

Elevated Atmospheric Carbon Dioxide Concentration

Increases in atmospheric [CO_2] are unlikely to directly affect pests, but indirect effects on pests caused by changes in plant chemistry (tissue quality) are nearly inevitable. The best-documented chemical change resulting from crop growth in CO_2-enriched atmospheres is the decrease in tissue amino acid and total N concentrations associated with higher C/N ratios (Cotrufo, Ineson, and Scott, 1998; Whittaker, 2001). This is an important chemical shift because insects feeding on these tissues must increase their consumption, sometimes as much as 40 percent, to meet their dietary N requirements (Whittaker, 2001). In spite of increased consumption, many chewing insects (particularly Lepidoptera larvae) appear unable to fully compensate for reduced plant N concentration and consequently their growth rates, survival, and fecundity suffer when feeding on plants grown in elevated [CO_2]s. This indicates that tissues of crop plants might become less nutritious in a higher-CO_2 world.

Although not proven in crops, CO_2 enrichment stimulates production of secondary compounds in many noncrop plant species. According to a literature survey, the concentration of phenolic compounds increased an average of 31 percent in leaves of plants grown in doubled-[CO_2] conditions (Bezemer and Jones, 1998). Such an effect of elevated [CO_2] on plant tissue composition might explain the consistent reduction in performance of insects fed plants grown in elevated [CO_2]s. How rising atmospheric [CO_2] will affect production of other types of secondary compounds, such as terpenoids and alkaloids, or inducible defense systems, is unknown.

In contrast to chewing insects, phloem feeders such as aphids sometimes perform better when allowed to feed from plants growing in elevated compared to ambient [CO_2] (Awmack, Harrington, and Leather, 1997). Since aphids tap into phloem and feed directly from the assimilate stream between source and sink, they are probably less sensitive to the changes in tissue chemistry caused by exposure to elevated [CO_2]. Whatever the cause, enhanced performance of aphids on plants grown in elevated [CO_2] indicates that direct aphid damage, as well as indirect damage caused by the viruses that they transmit to crops, may increase as atmospheric [CO_2] rises (Awmack, Harrington, and Leather, 1997).

Research to date has shown only small effects of atmospheric [CO_2] on disease incidence and severity, although little research has been done on this topic (Table 10.2). From a theoretical perspective, it might be expected that CO_2 enrichment would increase crop resistance to pathogens because elevated [CO_2] reduces crop N concentrations, which are almost always positively related to disease susceptibility (Thompson, Brown and Woodward, 1993). Alternatively, CO_2 enrichment might benefit soil pathogens by stimulating root growth and root exudation of sugars into soil. The limited data presently available indicate that crops become more tolerant of pathogens as atmospheric [CO_2] is increased (Table 10.2). Populations of bacteria and actinomycetes collected per unit area of cotton leaves were unaffected by CO_2 enrichment (550 ppm) in the field (Runion et al., 1994). Populations of some fungi were lower, some were unchanged, and others were higher as a result of elevated [CO_2] in the same cotton crop (Runion et al., 1994). It remains unclear if production of plant defensive chemicals and induction of defense systems such as the HR and SAR are affected by elevated [CO_2].

TABLE 10.2. Effects of elevated [CO_2] on some crop pathogens.

Pathogen	Host crop	Effect on disease or crop	Source
Fungi			
Erysiphe graminis	Barley	Crop growth reduction greater in elevated [CO_2]	A
	Wheat	Pathogen growth reduced in elevated [CO_2]	B
Phytophthora parasitica	Tomato	Crop tolerance to disease increased by elevated [CO_2]	C
Puccinia recondita	Wheat	None	D
Rhizoctonia solani	Cotton[a]	Trend for greater infestation of *soil* with elevated [CO_2]	E
Virus			
Barley yellow dwarf virus	Oat	Crop survival increased by elevated [CO_2]	F

Sources: (A) Hibberd, Whitbread, and Farrar (1996); (B) Thompson, Brown, and Woodward (1993); (C) Jwa and Walling (2002); (D) Tiedemann and Firsching (2000); (E) Runion et al. (1994); and (F) Malmström and Field (1997).

[a]Field experiment (FACE). A soil bioassay indicated no increase in damping-off potential.

Elevated Atmospheric Ozone Concentration

Crops exposed to elevated atmospheric [O_3] can become more sensitive to feeding by chewing insects, and when given a choice, insects may preferentially feed on leaves that have been exposed to elevated [O_3] rather than feeding on leaves grown in ambient [O_3] (Endress and Post, 1985). Diets consisting of plants grown in high [O_3]s can stimulate insect growth and decrease mortality (Jackson, Heagle, and Eckel, 1999). Tobacco hornworm moths even oviposited more eggs on tobacco plants growing in a high [O_3] (Jackson, Heagle, and Eckel, 1999). Stimulatory effects on insect growth of exposing plants to elevated [O_3] are probably related to a favorable shift (from the insect's perspective) in the plant's C/N ratio or amino acid content rather than a direct effect of [O_3] on the insect itself (Brown, McNeill, and Ashmore, 1992). But insect responses to plants grown in elevated [O_3] are variable; depending on the host plant and insect species, herbivory can be stimulated, inhibited, or unaffected by elevated [O_3] (Pfleeger, da Luz, and Mundt, 1999; Costa, Kennedy, and Heagle, 2001). This variability could be related to differential effects of [O_3] on secondary compounds, inducible defense reactions, and the sensitivity of a given pest to such changes. Furthermore, the timing of both elevated [O_3] and the presence of specific insect pests may affect experimental outcomes of specific plant-insect interactions. Little is known about the effects of elevated [O_3] on crop-insect interactions for most crops growing in the field.

It is well documented that high [O_3] episodes can stimulate the synthesis of secondary compounds, signal molecules, PR proteins (i.e., SAR), and antioxidative systems in some plants (Table 10.3). Exposure to high [O_3] even elicits production of cellular barriers such as callose, and enhanced lignification of cell walls. As a result, plants exposed to elevated [O_3] are sometimes, but not always, more resistant to pathogen attack (Table 10.4). There is some indication that exposure to elevated [O_3] increases resistance to obligate pathogens, but decreases resistance to facultative pathogens (Krupa et al., 2000). Exposure to high [O_3] also tends to increase host plant susceptibility to necrotrophic and root-rot fungi (Sandermann, 2000).

In addition to prior O_3 exposure affecting disease development, prior infection of crops by pathogens can affect a crop's response to

TABLE 10.3. Plant defense reactions induced by elevated [O_3] episodes.

Antioxidants	Ascorbate peroxidase (*Arabidopsis,* tobacco), catalases, glutathione peroxidase, glutathione S-transferase, and superoxide dismutases
Cellular barriers	Callose (tobacco), extensins (beech, parsley, pine, spruce), and lignins and fluorescent cell-wall components
PR proteins	Acidic and basic PR proteins (*Arabidopsis,* birch, parsley, spruce, tobacco)
Secondary compounds	Furanocoumarins (parsley), isoflavonoids (soybean), and stilbenes (grapevine, pine)
Signal molecules	*ACC (potato, spruce, tobacco, tomato), ethylene, and salicylic acid (*Arabidopsis,* tobacco)

Source: Modified from Sandermann (2000).

*See Chapter 2 for some details about the influence of O_3 exposure on ACC, ethylene, and antioxidative systems.

TABLE 10.4. Effects of elevated [O_3] on some crop diseases.

Pathogen	Host crop	Effect of O_3 on disease	Source
Bacteria			
Pseudomonas glycinea	Soybean	Reduced lesion numbers	A
Xanthomonas alfalfae	Alfalfa	Reduced infection severity	B
Fungi			
Bortytis cinerea	Potato	Only O_3-injured leaves infected	C
Drechslera sorokiniana	Wheat	Increased area of infection	D
Puccinia coronata	Oat	Reduced growth of uredia	E
P. graminis	Wheat	Reduced hyphal growth and urediospore production on O_3-injured leaves	E
P. recondita	Wheat	None	F

Sources: (A) Laurence and Wood (1978); (B) Howell and Graham (1977); (C) Manning et al. (1969); (D) Fehrmann, von Tiedemann, and Fabian (1986); (E) Pfleeger, de Luz, and Mundt (1999); and (F) Manning and von Tiedemann (1995).

Notes: Effects depended on sufficient doses of O_3, either before or after inoculation. Those doses differed between experiments, and some doses were unrepresentative of [O_3]s in the field. Details of experiments are in the source publications.

elevated [O_3]. For example, infection of alfalfa and bean by *Xanthomonas alfalfae* and *X. phaseoli,* respectively, increased crop tolerance to elevated [O_3] (Howell and Graham, 1977; Temple and Bisessar, 1979). Infection of wheat by *Puccinia graminis* reduced injury to substomatal mesophyll cells by elevated [O_3] (Heagle and Key, 1973).

Salinity

The effects of salinization on the susceptibility of crops to pests are poorly understood. In rice, soil salinity can increase tissue N content and decrease production of secondary compounds, and as a consequence, rice susceptibility to the whitebacked planthopper, a major pest in Asia, can be significantly increased when the crop is grown in salinized soil. This particular insect grew faster, had higher fecundity, lived longer, and had larger populations when grown on high-salt compared to low-salt (control) treatments (Salim and Akbar, 1995). In many other crops, however, tissue nitrogen concentration is reduced by soil salinity (Chapter 7), so these observations may be limited to rice.

Little is known about effects of soil salinity on crop diseases, though the severity of root diseases caused by pathogenic fungi has been reported to increase in tomato and alfalfa (Blaker and MacDonald, 1985; Pankhurst et al., 2001).

SYMBIOTIC RELATIONSHIPS WITH SOIL ORGANISMS

Many soil organisms form important symbiotic relationships with crops. For example, mycorrhizae are the symbiotic associations of the mycelia of fungi with plant roots. The fungi enhance plant water uptake, protect roots against some pathogens, and increase mineral nutrient uptake, especially phosphate. In return, the fungi receive their energy (in the form of organic materials such as sugars) from the plants (as much as 10 to 20 percent of the carbon assimilated in plant photosynthesis can be transported to mycorrhizal fungi [Treseder and Allen, 2000]). The predominant form of mycorrhizae found in crops is called endomycorrhizae (also called arbuscular mycorrhizal, or AM, fungi). Endomycorrhizal fungi penetrate root cells, living both inside and outside roots. In contrast, ectomycorrhizal fungi do not penetrate root cells, are unimportant in most crops, and will not be considered further herein.

Because mycorrhizal fungi depend on the host plant for their carbon and energy, any environmental change that influences crop photosynthesis or subsequent partitioning of carbon to crop roots might be expected to affect the fungi. For example, it is well-known that plants grown in low light (reduced photosynthesis) are colonized by fewer mycorrhizal fungi than plants grown in high light.

Some bacteria and actinomycetes also form symbiotic relationships with the roots of some crops. The most important of these groups are *Rhizobium* sp. bacteria which form a symbiosis with legumes (e.g., soybean, lupin, bean, pea, and cowpea). *Rhizobium* bacteria chemically reduce atmospheric nitrogen (N_2), forming NH_3, in a reaction catalyzed by the enzyme nitrogenase. This is important because plants cannot make direct use of N_2 in their metabolism. Rhizobia (plural for any of the *Rhizobium* bacterial species that form N_2-fixing symbioses in plant roots) live in nodules formed in/on roots of the host plant. The amount of reduced nitrogen supplied to crops in such nodules can be significant; for example, 100 to 200 kg N ha^{-1} $year^{-1}$ is probably a realistic average range for soybean during a good year (Loomis and Connor, 1992). In return for supplying the host plant with reduced nitrogen, rhizobia receive their carbon (mainly carbohydrates) and minerals from the plant. The nature of this exchange of nitrogen for carbon and minerals depends on plant metabolic state as well as soil physical and chemical conditions. Lack of water or an abundance of NH_4 or NO_3 in the soil solution both retard nitrogenase activity. A rapid supply of carbohydrate from the plant to the nodule can stimulate nitrogenase activity. As with mycorrhizal fungi, any environmental change that affects the carbon balance of the crop or the partitioning of carbon to roots might affect the success of rhizobia and the activity of nitrogenase.

Whether effects of environmental change on N_2 fixation will prove to be of much practical importance is an open question. Global production of legumes is decreasing in favor of higher-yielding cereal crops (Graham and Vance, 2000), and because native soil rhizobia often outcompete domesticated strains, which reduces the benefit of inoculation, the number of farmers inoculating seed with N_2-fixing bacteria before sowing is declining (Loomis and Connor, 1992; Trotman and Weaver, 1995). As a consequence, farming systems are relying more heavily on nitrogen derived industrially (in a process that emits CO_2 to the atmosphere), and less on that supplied by symbiotic bacteria.

For environmental changes that directly affect the crop but not the symbiont, such as atmospheric [CO_2] and [O_3], it is expected that changes in the symbiont will be a result of changes in the crop. Thus, the effects of changes in atmospheric composition on the symbionts may generally be incidental, rather than of primary importance to the crop and its yield. Conversely, changes in the soil environment, namely warming and increased salinity, may have primary effects on the symbiont, which may in turn affect the crop.

Warming

Effects of warming on AM fungi associated with a major crop do not appear to have been well studied (Fitter, Heinenmeyer, and Staddon, 2000). In one experiment, barley roots were colonized by AM fungi when soil temperature was 10°C, but not when soil temperature was 15°C. For the weed *Plantago lanceolata,* however, colonization by AM fungi was greater at 20°C than at 12°C (Fitter, Heinemeyer, and Staddon, 2000). In the forage crop western wheatgrass, a 4°C increase in air temperature decreased the root length colonized by AM fungi by 15 percent (Monz et al., 1994). It is unknown if these experimental results were due to an effect of temperature on the fungus, the plant, or both.

Temperature can influence N_2 fixation by rhizobia. High root temperature can inhibit N_2 fixation. For example, reductions in N_2 fixation were observed when temperature was above 30°C for clover, pea, and bean, and when temperature was above 35 to 40°C for soybean, guar, peanut, cowpea, and mung bean (Hafeez, Asad, and Malik, 1991; Michiels, Verreth, and Vanderleyden, 1994). However, low temperature can also inhibit N_2 fixation. Warming, therefore, may have negative effects on N_2 fixation in tropical areas with presently supraoptimal temperature, but could have positive effects in temperate regions where temperatures are now suboptimal.

Elevated Atmospheric Carbon Dioxide Concentration

Because elevated atmospheric [CO_2] generally stimulates crop growth, at least in C_3 crops, rising [CO_2] may be expected to stimulate AM fungi growth by enhancing the flow of carbohydrates to roots and then on to fungi. Although the percentage of roots colonized by mycorrhizal fungi is usually unchanged when plants are grown in an elevated [CO_2], fungal biomass generally increases in proportion to the increase in total root biomass (Treseder and Allen, 2000).

Effects of CO_2 enrichment on mycorrhizal fungi can vary greatly with crop species, soil physical and chemical conditions, and time of the year. For example, total AM fungal hyphal length on grain sorghum roots was enhanced 109 percent by elevated [CO_2] (i.e., 550 compared to 370 ppm) under well-watered conditions, but 267 percent under water-stress conditions (Rillig et al., 2001). Although it is likely that rising atmospheric [CO_2] will stimulate mycorrhizae in many cases, the extent of the enhancement will probably depend on many factors and cannot be reliably predicted at present.

Elevated atmospheric [CO_2] stimulates N_2 fixation in a number of legumes, including soybean, pea, and clover, primarily by increasing nodule number or size (Phillips et al., 1976; Williams, Dejong, and Phillips, 1981; Finn and Brunn, 1982; Reddy, Acock, and Acock, 1989; Serraj, Sinclair, and Allen, 1998). The proportion of plant nitrogen derived from N_2 fixation sometimes goes up as well (Lilley et al., 2001), particularly when concentrations of NO_3 and NH_4 in the soil are limited (Zanetti, Hartwig, and Nösberger, 1998). Effects of CO_2 enrichment on specific nodule activity (i.e., mass of nitrogen assimilated per mass of nodule) are inconsistent (Rogers, Runion, and Krupa, 1994).

Compared to bacteria living within nodules, populations of free-living rhizobia in soil may be more sensitive to elevated atmospheric [CO_2]. Under clover, for example, a doubling in the population of rhizosphere *Rhizobium* bacteria was observed (Schortenmeyer et al., 1996), and the amount of N_2-fixing bacteria in the bulk soil was also stimulated in a rice paddy exposed to elevated [CO_2] (Allen, 1990).

Effects of CO_2 enrichment on the quantity or activity of N_2-fixing bacteria, whether located in nodules, the rhizosphere, or the bulk soil, are probably caused by greater carbohydrate flow from shoots to roots (Stulen and den Hertog, 1993). Greater demand for nitrogen by plants in CO_2-enriched atmospheres (the plants are bigger, so require more nitrogen to maintain a given enzyme complement) could also play a role in stimulating N_2 fixation (Zanetti, Hartwig, and Nösberger, 1998).

Elevated Atmospheric Ozone Concentration

Because elevated atmospheric [O_3] can inhibit crop photosynthesis, growth, and partitioning of carbon belowground, it is expected that elevated [O_3] would inhibit mycorrhizal formation. In keeping with this notion, negative effects of elevated [O_3] on mycorrhizae have been reported for tomato and soybean (reviewed in Shafer and Schoeneberger, 1994).

The effect of mycorrhizae on crop's sensitivity to elevated [O_3] may also be important. For example, mycorrhizal crops are sometimes more resistant to elevated [O_3] than are nonmycorrhizal crops (Shafer and Schoeneberger, 1994). The mechanisms are unclear.

Exposure of legumes to elevated [O_3] was reported to decrease nodulation and N_2 fixation in soybean and pinto bean (Tingey and Blum, 1973; Blum and Tingey, 1977; Flagler et al., 1987). Negative effects of elevated [O_3] on N_2 fixation are not directly on nitrogenase or the bacteria because O_3 does not penetrate into the soil. Rather, O_3 inhibits both photosynthesis and assimilate transport to roots, and this reduces the supply of energy (i.e., reduced carbon) to rhizobia (Blum and Tingey, 1977).

Salinity

Relatively little attention has been paid to the effects of salinity on mycorrhizae (Bethlenfalvay, 1992; Azcón and El-Atrash, 1997), though it is known that most salt-tolerant species, including those native to salt marshes, are mycorrhizal (Estaun, 1989). Moreover, a few studies indicate a positive effect of mycorrhizae on the salt tolerance of some crops (reviewed in Bethlenfalvay, 1992). However, as is the case for crop species and varieties, fungal species differ substantially in their ability to withstand high-salt environs. Although understanding is limited, development of compatible mycorrhizal relationships involving both a salt-tolerant plant and a salt-tolerant fungus might prove to be an important aspect of crop adaptation to salinity.

Rhizobia are both directly and indirectly sensitive to soil salinity. The direct effects on rhizobia are generally caused by ion (i.e., Na^+ and Cl^-) toxicity, not osmotic effects (Elsheikh, 1998). Osmotic problems can be overcome by rhizobia, albeit with an energy cost, through osmotic adjustment. Indirect effects of salinity on N_2 fixation are caused by salt stress imposed on the host plant, or by changes in soil pH. Direct and indirect effects combine to inhibit multiplication of rhizobia, nodulation of roots, and rate of N_2 fixation. This may explain why legumes are more tolerant of soil salinity when supplied with NO_3 and/or NH_4 compared to an obligate dependence on N_2 assimilation (Serraj and Drevon, 1998). Some have even suggested that the sensitivity of the *Rhizobium*-legume symbiosis contributes largely to the overall salt sensitivity of legumes (Serraj and Drevon, 1998).

Some rhizobia are more tolerant of salt than others. Identification of resistant strains, and incorporating them into legume cropping systems, could increase crop productivity on salinized soils. Although progress in the field is lagging, salt tolerance of many N_2-fixing bacteria has already been characterized in the laboratory (Elsheikh, 1998; van Hoorn et al., 2001). It is important to note, however, that rhizobia are generally more salt-tolerant than their plant hosts, so identifying or developing salt-resistant rhizobia may not lead to yield improvements unless salt-resistant legume varieties are developed in parallel (Elsheikh, 1998; Serraj and Drevon, 1998).

MULTIPLE ENVIRONMENTAL CHANGES AND IMPLICATIONS FOR BIOTIC INTERACTIONS

What effects the balance of environmental changes will have on biotic interactions is unclear. With respect to soil organisms, for instance, rising atmospheric [CO_2] should enhance the growth and activity of both mycorrhizae and rhizobia, but the detrimental effects of rising [O_3] might counteract these benefits. Furthermore, rising [CO_2] and [O_3] might also have counteracting effects on relationships between crops and pests. Similarly, warming will hasten, while rising [CO_2] is likely to slow, insect development. Interactive effects of multiple changes in crop/weed interactions are completely unknown.

It is clear that changes in temperature, precipitation (soil moisture), and salinity can be expected to interact with rising atmospheric [CO_2] and [O_3] to affect biotic interactions. But generalizations about the magnitudes of such effects, or even estimations of whether the effects will be positive or negative for the crop, cannot be made due to the small amount of relevant experimental data and lack of experimentally tested theory. Experiments controlling temperature, [CO_2], and [O_3] can help untangle the net benefits or harm of changes in multiple components of the atmosphere, but such research is only beginning.

SUMMARY

Crops must contend with insects, diseases, and weeds. Changes in the physical or chemical environments (e.g., rising atmospheric

$[CO_2]$ and $[O_3]$, warming, and soil salinization) will certainly alter some of the interactions between crop plants and the other organisms that share their locales. Such alterations could be beneficial or detrimental to the crop and its yield. Unfortunately, because of the large number of interactions between crops and other organisms, and the confounding nature of multiple environmental changes on both crops and their coexisting organisms, few hard generalizations can be made. A few speculative examples are put forward as issues to be considered.

1. Warming may favor many weeds compared to most crops, though rising atmospheric $[CO_2]$ may favor C_3 crops competing with C_4 weeds. Because many weeds are especially successful at exploiting new resources, rising $[CO_2]$ is also likely to enhance C_3 weed growth. With respect to increasing stresses, such as rising atmospheric $[O_3]$ and soil salinization, it may be significant that there are many more weed species than crop species and as a result weed populations may adapt more quickly to environmental changes than the crops they compete with.
2. Warming is likely to increase the abundance and growth rate, and extend the geographic range, of many important crop-eating insects. Warming is also likely to stimulate microbial pathogens, but this effect will depend largely on associated changes in rainfall patterns, soil moisture, and air humidity.
3. Environmental changes are likely to alter crop tissue chemistry, including nitrogen and water content as well as metabolism of inducible defensive mechanisms. These effects, should they occur, will make crop tissues more or less palatable/nutritious to insects and more or less susceptible to microbial pathogens. For instance, rising $[CO_2]$ might lower leaf nitrogen content, leading to slower growth and reduced fecundity of insect pests. Exposure to high $[O_3]$ often stimulates inducible plant defense reactions, sometimes leading to enhanced crop resistance to microbial pathogens. However, crops exposed to O_3 have elevated tissue nitrogen levels, which tends to make them more attractive and nutritious to herbivores. Complex interactive effects of warming with CO_2 and O_3 are likely but have not been studied.
4. The effects of rising atmospheric $[CO_2]$ and $[O_3]$ on crop symbionts found in the soil will be through effects on the crop, i.e.,

the effects on the symbionts will follow the primary effect on the crop. The effects of warming and salinization may be primarily on the crop, the symbiont, or both.

5. Rising CO_2 levels often stimulate mycorrhizae and rhizobia, to the benefit of both crops and symbionts. Elevated $[O_3]$, however, reduces the supply of carbohydrates to belowground tissues, which decreases colonization of roots by beneficial fungi and reduces nodulation and N_2-fixation by rhizobia. High soil salinity also inhibits colonization of roots by symbionts.
6. Soil warming will likely benefit rhizobia and stimulate N_2-fixation in temperature regions, but might harm this symbiosis in tropical regions where soil temperatures are already optimal or supraoptimal. Little is known regarding the influence of warming on soil fungi.

Future research will clarify, quantify, or correct these examples. It will also add to the list of known changes in the biotic environment that will be important to crops as their physical and chemical environments continue to change. The key point is that crop plants are not the only organisms determining yield that are likely to be affected by environmental changes.

Appendix A

List of Plant Species

Common name	Latin binomial	Primary photosynthetic pathway used
Alfalfa	*Medicago sativa* L.	C_3
Almond	*Prunus amygdalus* L.	C_3
Amaranth	*Amaranthus* spp.	C_4
Amaranth, Palmer	*Amaranthus palmeri* S. Wats	C_4
Apple	*Malus* spp.	C_3
Apricot	*Prunus armeniaca* L.	C_3
Asparagus	*Asparagus officinalis* L.	C_3
Avocado	*Persea americana* P. Mill.	
Banana	*Musa sapientum* L.	
Barley	*Hordeum vulgare* L.	C_3
Barnyard grass	*Echinochloa glabrescens* L.	C_4
Bean	*Phaseolus vulgaris* L.	C_3
Bean (field)	*Vicia faba* L.	C_3
Beech	*Fagus* spp.	C_3
Bentgrass	*Agrostis capillaris* L.	C_3
Birch	*Betula* spp.	C_3
Brazilnut	*Bertholletia excelsa* Humb. & Bonpl.	
Broad bean	*See* faba bean	
Buckwheat	*Fagopyrum esculentum* Moench	
Butternut	*Juglans cinerea* L.	
Cabbage	*Brassica oleracea* L.	C_3
Canada thistle	*Cirsium arvense* (L.) Scop.	C_3
Canola	*See* rape	
Carob	*Ceratonia siliqua* L.	
Carrot	*Daucus carota* L.	C_3
Cashew	*Anacardium occidentale* L.	
Cassava	*Manihot esculenta* Crantz	

Common name	Latin binomial	Primary photosynthetic pathway used
Castor bean	*Ricinus communis* L.	C_3
Chickpea	*Cicer arietinum* L.	C_3
Clover	*Trofolium* spp.	C_3
Cocklebur (common or rough)	*Xanthium strumarium* L.	C_3
Cocoa	*Theobroma cacao* L.	
Coconut	*Cocus nucifera* L.	
Coffee	*Coffea* spp.	
Corn	*See* maize	
Cotton	*Gossypium hirsutum* L.	C_3
Cotton (Pima)	*Gossypium barbadense* L.	C_3
Cowpea	*Vigna unguiculata* (L.) Walp.	
Cranberry	*Vaccinium macrocarpon* Ait.	C_3
Cucumber	*Cucumis sativus* L.	C_3
Dandelion	*Taraxacum officinale* Wigg.	C_3
Date (date palm)	*Phoenix dactylifera* L.	
Eggplant	*Solanum melongena* L.	
Faba bean	*Vicia faba* L.	C_3
Fava bean	*See* faba bean	
Fig	*Ficus carica* L.	
Flax	*Linum usitatissimum* L.	C_3
Flaxseed	*See* flax	
Foxtail millet	*Setaria italica* L.	
Garbanzo bean	*See* chickpea	
Gingelly	*See* sesame	
Grape	*Vitis vinifera* L.	C_3
Grapevine	*See* grape	
Guar	*Cyamopsis tetragonoloba* (L.) Taubert	
Guava	*Psidium guajava* L.	
Hazelnut	*Corylus* spp.	
Johnsongrass	*Sorghum halapense* (L.) Pers.	C_4
Jojoba	*Simmondsia chinensis* (Link) Schneid.	
Kumquat	*Fortunella* spp.	
Lambsquarters	*Chenopodium album* L.	C_3
Lentil	*Lens culinaris* Medik.	

Common name	Latin binomial	Primary photosynthetic pathway used
Lettuce	*Lactuca* spp.	C_3
Lima bean	*Phaseolus lunatus* L.	C_3
Loquat	*Eriobotrya japonica* (Thunb.) Lindl.	
Lupine	*Lupinus* spp.	
Maize	*Zea mays* L.	C_4
Mangels	*Beta vulgaris* L.	C_3
Mango	*Mangifera indica* L.	
Millet (pearl millet)	*Pennisetum typhoides* S. & H.	C_4
Millet (wild proso millet)	*Panicum miliaceum* L.	C_4
Mung bean	*Vigna radiata* (L.) R. Wilczek	C_3
Oat	*Avena* spp.	C_3
Okra	*Abelmoschus esculentus* (L.) Moench	
Olive	*Olea europaea* L.	C_3
Onion	*Allium* spp.	C_3
Orange	*Citrus sinensis* (L.) Osbeck	C_3
Palm (oil)	*Elaesis guineensis* Jacq.	
Papaya	*Carica papaya* L.	
Parsley	*Petroselium crispum* (P. Mill.) Nyman ex A.W. Hill	C_3
Pea	*Pisum* spp.	C_3
Peach	*Prunus persica* (L.) Batsch	C_3
Peanut	*Arachis hypogaea* L.	C_3
Pear	*Pyrus communis* L.	C_3
Pecan	*Carya illinoinensis* (Wangenh.) K. Koch	C_3
Pepper	*Capsicum* spp.	C_3
Pigeonpea	*Cajanus cajan* (L.) Millsp.	
Pigweed, redroot	*Amaranthus retroflexus* L.	C_4
Pilinut	*Canarium ovatum* Engler	
Pine	*Pinus* spp.	C_3
Pine, loblolly	*Pinus taeda* L.	C_3
Pistachio	*Pistacia vera* L.	
Plantain	*Musa* × *paradisiaca* L. (pro sp.)	
Popcorn	*See* maize	
Potato	*Solanum tuberosum* L.	C_3
Plum	*Prunus* spp.	C_3

Common name	Latin binomial	Primary photosynthetic pathway used
Prune	*Prunus* spp.	C_3
Pumpkin	*Cucurbita pepo* L.	
Purslane	*Portulaca* spp. L.	
Quackgrass	*Elytrigia repens* (L.) Nevski	C_3
Radish	*Raphanus sativus* L.	C_3
Raisin	*See* grape	
Rape	*Brassica napus* L.	
Rapeseed	*See* rape	
Red beet	*Beta vulgaris* L.	C_3
Rice	*Oryza sativa* L.	C_3
Rye	*Secale cereale* L.	C_3
Ryegrass	*Lolium* spp.	C_3
Safflower	*Carthamus tinctorius* L.	
Sesame	*Sesamum indicum* L.	
Sicklepod	*Senna obtusifolia* (L.) Irwin & Barneby	
Small broomrape	*Orobanche minor* Smith	
Sorghum (grain sorghum)	*Sorghum bicolor* (L.) Moench	C_4
Soybean	*Glycine max* (L.) Merrill	C_3
Spinach	*Spinacia oleracea* L.	C_3
Spruce	*Picea* spp.	C_3
Squash	*Cucurbita* spp.	C_3
Sugarbeet	*Beta vulgaris* L.	C_3
Sugarcane	*Saccharum* hybrids	C_4
Sunflower	*Helianthus annuus* L.	C_3
Sweet corn	*See* maize	
Sweetpotato	*Ipomoea batatas* (L.) Lam.	C_3
Swiss chard	*Beta vulgaris* L.	C_3
Tapioca	*See* cassava	
Tea	*Camellia sinensis* (L.) O.Kuntze	C_3
Texas panicum	*Panicum texanum* Buckl.	C_4
Tobacco	*Nicotiana tabacum* L.	C_3
Tomato	*Lycopersicon esculentum* (L.) Mill.	C_3
Turnip	*Brassica* rapa	
Velvetleaf	*Abutilon theophrasti* Medicus	C_3
Walnut (black)	*Juglans nigra* L.	

Common name	Latin binomial	Primary photosynthetic pathway used
Western wheatgrass	*Pascopyrum smithii* (rydb.) A. Love	C_3
Wheat	*Triticum* spp., often *Triticum aestivum* L.	C_3
Winged bean	*Psophocarpus tetragonolobus* (L.) DC.	
Yam	*Dioscorea* spp.	

Appendix B

Definition of U.S. "Principal Crops"

Annual statistics (estimates) are available for the area of "principal crops" harvested (1909 to present) and/or planted (1929 to present) in the United States. The difference between area planted and area harvested is largely a reflection of crop failure. Data are available at <http://usda.mannlib.cornell.edu/data-sets/crops/96120/>.

Principal U.S. crops during the years 1909 to 1990 were barley, dry edible beans, maize (called "corn" in USDA databases), cotton, flaxseed, dry pea (1963 to 1973 only), oat, peanut, popcorn (1963 to 1973 only), potato, sweetpotato (beginning in 1963), rice, rye, sorghum, soybean, sugarbeet, sunflower (beginning in 1975), winter wheat, durum wheat, and other spring wheat. The area of all hay, tobacco, and sugarcane harvested was used to compute area planted and area harvested. Thus, the area of hay (about one fifth of the area of all principal crops in recent years), tobacco, and sugarcane harvested was the same as the area planted when computing area of principal crops (i.e., no hay, tobacco, or sugarcane "failures" are accounted for).

A new series (set of crops) was begun in 1983 (data for the old series overlapped with the new series through 1990). New series crops are maize (for grain and silage), sorghum (for grain and silage), oat, barley, winter wheat, durum wheat, other spring wheat, rice, rye, soybean (for beans), peanut (for nuts), sunflower, cotton, dry edible bean, canola (rape), potato, and sugarbeet. The area of all hay, tobacco, and sugarcane harvested is used to compute area planted and area harvested. Area planted also includes the area of unharvested small grains planted as cover crops.

The area of principal crops planted accounted for 79 to 89 percent of total U.S. cropland (defined as "cropland used for crops" plus "idle cropland," but excluding "cropland used only for pasture"), according to a comparison of the area of principal crops planted to the area of total U.S. cropland as estimated during the 1940, 1950, 1959, 1969, 1978, 1982, 1987, 1992, and 1997 Censuses of Agriculture.

References

Abdullah, Z., R. Ahmad, and J. Ahmed (1978). Salinity induced changes in the reproductive physiology of wheat plants. *Plant and Cell Physiology* 19: 99-106.

Aben, J.M.M., M. Janssen-Jurkoviková, and E.H. Adema (1990). Effects of low-level ozone exposure under ambient conditions on photosynthesis and stomatal control of *Vicia faba* L. *Plant, Cell and Environment* 13: 463-469.

Ackerly, D.D., J.S. Coleman, S.R. Morse, and F.A. Bazzaz (1992). CO_2 and temperature effects on leaf area production in two annual plant species. *Ecology* 73: 1260-1269.

Adams, C.A. and R.W. Rinne (1980). Moisture content as a controlling factor in seed development and germination. *International Review of Cytology* 68: 1-8.

Adaros, G., H.-J. Weigel, and H.-J. Jäger (1991). Concurrent exposure to SO_2 and/or NO_2 alters growth and yield responses of wheat and barley to low concentrations of O_3. *New Phytologist* 118: 581-591.

Adedipe, N.O., D. Hofstra, and D.P. Ormrod (1972). Effects of sulfur nutrition on phytotoxicity and growth responses of bean plants to ozone. *Phytopathology* 50: 1789-1793.

Agamathu, P. and W.J. Broughton (1986). Factors affecting the development of the rooting system in young oil palms (*Elaesis guineensis* Jacq.). *Agriculture Ecosystems and Environment* 17: 173-179.

Ahmadi, A. and D.A. Baker (2001). The effect of water stress on grain filling processes in wheat. *Journal of Agricultural Science* 136: 257-269.

Ahmed, F.E., A.E. Hall, and D.A. DeMason (1992). Heat injury during floral development in cowpea (*Vigna unguiculata,* Fabaceae). *American Journal of Botany* 79: 784-791.

Ahmed, F.E., A.E. Hall, and M.A. Madore (1993). Interactive effects of high temperature and elevated carbon dioxide concentration on cowpea (*Vigna unguiculata* [L.] Walp.). *Plant, Cell and Environment* 16: 835-842.

Aiken, R.M. and A.J.M. Smucker (1996). Root system regulation of whole plant growth. *Annual Review of Phytopathology* 34: 325-346.

Ainsworth, E.A., P.A. Davey, C.J. Bernacchi, O.C. Dermody, E.A. Heaton, D.J. Moore, P.B. Morgan, S.L. Naidu, H.-S. Yoo Ra, X.-G. Xu, et al. (2002). A meta-analysis of elevated [CO_2] effects on soybean (*Glycine max*) physiology, growth and yield. *Global Change Biology* 8: 695-709.

Ajtay, G.L., P. Ketner, and P. Duvigneaud (1979). Terrestrial primary production and phytomass. In *The Global Carbon Cycle*, eds. B. Bolin, E.T. Degens, S. Kempe, and P. Ketner. Chichester, UK: John Wiley and Sons, pp. 129-181.

Alberto, A.M.P., L.H. Ziska, C.R. Cervancia, and P.A. Manalo (1996). The influence of increasing carbon dioxide and temperature on competitive interactions between a C_3 crop, rice *(Oryza sativa)* and a C_4 weed *(Echinochloa glabrescens)*. *Australian Journal of Plant Physiology* 23: 795-802.

Ali, I.A., U. Kafkafi, Y. Sugimoto, and S. Inanaga (1994). Response of sand-grown tomato supplied with varying ratios of nitrate/ammonium to constant and variable root temperature. *Journal of Plant Nutrition* 17: 2001-2024.

Ali, I.E.A., U. Kafkafi, I. Yamaguchi, Y. Sugimoto, and S. Inanaga (1998). Response of oilseed rape plant to low root temperature and nitrate-ammonium ratios. *Journal of Plant Nutrition* 21: 1463-1481.

Ali, I.E.A., U. Kafkafi, I. Yamaguchi, Y. Sugimoto, and S. Inanaga (2000). Growth, transpiration, root-born cytokinins and gibberellins, and nutrient compositional changes in sesame exposed to low root-zone temperature under different ratios of nitrate:ammonium supply. *Journal of Plant Nutrition* 23: 123-140.

Al-Khatib, K. and G.M. Paulsen (1999). High-temperature effects on photosynthetic processes in temperate and tropical cereals. *Crop Science* 39:119-125.

Allen, L.H. (1990). Plant responses to rising carbon dioxide and potential interactions with air pollutants. *Journal of Environmental Quality* 19: 15-34.

Allen, L.H., Jr. (1992). Free-air CO_2 enrichment field experiments: An historical overview. *Critical Reviews in Plant Sciences* 11: 121-134.

Allen, L.H., J.T. Baker, and K.J. Boote (1996). The CO_2 fertilization effect: Higher carbohydrate production and retention as biomass and seed yield. In *Global Climate Change and Agricultural Production,* eds. F. Bazzaz and W. Sombroek. Rome: Food and Agriculture Organization, pp. 65-100.

Allen, L.H. and K.J. Boote (2002). Crop ecosystem responses to climatic change: Soybean. In *Climate Change and Global Crop Productivity,* eds. K.R. Reddy and H.F. Hodges. Wallingford, UK: CABI Publishing, pp. 133-160.

Allen, L.H., Jr., B.G. Drake, H.H. Rogers, and J.H. Shinn (1992). Field techniques for exposure of plants and ecosystems to elevated CO_2 and other trace gases. *Critical Reviews in Plant Sciences* 11: 85-119.

Aloni, B., T. Pashkar, and L. Karni (1991). Partitioning of [^{14}C]sucrose and acid invertase activity in reproductive organs of pepper plants in relation to their abscission under heat stress. *Annals of Botany* 67: 371-377.

Aloni, B., M. Peet, M. Pharr, and L. Karni (2001). The effect of high temperature and high atmospheric CO_2 on carbohydrate changes in bell pepper *(Capsicum annuum)* pollen in relation to its germination. *Physiologia Plantarum* 112: 505-512.

Amthor, J.S. (1988). Growth and maintenance respiration in leaves of bean (*Phaseolus vulgaris* L.) exposed to ozone in open-top chambers in the field. *New Phytologist* 110: 319-325.

Amthor, J.S. (1989). *Respiration and Crop Productivity.* New York: Springer-Verlag.

Amthor, J.S. (1991). Respiration in a future, higher-CO_2 world. *Plant, Cell and Environment* 14: 13-20.

Amthor, J.S. (1994a). Plant respiratory responses to the environment and their effects on the carbon balance. In *Plant–Environment Interactions,* ed. E Wilkinson. New York: Marcel Dekker, pp. 501-554.

Amthor, J.S. (1994b). Respiration and carbon assimilate use. In *Physiology and Determination of Crop Yield,* eds. K.J. Boote, J.M. Bennett, T.R. Sinclair, and G.M. Paulsen. Madison, WI: American Society of Agronomy, pp. 221-250.

Amthor, J.S. (1997). Plant respiratory responses to elevated carbon dioxide partial pressure. In *Advances in Carbon Dioxide Effects Research,* eds. L.H. Allen Jr., M.B. Kirkham, D.M. Olszyk, and C.E. Whitman. Madison, WI: American Society of Agronomy, pp. 35-77.

Amthor, J.S. (1999). Increasing atmospheric CO_2 concentration, water use, and water stress: Scaling up from the plant to the landscape. In *Carbon Dioxide and Environmental Stress,* eds. Y. Luo and H.A. Mooney. San Diego: Academic Press, pp. 33-59.

Amthor, J.S. (2000a). Direct effect of elevated CO_2 on nocturnal in situ leaf respiration in nine temperate deciduous tree species is small. *Tree Physiology* 20: 139-144.

Amthor, J.S. (2000b). The McCree—de Wit—Penning de Vries—Thornley respiration paradigms: 30 years later. *Annals of Botany* 86: 1-20.

Amthor, J.S. (2001). Effects of atmospheric CO_2 concentration on wheat yield: Review of results from experiments using various approaches to control CO_2 concentration. *Field Crops Research* 37: 1-34.

Amthor, J.S. (2003). Efficiency of lignin biosynthesis: A quantitative analysis. *Annals of Botany* 91: 673-695.

Amthor, J.S. and J.R. Cumming (1988). Low levels of ozone increase bean leaf maintenance respiration. *Canadian Journal of Botany* 66: 724-726.

Amthor, J.S., G.W. Koch, J.R. Willms, and D.B. Layzell (2001). Leaf O_2 uptake in the dark is independent of coincident CO_2 partial pressure. *Journal of Experimental Botany* 52: 2235-2238.

Amthor, J.S. and R.S. Loomis (1996). Integrating knowledge of crop responses to elevated CO_2 and temperature with mechanistic simulation models: Model components and research needs. In *Carbon Dioxide and Terrestrial Ecosystems,* eds. G.W. Koch and H.A. Mooney. San Diego: Academic Press, pp. 317-345.

Amthor, J.S., R.J. Mitchell, G.B. Runion, H.H. Rogers, S.A. Prior, and C.W. Wood (1994). Energy content, construction cost and phytomass accumulation of *Glycine max* (L.) Merr. and *Sorghum bicolor* (L.) Moench grown in elevated CO_2 in the field. *New Phytologist* 128: 443-450.

Amtmann, A. and D. Sanders (1999). Mechanisms of Na^+ uptake by plant cells. *Advances in Botanical Research* 29: 76-112.

Amundson, R.G., R.J. Kohut, A.W. Schoettle, R.M. Raba, and P.B. Reich (1987). Correlative reductions in whole-plant photosynthesis and yield of winter wheat caused by ozone. *Phytopathology* 77: 75-79.

Amundson, R.G., R.M. Raba, A.W. Schoettle, and P.B. Reich (1986). Response of soybean to low concentrations of ozone: II. Effects on growth, biomass allocation, and flowering. *Journal of Environmental Quality* 15: 161-167.

An, L.-Z. and X.-L. Wang (1997). Changes in polyamine contents and arginine decarboxylase activity in wheat leaves exposed to ozone and hydrogen fluoride. *Journal of Plant Physiology* 150: 184-187.

Andalo, C., C. Raquin, N. Machon, B. Godelle, and M. Mousseau (1998). Direct and maternal effects of elevated CO_2 on early root growth of germinating *Arabidopsis thaliana* seedlings. *Annals of Botany* 81: 405-411.

Andersen, C.P. (2003). Source-sink balance and carbon allocation below ground in plants exposed to ozone. *New Phytologist* 157: 213-228.

Anderson, W.B. and W.D. Kemper (1964). Corn growth as affected by aggregate stability, soil temperature, and soil moisture. *Agronomy Journal* 5: 453-456.

Arndt, C.H. (1945). Temperature-growth relations of the root and hypocotyls of cotton seedlings. *Plant Physiology* 20: 200-220.

Arp, W.J. (1991). Effects of source sink relationships on photosynthetic acclimation to elevated carbon dioxide. *Plant, Cell and Environment* 14: 869-876.

Ashmore, M.R. and F.M. Marshall (1999). Ozone impacts on agriculture: An issue of global concern. *Advances in Botanical Research* 29: 31-52.

Ashraf, M. and S. Mehmood (1990). Response of four *Brassica* species to drought stress. *Environmental and Experimental Botany* 30: 93-100.

Aslam, Z., W.D. Jeschke, W.G. Barrett-Lennard, T.L. Setter, E. Watkin, and H. Greenway (1986). Effects of external NaCl on the growth of *Atriplex amnicola* and the ion relations and carbohydrate status of the leaves. *Plant, Cell and Environment* 9: 571-580.

Atkin, O.K., E.J. Edwards, and B.R. Loveys (2000). Response of root respiration to changes in temperature and its relevance to global warming. *New Phytologist* 147: 141-154.

Austin, R.B., J.A. Edrich, M.A. Ford, and R.D. Blackwell (1977). The fate of the dry matter, carbohydrates and ^{14}C lost from leaves and stems of wheat during grain filling. *Annals of Botany* 41: 1309-1321.

Awmack, C.S., R. Harrington, and S.R. Leather (1997). Host plant effects on the performance of the aphid *Aulacorthum solani* (Kalt.) (Homoptera: Aphididae) at ambient and elevated CO_2. *Global Change Biology* 3: 545-549.

Ayers, A.D., J.W. Brown, and C.H. Wadleigh (1952). Salt tolerance of barley and wheat in soil plots receiving several salinization regimes. *Agronomy Journal* 44: 307-310.

Azcón, R. and F. El-Atrash (1997). Influence of arbuscular mycorrhizae and phosphorus fertilization on growth, nodulation and N_2 fixation (^{15}N) in *Medicago sativa* at four salinity levels. *Biology and Fertility of Soils* 24: 81-86.

Azcón-Bieto, J. and C.B. Osmond (1983). Relationship between photosynthesis and respiration. *Plant Physiology* 71: 574-581.

Azevedo, R.A., R.M. Alas, R.J. Smith, and P.J. Lea (1998). Response of antioxidant enzymes to transfer from elevated carbon dioxide to air and ozone fumigation, in the leaves and roots of wild-type and catalase-deficient mutant of barley. *Physiologia Plantarum* 104: 280-292.

Badiani, M., A. D'Annibale, F. Miglietta, and A. Raschi (1993). The antioxidant status of soybean (*Glycine max*) leaves grown under natural CO_2 enrichment in the field. *Australian Journal of Plant Physiology* 20: 275-284.

Baker, D.A. (2000). Long-distance vascular transport of endogenous hormones in plants and their role in source:sink regulation. *Israel Journal of Plant Sciences* 48: 199-203.

Baker, J.T., L.H. Allen, K.J. Boote, and N.B. Pickering (1996). Assessment of rice responses to global climate change: CO_2 and temperature. In *Carbon Dioxide and Terrestrial Ecosystems,* eds. G.W. Koch and H.A. Mooney. San Diego: Academic Press, pp. 265-282.

Baker, J.T., L.H. Allen, K.J. Boote, and N.B. Pickering (2000). Direct effects of atmospheric carbon dioxide concentration on whole canopy dark respiration of rice. *Global Change Biology* 6: 275-286.

Bale, J.S., G.J. Masters, I.D. Hodkinson, C. Awmack, T.M. Bezemer, V.K. Brown, J. Butterfield, A. Buse, J.C. Coulson, J. Farrar, et al. (2002). Herbivory in global climate change research: Direct effects of rising temperature on insect herbivores. *Global Change Biology* 8: 1-16.

Balibrea, M.E., J. Dell-Amico, M.C. Bolarin, and F. Pérez-Alfocea (2000). Carbon partitioning and sucrose metabolism in tomato plants growing under salinity. *Physiologia Plantarum* 110: 503-511.

Bambridge, L., R. MacLeod, and R. Harmer (1995). Plant growth and cell division in *Pisum sativum* L. and *Picea sitchensis* (Bong.) Carr. exposed to ozone. *New Phytologist* 130: 75-80.

Bancal, P. and F. Soltani (2002). Source-sink partitioning. Do we need Munch? *Journal of Experimental Botany* 53: 1919-1928.

Barber, S.A., A.D. Mackay, R.O. Kuchenbuck, and P.B. Barraclough (1989). Effects of soil temperature and water on maize root growth. *Developments in Plant and Soil Science* 36: 231-233.

Barber, S.A. and R.A. Olson (1968). Fertilizer use in corn. In *Changing Patterns in Fertilizer Use,* eds. L.B. Nelson, M.H. McVickar, R.D. Munson, L.F. Seatz, S.L. Tisdale, and W.C. White. Madison, WI: Soil Science Society of America, pp. 165-175.

Barnett, T.P., D.W. Pierce, and R. Schnur (2001). Detection of anthropogenic climate change in the World's oceans. *Science* 292: 270-274.

Barnett, V. (1994). Statistics and the long-term experiments: Past achievements and future challenges. In *Long-Term Experiments in Agricultural and Ecological*

Sciences, eds. R.A. Leigh and A.E. Johnston. Wallingford, UK: CAB International, pp. 165-183.

Barnola, J.M., D. Raynaud, C. Lorius, and N.I. Barkov (1999). Historical CO_2 record from the Vostok ice core. In *Trends: A Compendium of Data on Global Change.* Oak Ridge, TN: Carbon Dioxide Information Analysis Center, Oak Ridge National Laboratory, U.S. Department of Energy, <cdiac.esd.ornl.gov> accessed March 1, 2002.

Bassil, E.S. and S.R. Kaffka (2002). Response of safflower (*Carthamus tinctorius* L.) to saline soils and irrigation. I. Consumptive water use. *Agricultural Water Management* 54: 67-80.

BassiriRad, H. (2000). Kinetics of nutrient uptake by roots: Responses to global change. *New Phytologist* 147: 155-169.

BassiriRad, H., M.M. Caldwell, and C. Bilbrough (1993). Effects of soil temperature and nitrogen status on kinetics of NO_3^- uptake by roots of field grown *Agropyron desertorum* (Fisch. Ex link) Schult. *New Phytologist* 123: 485-489.

BassiriRad, H., S.A. Prior, R.J. Norby, and H.H. Rogers (1999). A field method of determining NH_4^+ and NO_3^- uptake kinetics in intact roots: Effects of CO_2 enrichment on trees and crop species. *Plant and Soil* 217: 195-204.

BassiriRad, H., J.W. Radin, and K. Matsuda (1991). Temperature-dependent water and ion transport properties of barley and sorghum roots. I. Relationship to leaf growth. *Plant Physiology* 97: 426-432.

BassiriRad, H., D.T. Tissue, J.F. Reynolds, and F.S. Chapin III (1996). Response of *Eriophorum vaginatum* to CO_2 enrichment at different soil temperatures: Effects on growth, root respiration and PO_4^{3-} uptake kinetics. *New Phytologist* 133: 423-430.

Batts, G.R., R.H. Ellis, J.I.L. Morison, P.N. Nkemka, P.J. Gregory, and P. Hadley (1998). Yield and partitioning in crops of contrasting cultivars of winter wheat in response to CO_2 and temperature in field studies using temperature gradient tunnels. *Journal of Agricultural Science* 130: 17-27.

Batts, G.R., J.I.L. Morison, R.H. Ellis, P. Hadley, and T.R. Wheeler (1997). Effects of CO_2 and temperature on growth and yield of crops of winter wheat over four seasons. *European Journal of Agronomy* 7: 43-52.

Baxter, R. and J.F. Farrar (1999). Export of carbon from leaf blades of *Poa alpina* L. at elevated CO_2 and two nutrient regimes. *Journal of Experimental Botany* 50: 1215-1221.

Beerling, D.J. and F.I. Woodward (1995). Stomatal responses of variegated leaves to CO_2 enrichment. *Annals of Botany* 75: 507-511.

Benavídes, M.P., P.L. Marconi, S.M. Gallego, M.E. Comba, and M.L. Tomaro (2000). Relationship between antioxidant defence systems and salt tolerance in *Solanum tuberosum. Australian Journal of Plant Physiology* 27: 273-278.

Bergweiler, C.J. and W.J. Manning (1999). Inhibition of flowering and reproductive success in spreading dogbane *(Apocynum androsaemifolium)* by exposure to ambient ozone. *Environmental Pollution* 105: 333-339.

Bernstein, L., L.E. Francois, and R.A. Clark (1974). Interactive effects of salinity and fertility on yields of grains and vegetables. *Agronomy Journal* 66: 412-421.

Bernstein, N., W.K. Silk, and A. Läuchli (1993). Growth and development of sorghum leaves under conditions of NaCl stress: Spatial and temporal aspects of leaf growth inhibition. *Planta* 191: 433-439.

Berry, J. and O. Björkman (1980). Photosynthetic response and adaptation to temperature in higher plants. *Annual Review of Plant Physiology* 31: 491-543.

Berry, J.A. and J.S. Downton (1982). Environmental regulation of photosynthesis. In *Photosynthesis: Development, Carbon Metabolism and Plant Productivity,* Volume II, ed. Govingee. New York: Academic Press, pp. 263-343.

Bethlenfalvay, G.J. (1992). Mycorrhizae and crop productivity. In *Mycorrhizae in Sustainable Agriculture,* eds. G.J. Bethlenfalvay and R.G. Linderman. Madison, WI: American Society of Agronomy, pp. 1-27.

Beveridge, C. (2000). The ups and downs of signaling between root and shoot. *New Phytologist* 147: 413-416.

Bezemer, T.M. and T.H. Jones (1998). Plant-insect herbivore interactions in elevated atmospheric CO_2: Quantitative analyses and guild effects. *Oikos* 82: 212-222.

Bindi, M., L. Fibbi, A. Frabotta, M. Chiesi, G. Selvaggi, and V. Magliulo (1999). Free air CO_2 enrichment of potato (*Solanum tuberosum* L.). In *Annual Report for Changing Climate and Potential Impacts on Potato Yield and Quality (CHIP), Contract ENV4-CT97-0489.* Brussels, Belgium: Commission of the European Union, pp. 160-196.

Bindi, M., L. Fibbi, A. Frabotta, M.G. Ottaviani, and V. Magliulo (1998). Free air CO_2 enrichment of potato (*Solanum tuberosum* L.). In *Annual Report for Changing Climate and Potential Impacts on Potato Yield and Quality (CHIP), Contract ENV4-CT97-0489.* Brussels, Belgium: Commission of the European Union, pp. 133-163.

Bindi, M., L. Fibbi, M. Lanini, and F. Miglietta (2001). Free air CO_2 enrichment (FACE) of grapevine (*Vitis vinifera* L.): I. Development and testing of the system for CO_2 enrichment. *European Journal of Agronomy* 14: 135-143.

Biscoe, P.V., R.K. Scott, and J.L. Monteith (1975). Barley and its environment. III. Carbon budget of the stand. *Journal of Applied Ecology* 12: 269-293.

Biswas, P.K., D.R. Hileman, P.P. Ghosh, N.C. Bhattacharya, and J.N. McCrimmon (1996). Growth and yield responses of field-grown sweet potato to elevated carbon dioxide. *Crop Science* 36: 1234-1239.

Black, V.J., C.R. Black, J.A. Roberts, and C.A. Stewart (2000). Impact of ozone on the reproductive development of plants. *New Phytologist* 147: 421-447.

Blackwell, P.S. and E.A. Wells (1983). Limiting oxygen flux densities for oat root extension. *Plant and Soil* 73: 129-139.

Blaker, N.S. and J.D. MacDonald (1985). Effect of soil salinity on the formation of sporangia and zoospores by three isolates of Phytophthora. *Phytopathology* 75: 270-274.

Blum, A., N. Klueva, and H.T. Nguyen (2001). Wheat cellular thermotolerance is related to yield under heat stress. *Euphytica* 117: 117-123.

Blum, U. and D.T. Tingey (1977). A study of the potential ways in which ozone could reduce root growth and nodulation of soybean. *Atmospheric Environment* 11: 737-739.

Blumenthal, C., P.J. Stone, P.W. Gras, F. Bekes, B. Clarke, E.W.R. Barlow, R. Appels, and C.W. Wrigley (1998). Heat-shock protein 70 and dough-quality changes resulting from heat stress during grain filling in wheat. *Cereal Chemistry* 75: 43-50.

Blumwald, E., G.S. Aharon, and M.P. Apse (2000). Sodium transport in plant cells. *Biochimica et Biophysica Acta* 1465: 140-151.

Bormann, F.H. and G.E. Likens (1979). *Pattern and Process in a Forested Ecosystem*. New York: Springer-Verlag.

Bosac, C., V.J. Black, J.A. Roberts, and C.R. Black (1998). Impact of ozone on seed yield and quality and seedling vigour in oilseed rape (*Brassica napus* L.). *Journal of Plant Physiology* 153: 127-134.

Botella, M.A., A.C. Cerdá, and S.H. Lips (1993). Dry matter production, yield, and allocation of carbon-14 assimilates by wheat as affected by nitrogen source and salinity. *Agronomy Journal* 85: 1044-1049.

Bouchereau, A., A. Aziz, F. Larher, and J. Martin-Tanguy (1999). Polyamines and environmental challenges: Recent development. *Plant Science* 140: 103-125.

Bouma, T.J., R. De Visser, P.H. Van Leeuwen, M.J. De Kock, and H. Lambers (1995). The respiratory energy requirements involved in nocturnal carbohydrate export from starch-storing mature source leaves and their contribution to leaf dark respiration. *Journal of Experimental Botany* 46: 1185-1194.

Bouma, T.J., K.L. Nielsen, D.M. Eissenstat, and J.P. Lynch (1997). Soil CO_2 concentration does not affect growth or root respiration in bean or citrus. *Plant, Cell and Environment* 20: 1495-1505.

Bowen, G.D. (1991). Soil temperature, root growth, and plant function. In *Plant roots: The hidden half,* eds. Y. Waisel, A. Eshel, and U. Kafkafi. New York: Marcel Dekker, pp. 309-330.

Bowes, G. (1991). Growth at elevated CO_2: Photosynthetic responses mediated through Rubisco. *Plant, Cell and Environment* 14: 795-806.

Bowes, G. (1993). Facing the inevitable: Plants and increasing atmospheric CO_2. *Annual Review of Plant Physiology and Plant Molecular Biology* 44: 309-332.

Bowler, C. and R. Fluhr (2000). The molecular basis of cross tolerance. *Trends in Plant Sciences* 5: 241-246.

Bowyer, J.R. and R.C. Leegood (1997). Photosynthesis. In *Plant Biochemistry*, eds. P.M. Dey and J.B. Harborne. San Diego: Academic Press, pp. 49-110.

Boyer, J.S. (1968). Relationship of water potential to growth of leaves. *Plant Physiology* 43: 1056-1062.

Bradshaw, W.E. and C.M. Holzapfel (2001). Genetic shift in photoperiodic response correlated with global warming. *Proceedings of the National Academy of Sciences USA* 98: 14509-14511.

Bravo-F, P. and E.G., Uribe (1981). Temperature dependence on concentration kinetics of absorption of phosphate and potassium in corn roots. *Plant Physiology* 67: 815-819.

Brennan, J.P. and K. Quade (2000). Longer-term changes in Australian wheat yields. *Agricultural Science* NS 13(3): 37-41.

Bressan, R.A., P.M. Hasegawa, and J.M. Pardo (1998). Plants use calcium to resolve salt stress. *Trends in Plant Science* 3: 411-412.

Brewer, L.K., F.B. Guillemet, and R.K. Creveling (1961). Influence of N-P-K fertilization on incidence and severity of oxidant injury to mangels and spinach. *Soil Science* 92: 298-301.

Broecker, W.S. and T.-H. Peng (1982). *Tracers in the Sea*. Palisades, NY: Eldigio Press, Lamont-Doherty Geological Observatory of Columbia University.

Brooks, A. and G.D. Farquhar (1985). Effect of temperature on the CO_2/O_2 specificity of ribulose-1,5-bisphosphate carboxylase/oxygenase and the rate of respiration in the light. *Planta* 165: 397-406.

Brooks, T.J., G.W. Wall, P.J. Pinter, B.A. Kimball, R.L. LaMorte, S.W. Leavitt, A.D. Matthias, F.J. Adamsen, D.J. Hunsaker, and N.N. Webber (2001). Acclimation response of spring wheat in a free-air CO_2 enrichment (FACE) atmosphere with variable soil nitrogen regimes. 3. Canopy architecture and gas exchange. *Photosysnthesis Research* 66: 97-108.

Brouquisse, R., F. James, P. Raymond, and A. Pradet (1991). Study of glucose starvation in excised maize root tips. *Plant Physiology* 96: 619-626.

Brown, V.C., S. McNeill, and M.R. Ashmore (1992). The effects of ozone fumigation on the performance of the black bean aphid, *Aphis fabae* Scop., feeding on broad beans, *Vicia faba* L. *Agriculture, Ecosystems and Environment* 38: 71-78.

Brüggemann, L.I., I.I. Pottosin, and G. Schönknecht (1998). Cytoplasmic polyamines block the fast-activating vacuolar cation channel. *The Plant Journal* 16: 101-105.

Brugnoli, E. and O. Björkman (1992). Growth of cotton under continuous salinity stress: Influence on allocation pattern, stomatal and non-stomatal components of photosynthesis and dissipation of excess light energy. *Planta* 187: 335-347.

Bruhn, D., T.N. Mikkelsen, and O.K. Atkin (2002). Does the direct effect of atmospheric CO_2 concentration on leaf respiration vary with temperature? Responses in two species of *Plantago* that differ in relative growth rate. *Physiologia Plantarum* 114: 57-64.

Bunce, J.A. (1990). Short- and long-term inhibition of respiratory carbon dioxide efflux by elevated carbon dioxide. *Annals of Botany* 65: 637-642.

Bunce, J.A. (1993). Growth, survival, competition, and canopy carbon dioxide and water vapor exchange of first year alfalfa at an elevated CO_2 concentration. *Photosynthetica* 29: 557-565.

Bunce, J.A. (1995a). Effects of elevated carbon dioxide concentration in the dark on the growth of soybean seedlings. *Annals of Botany* 75: 365-368.

Bunce, J.A. (1995b). Long-term growth of alfalfa and orchard grass plots at elevated carbon dioxide. *Journal of Biogeography* 22: 341-348.

Bunce, J.A. (1996). Growth at elevated carbon dioxide concentration reduces hydraulic conductance in alfalfa and soybean. *Global Change Biology* 2: 155-158.

Bunce, J.A. (2002). Carbon dioxide concentration at night affects translocation from soybean leaves. *Annals of Botany* 90: 399-403.

Bunce, J.A. and L.H. Ziska (1996). Responses of respiration to increases in carbon dioxide concentration and temperature in three soybean cultivars. *Annals of Botany* 77: 507-514.

Bunce, J.A. and L.H. Ziska (2000). Crop ecosystem responses to climatic change: Crop/weed interactions. In *Climate Change and Global Crop Productivity,* eds. K.R. Reddy and H.F. Hodges. Wallingford, UK: CABI Publishing, pp. 333-352.

Buringh, P. and R. Dudal (1987). Agricultural land use in space and time. In *Land Transformation in Agriculture,* eds. M.G. Wolman and F.G.A. Fournier. Chichester, UK: John Wiley and Sons, pp. 9-43.

Burke, J.J. (1990). High temperature stress and adaptation in crops. In *Stress Responses in Plants: Adaptation and Acclimation Mechanisms,* eds. R.G. Alscher and J.R. Cumming. New York: Wiley-Liss, pp. 295-309.

Burstrom, H. (1956). Temperature and root cell elongation. *Physiologia Plantarum* 9: 682-692.

Burton, A.J. and K.S. Pregitzer (2002). Measurement carbon dioxide concentration does not affect root respiration of nine tree species in the field. *Tree Physiology* 22: 67-72.

Butzen, S. and M. Cummings (1999). Corn grain protein: Understanding the nutritional profile of corn grain and the effects of management and growing conditions on nutritional quality. Part 2: Kernel characteristics and genetic effects in relation to protein content. *Crop Insights* 9(11), Pioneer Hi-Bred International, Inc. <www.pioneer.com>, accessed May 22, 2002.

Buyanovsky, G.A. and G.H. Wagner (1998). Changing role of cultivated land in the global carbon cycle. *Biology and Fertility of Soils* 27: 242-245.

Byrd, G.T. (1992). Dark respiration in C_3 and C_4 species. PhD Dissertation. Athens, GA: University of Georgia.

Cakmak, I. and C. Engels (1999). Role of mineral nutrients in photosynthesis and yield formation. In *Mineral Nutrition of Crops,* ed. Z. Rengel. Binghamton, NY: Food Products Press, pp. 141-169.

Calderini, D.F. and G.A. Slafer (1998). Changes in yield and yield stability in wheat during the 20th century. *Field Crops Research* 57: 335-347.

Caldwell, M.M., C.L. Ballare, J.F. Bornman, S.D. Flint, L.O. Bjorn, A.H. Teramura, G. Kulandaivelu, and M. Tevini (2003). Terrestrial ecosystems increased solar ultraviolet radiation and interactions with other climatic change factors. *Photochemical and Photobiological Sciences* 2: 29-38.

Caldwell, M.M., L.O. Björn, J.F. Bornman, S.D. Flint, G. Kulandaivelu, A.H. Teramura, and M. Tevini (1998). Effects of increased solar ultraviolet radiation on terrestrial ecosystems. *Journal of Photochemistry and Photobiology B: Biology* 46: 40-52.

Caldwell, M.M. and S.D. Flint (1994). Solar ultraviolet radiation and ozone layer change: Implications for crop plants. In *Physiology and Determination of Crop Yield*, eds. K.J. Boote, J.M. Bennett, T.R. Sinclair, and G.M. Paulsen. Madison, WI: American Society of Agronomy, pp. 487-507.

Cannon, R.J.C. (1998). The implications of predicted climate change for insect pests in the UK, with emphasis on non-indigenous species. *Global Change Biology* 4: 785-796.

Carrasco-Rodriguez, J.L. and S. del Valle-Tascon (2001). Impact of elevated ozone on chlorophyll *a* fluorescence in field-grown oat *(Avena sativa). Environmental and Experimental Botany* 45: 133-142.

Carvajal, M., D.T. Cooke, and D.T. Clarkson (1996). Plasma membrane fluidity and hydraulic conductance in wheat roots: Interactions between root temperature and nitrate or phosphate deprivation. *Plant, Cell and Environment* 19: 1110-1114.

Casperson, J.P., S.W. Pacala, J.C. Jenkins, G.C. Hurtt, P.R. Moorcroft, and R.A. Birdsey (2000). Contributions of land-use history to carbon accumulation in U.S. forests. *Science* 290: 1148-1151.

Chameides, W.L., P.S. Kasibhatia, J. Yienger, and H. Levy II (1994). Growth of continental-scale metro-agro-plexes, regional ozone pollution, and world food production. *Science* 264: 74-77.

Chang, P.-F.L., B. Damsz, A.K. Kononowicz, M. Reuveni, Z. Chen, Y. Xu, K. Hedges, C.C. Tseng, N.K. Singh, M.L. Binzel, et al. (1996). Alterations in cell membrane structure and expression of a membrane-associated protein after adaptation to osmotic stress. *Physiologia Plantarum* 98: 505-516.

Chapin, F.S. (1974). Morphological and physiological mechanisms of temperature compensation in phosphate absorption along a latitudinal gradient. *Ecology* 55: 1180-1198.

Chapman, H.W., L.S. Gleason, and W.E. Loomis (1954). The carbon dioxide content of field air. *Plant Physiology* 29: 500-503.

Chaudhuri, U.N., R.B. Burnett, M.B. Kirkham, and E.T. Kanemasu (1986). Effect of carbon dioxide on sorghum yield, root growth, and water use. *Agricultural and Forest Meteorology* 37: 109-122.

Chaudhuri, U.N., M.B. Kirkham, and E.T. Kanemasu (1990). Root growth of winter wheat under elevated carbon dioxide and drought. *Crop Science* 30: 853-857.

Cheikh, N. and R.J. Jones (1994). Disruption of maize kernel growth and development by heat stress. *Plant Physiology* 106: 45-51.

Chernikova, T., M.J. Robinson, E.H. Lee, and C.L. Mulchi (2000). Ozone tolerance and antioxidant enzyme activity in soybean cultivars. *Photosynthesis Research* 64: 15-26.

Cherry, J.H., Heuss-LaRosa, K., and Mayer, R.R. (1989). Adaptation of thermotolerance in cowpea suspension cultures. In *Environmental Stress in Plants,* ed. J.H. Cherry. NATO ASI Series. Series G: Ecological Sciences, Volume 19. Berlin: Springer Verlag, pp. 355-369.

Chiang, H. and D.M. Norris (1983). Phenolic and tannin contents as related to anatomical parameters of soybean resistance to agromyzid bean flies. *Journal of Agricultural and Food Chemistry* 31: 726-730.

Chiou, T.-J. and D.R. Bush (1998). Sucrose is a signal molecule in assimilate partitioning. *Proceedings of the National Academy of Science USA* 95: 4784-4788.

Chmielewski, F.-M. and J.M. Potts (1995). The relationship between crop yields from an experiment in southern England and long-term climate variations. *Agricultural and Forest Meteorology* 73: 43-66.

Chowdhury, S.I. and I.F. Wardlaw (1978). The effect of temperature on kernel development in cereals. *Australian Journal of Agricultural Research* 29: 205-223.

Clarkson, D.T. (1977). Membrane structure and transport. In *The Molecular Biology of Plant Cells,* ed. H. Smith. Oxford: Blackwell, pp. 24-63.

Clarkson, D.T., M.J. Earnshaw, P.J. White, and H.D. Cooper (1988). Temperature dependent factors influencing nutrient uptake: An analysis of responses at different levels of organization. In *Plants and Temperature,* Volume 42, eds. S.P. Long and F.I. Woodward. Cambridge, UK: Society for Experimental Biology, pp. 281-330.

Clarkson, D.T., M.J. Hopper, and L.H.P. Jones (1986). The effect of root temperature on the uptake of nitrogen and the relative size of the root system in *Lolium perenne* L. 1. Solutions containing both NH_4 and NO_3. *Plant, Cell and Environment* 9: 535-545.

Clarkson, D.T. and A.J. Warner (1979). Relationships between root temperature and transport of ammonium and nitrate ions by Italian and perennial ryegrass (*Lolium multiflorum* and *Lolium perenne*). *Plant Physiology* 64: 557-561.

Clayton, H., M.R. Knight, H. Knight, M.R. McAinsh, and A.M. Hetherington (1999). Dissection of the ozone-induced calcium signature. *Plant Journal* 17: 575-579.

Cleland, R.E. (1980). Auxin and H^+-excretion: The state of our knowledge. In *Plant Growth Substances,* ed. F. Scoog. New York: Academic Press, pp. 71-78.

Clowes, F.A.L. and R. Wadekar (1988). Modelling of the root cap of Zea mays L. in relation to temperature. *New Phytologist* 108: 259-262.

Cohen, J.E. (1995). *How Many People Can the Earth Support?* New York and London: W.W. Norton and Company.

Colby, W.H., F.W. Crook, and S.-E.H. Webb (1992). *Agricultural Statistics of the People's Republic of China, 1949-90.* Washington, DC: U.S. Department of Agriculture, Agriculture and Trade Analysis Division, Economic Research Service, Statistical Bulletin No. 844.

Collier, R.H., S. Finch, K. Phelps, and A.R. Thompson (1991). Possible impact of global warming on cabbage root fly *(Delia radicum)* activity in the UK. *Annals of Applied Biology* 118: 261-271.

Conley, M.M., B.A. Kimball, T.J. Brooks, P.J. Pinter Jr., D.J. Hunsaker, G.W. Wall, N.R. Adam, R.L. LaMorte, A.D. Matthias, T.L. Thompson, et al. (2001). CO_2 enrichment increases water-use efficiency in sorghum. *New Phytologist* 151: 407-412.

Conroy, J.P., S. Seneweera, A.S. Basra, G. Rogers, and B. Nissen-Wooller (1994). Influence of rising atmospheric CO_2 concentrations and temperature on growth, yield and grain quality of cereal crops. *Australian Journal of Plant Physiology* 21: 741-758.

Conway, T.J. and P.P. Tans (1996). *Atmospheric Carbon Dioxide Mixing Ratios from the NOAA Climate Monitoring and Diagnostics Laboratory Cooperative Flask Sampling Network, 1997-1993.* ORNL/CDIAC-73, NDP-005/R3. Oak Ridge, TN: Carbon Dioxide Information Analysis Center, Oak Ridge National Laboratory, U.S. Department of Energy.

Cooley, D.R. and W.J. Manning (1987). The impact of ozone on assimilate partitioning in plants: A review. *Environmental Pollution* 47: 95-113.

Cornic, G. and P.G. Jarvis (1972). Effects of oxygen on CO_2 exchange and stomatal resistance in Sitka spruce and maize at low irradiances. *Photosynthetica* 6: 225-239.

Cosgrove, D.J. (1993). Wall extensibility: Its nature, measurement and relationship to plant cell growth. *New Phytologist* 124: 1-23.

Cosgrove, D.J. (1997). Relaxation in a high-stress environment: The molecular basis of extensible cell walls and cell enlargement. *Plant Cell* 9: 1031-1041.

Costa, S.D., G.G. Kennedy, and A.S. Heagle (2001). Effect of host plant ozone stress on Colorado potato beetles. *Physiological and Chemical Ecology* 30: 824-831.

Cotrufo, M.F., P. Ineson, and A. Scott (1998). Elevated CO_2 reduces the nitrogen concentration of plant tissues. *Global Change Biology* 4: 43-54.

Council for Agricultural Science and Technology (1999). *Animal Agriculture and Global Food Supply.* Report No. 135. Ames, IA: Council for Agricultural Science and Technology.

Cousins, A.B., N.R. Adam, G.W. Wall, B.A. Kimball, P.J. Pinter Jr., S.W. Leavitt, R.L. LaMorte, A.D. Matthias, M.J. Ottman, T.L. Thompson, and A.N. Webber (2001). Reduced photorespiration and increased energy-use efficiency in young CO_2-enriched sorghum leaves. *New Phytologist* 150: 275-284.

Covell, S., R.H. Ellis, E.H. Roberts, and R.J. Summerfield (1986). The influence of temperature on seed germination rate in grain legumes. I. A comparison of chickpea, lentil, soyabean and cowpea at constant temperatures. *Journal of Experimental Botany* 37: 705-715.

Cowling, D.W. and M.J. Koziol (1982). Mineral nutrition and plant response to air pollutants. In *Effects of Gaseous Air Pollution in Agriculture and Horticulture,* eds. M.H. Unsworth and D.P. Ormrod. London: Butterworths, pp. 349-375.

Cox, P.M., R.A. Betts, C.D. Jones, S.A. Spall, and I.J. Totterdell (2000). Acceleration of global warming due to carbon-cycle feedbacks in a coupled climate model. *Nature* 408: 184-187.

Cramer, G.R. and D.C. Bowman (1992). Kinetics of maize leaf elongation. 1. Response of a sodium excluding cultivar and a Na-including cultivar to varying Na/Ca salinities. *Journal of Experimental Botany* 43: 857-864.

Cramer, G.R., A. Läuchli, and V.S. Polito (1985). Displacement of Ca^{2+} by Na^{+} from the plasmalemma of root cells. A primary response to salt stress? *Plant Physiology* 79: 207-211.

Cramer, G.R., C.L. Schmidt, and C. Bidart (2001). Analysis of cell wall hardening and cell wall enzymes of salt-stressed maize *(Zea mays)* leaves. *Australian Journal of Plant Physiology* 28: 101-109.

Crawford, R.M.M. and D.W. Wolfe (1999). Temperature: Cellular to whole plant and population responses. In *Carbon Dioxide and Environmental Stress,* eds. Y. Luo and H.A. Mooney. San Diego: Academic Press, pp. 61-106.

Cruz, C., S.H. Lips, and M.A. Martins-Loucâo (1993). The effect of nitrogen source on photosynthesis of carob at high CO_2 concentrations. *Physiologia Plantarum* 89: 552-556.

Cubasch, U., G.A. Meehl, G.J. Boer, R.J. Stouffer, M. Dix, A. Noda, C.A. Senior, S. Raper, K.S. Yap, et al. (2001). Projections of future climate change. In *Climate Change 2001: The Scientific Basis,* eds. J.T. Houghton, Y. Ding, D.J. Griggs, M. Noguer, P.J. van der Linden, X. Dai, K. Maskell, and C.A. Johnson. Cambridge, UK: Cambridge University Press, pp. 525-582.

Cumbus, I.P. and P.H. Nye (1982). Root zone temperature effects on growth and nitrate absorption in rape (*Brassica napus* cv. Emerald). *Journal of Experimental Botany* 33: 1138-1146.

Cumbus, I.P. and P.H. Nye (1985). Root zone temperature effects on growth and phosphate absorption in rape, *Brassica napus* cv. Emerald. *Journal of Experimental Botany* 36: 219-227.

Daepp, M., D. Suter, J.P.F. Almeida, H. Isopp, U. Hartwig, M. Frehner, H. Blum, J. Nösberger, and A. Lüscher (2000). Yield response of *Lolium perenne* swards to free-air CO_2 enrichment increased over six years in a high N input system on fertile soil. *Global Change Biology* 6: 805-816.

Dale, J.E. (1988). The control of leaf expansion. *Annual Review of Plant Physiology and Plant Molecular Biology* 39: 267-295.

Darall, N.M. (1989). The effect of air pollutants on physiological processes in plants. *Plant, Cell and Environment* 12: 1-30.

Davey, P.A., S. Hunt, G.L. Hymus, E.H. DeLucia, B.G. Drake, D.F. Karnosky, and S.P. Long (2004). Respiratory oxygen uptake is not decreased by an instantaneous elevation of [CO_2], but is increased by long-term growth in the field at elevated [CO_2]. *Plant Physiology* 134: 520-527.

Dekov, I., T. Tsonev, and I. Yordanov (2000). Effects of water stress and high-temperature stress on the structure and activity of photosynthetic apparatus of *Zea mays* and *Helianthus annuus*. *Photosynthetica* 38: 361-366.

Demmers-Derks, H., R.A.C. Mitchell, V.J. Mitchell, and D.W. Lawlor (1998). Response of sugarbeet (*Beta vulgaris* L.) yield and biochemical composition to elevated CO_2 and temperature at two nitrogen applications. *Plant, Cell and Environment* 21: 829-836.

Dhingra, H.R. and T.M. Varghese (1986). Effect of NaCl salinity on the activities of amylase and invertase in *Zea mays* L. pollen. *Annals of Botany* 57: 101-104.

Dhingra, H.R. and T.M. Varghese (1997). Flowering and sexual reproduction under salt stress. In *Strategies for Improving Salt Tolerance in Higher Plants,* eds. P.K. Jaiwal, R.P Singh, and A. Gulati. New Hampshire: Science Publishers, Inc., pp. 221-245.

Dijkstra, P., S. Nonhebel, C. Grashoff, J. Goudriaan, and S.C. van de Geijn (1996). Response of growth and CO_2 uptake of spring wheat and faba bean to CO_2 concentration under semifield conditions: Comparing results of field experiments and simulations. In *Carbon Dioxide and Terrestrial Ecosystems,* eds. G.W. Koch and H.A. Mooney. San Diego: Academic Press, pp. 251-264.

Dijkstra, P., A.H.M.C. Schapendonk, K. Groenwold, M. Jansen, and S.C. van de Geijn (1999). Seasonal changes in the response of winter wheat to elevated atmospheric CO_2 concentration grown in open-top chambers and field tracking enclosures. *Global Change Biology* 5: 563-576.

Dionisio-Sese, M.L. and S. Tobita (1998). Antioxidant response of rice seedlings to salinity stress. *Plant Science* 135: 1-9.

Dizengremel, P. (2001). Effects of ozone on the carbon metabolism of forest trees. *Plant Physiology and Biochemistry* 39: 729-742.

Doerner, P., J.E. Jorgensen, R. You, J. Steppuhn, and C. Lamb (1996). Control of root growth and development by cyclin expression. *Nature* 380: 520-523.

Donald, C.M. (1962). In search of yield. *Journal of the Australian Institute of Agricultural Science* 28: 171-178.

Donald, C.M. and J. Hamblin (1976). The biological yield and harvest index of cereals as agronomic and plant breeding criteria. *Advances in Agronomy* 28: 361-405.

Donnelly, A., J. Craigon, C.R. Black, J.J. Colls, and G. Landon (2001a). Does elevated CO_2 ameliorate the impact of O_3 on chlorophyll content and photosynthesis in potato *(Solanum tuberosum)*. *Physiologia Plantarum* 111: 501-511.

Donnelly, A., J. Craigon, C.R. Black, J.J. Colls, and G. Landon (2001b). Elevated CO_2 increases biomass and tuber yield in potato even at high ozone concentrations. *New Phytologist* 149: 265-274.

Dornbos, D.L. and R.E. Mullen (1992). Soybean seed protein and oil contents and fatty acid composition adjustments by drought and temperature. *Journal of the American Oil Chemists' Society* 69: 228-231.

Dow el-madina, I.M. and A.E. Hall (1986). Flowering of contrasting cowpea (*Vigna unguiculata* [L.] Walp.) genotypes under different temperatures and photoperiods. *Field Crops Research* 14: 87-104.

Drake, B.G., M.A. Gonzàlez-Meler, and S.P. Long (1997). More efficient plants: A consequence of rising atmospheric CO_2? *Annual Review of Plant Physiology and Molecular Biology* 48: 609-639.

Dupuis, I. and C. Dumas (1990). Influence of temperature stress on in vitro fertilization and heat shock protein synthesis in maize (*Zea mays* L.) reproductive tissues. *Plant Physiology* 94: 665-670.

Duvick, D.N. and K.G. Cassman (1999). Post-Green Revolution trends in yield potential of temperate mazie in the North-Central United States. *Crop Science* 39: 1622-1630.

Eastin, J.D. (1983). Sorghum. In *Potential Productivity of Field Crops Under Different Environments,* eds. W.H. Smith and S.J. Banta. Los Baños, Laguna, Philippines: International Rice Research Institute, pp. 181-204.

Easterling, D.R., B. Horton, P.D. Jones, T.C. Peterson, T.R. Karl, D.E. Parker, M.J. Salinger, V. Razuvayev, N. Plummer, P. Jamason, and C.K. Folland (1997). Maximum and minimum temperature trends for the globe. *Science* 277: 364-367.

Easterling, D.R., T.R. Karl, K.P. Gallo, D.A. Robinson, K.E. Trenberth, and A. Dai (2000). Observed climate variability and change of relevance to the biosphere. *Journal of Geophysical Research* 105: 20101-20114. <lwf.ncdc.noaa.gov/oa/climate/research/anomalies/annual_land.ts> accessed February 7, 2003.

Easterling, D.R., G.A. Meehl, C. Parmesan, S.A. Changnon, T.R. Karl, and L.O. Mearns (2000). Climate extremes: Observations, modeling, and impacts. *Science* 289: 2068-2074.

Edreva, A., I. Yordanov, R. Kardjieva, and E. Gesheva (1998). Heat shock responses of bean plants: Involvement of free radicals, antioxidants and free radical/active oxygen scavenging systems. *Biologia Plantarum* 41: 185-191.

Edwards, N.T. (1991). Root and soil respiration responses to ozone in *Pinus taeda* L. seedlings. *New Phytologist* 118: 315-321.

Egli, D.B. (1998). *Seed Biology and the Yield of Grain Crops.* New York: CAB International.

Egli, D.B. and D.M. TeKrony (1997). Species differences in seed water status during seed maturation and germination. *Seed Science Research* 7: 3-11.

Ehleringer, J.R. (1981). Leaf absorptances of Mojave and Sonoran Desert plants. *Oecologia* 49: 366-370.

Ehleringer, J.R. and O. Björkman (1977). Quantum yields for CO_2 uptake in C_3 and C_4 plants. Dependence on temperature, CO_2, and O_2 concentration. *Plant Physiology* 59: 86-90.

Ehleringer, J.R. and O. Björkman (1978). Pubescence and leaf spectral characteristics of a desert shrub *Encelia frinosa*. *Oecologia* 36: 151-162.

Eissenstat, D.M. (1992). Costs and benefits of constructing roots of small diameter. *Journal of Plant Nutrition* 15: 763-782.

Eissenstat, D.M., C.E. Wells, R.D. Yanai, and J.L. Whitbeck (2000). Building roots in a changing environment: Implications for root longevity. *New Phytologist* 147: 33-42.

Elkiey, T. and D.P. Ormrod (1980). Sorption of ozone and sulphur dioxide by petunia leaves. *Journal of Environmental Quality* 9: 93-95.

Ellis, R.H., S. Covell, R.H. Roberts, and R.J. Summerfield (1986). The influence of temperature on seed germination rate in grain legumes. II. Intraspecific variation in chickpea (*Cicer arietinum* L.) at constant temperatures. *Journal of Experimental Botany* 37: 1503-1515.

Ellis, R.H., G. Simon, and S. Covell (1987). The influence of temperature on seed germination rate in grain legumes. *Journal of Experimental Botany* 38: 1033-1043.

Elsheikh, E.A.E. (1998). Effects of salt on rhizobia and bradyrhizobia: A review. *Annals of Applied Biology* 132: 507-524.

Endress, A.G. and S.L. Post (1985). Altered feeding preference of Mexican bean beetle *Epilachna varivestis* for ozonated soybean foliage. *Environmental Pollution* 39: 9-16.

Engles, C., L. Munkle, and H. Marschner (1992). Effect of root zone temperature and shoot demand on uptake and xylem transport of macronutrients in maize (*Zea Mays* L.). *Journal of Experimental Botany* 43: 537-547.

Enoch, H.Z. and S.J. Honour (1993). Significance of increasing ambient CO_2 for plant growth and survival, and interactions with air pollutants. In *Interacting Stresses on Plants in a Changing Climate*, eds. M.B. Jackson and C.R. Black. Berlin: Springer-Verlag, pp. 51-75.

Enoch, H.Z. and Zieslin, N. (1988). Growth and development of plants in response to carbon dioxide concentrations. *Applied Agricultural Research* 3: 248-256.

Enstone, D.E. and D.E. Peterson (1992). The apoplastic permeability of root apices. *Canadian Journal of Botany* 70: 1502-1512.

Epstein, E. (1998). How calcium enhances plant salt tolerance. *Science* 280: 1906-1907.

Epstein, E. and C.E. Hagen (1952). A kinetic study of the absorption of alkali cations by barley roots. *Plant Physiology* 27: 457-474.

Estaun, M.V. (1989). Effect of sodium chloride and mannitol on germination and hyphal growth of the vescicular-arbuscular mycorrhizal fungus *Glomus mosseae*. *Agriculture, Ecosystems and Environment* 29: 123-129.

Etheridge, D.M., L.P. Steele, R.L. Langenfelds, R.J. Francey, J.-M. Barnola, and V.A. Morgan (1996). Natural and anthropogenic changes in atmospheric CO_2 over the last 1000 years from air in Antarctic ice and firn. *Journal of Geophysical Research* 101: 4115-4128.

Evans, J.R., T.D. Sharkey, J.A. Berry, and G.D. Farquhar (1986). Carbon isotope discrimination measured concurrently with gas exchange to investigate CO_2 diffusion in leaves of higher plants. *Australian Journal of Plant Physiology* 13: 281-292.

Evans, L.T. (1963). Extrapolation from controlled environments to the field. In *Environmental Control of Plant Growth,* ed. L.T. Evans. New York: Academic Press, pp. 421-435.

Evans, L.T. (1975). The physiological basis of crop yield. In *Crop Physiology: Some Case Histories,* ed. L.T. Evans. Cambridge, UK: Cambridge University Press, pp. 327-355.

Evans, L.T. (1993). *Crop Evolution, Adaptation and Yield.* Cambridge, UK: Cambridge University Press.

Evans, L.T. (1997). Adapting and improving crops: The endless task. *Philosophical Transactions of the Royal Society of London* B 352: 901-906.

Evans, L.T. (1998). *Feeding the Ten Billion: Plants and Population Growth.* Cambridge, UK: Cambridge University Press.

Everard, J.D., R. Gucci, S.C. Kann, J.A. Flore, and W.H. Loescher (1994). Gas exchange and carbon partitioning in the leaves of celery (*Apium graveolens* L.) at various levels of root zone salinity. *Plant Physiology* 106: 281-292.

Evlagon, D., I. Ravina, and P.M. Neumann (1992). Effects of salinity stress and calcium on hydraulic conductivity and growth in maize seedling roots. *Journal of Plant Nutrition* 15: 795-803.

Fangmeier, A., U. Brockerhoff, U. Grüters, and H.J. Jäger (1994). Growth and yield response of spring wheat (*Triticum aestivum* L. cv. Turbo) grown in open-top chambers to ozone and water stress. *Environmental Pollution* 83: 317-325.

Fangmeier, A., B. Chrost, P. Högy, and K. Krupinska (2000). CO_2-enrichment enhances flag leaf senescence in barley due to greater grain nitrogen sink capacity. *Environmental and Experimental Botany* 44: 151-164.

Fangmeier, A., U. Gruters, U. Hertstein, A. Sandhage-Hofmann, B. Vermehren, and H.J. Jager (1996). Effects of elevated CO_2, nitrogen supply and tropospheric ozone on spring wheat: I. Growth and yield. *Environmental Pollution* 91: 381-390.

FAO (1997). *Production Yearbook,* Volume 50. Rome: Author.

Farrar, J.F. (1988). Temperature and the partitioning and translocation of carbon. In *Plants and Temperature,* Volume 42, eds. S.P. Long and F.I. Woodward. Cambridge, UK: Society for Experimental Biology, pp. 203-235.

Farrar, J.F. (1991). Starch turnover: Its role in source-sink relations. In *Phloem Transport and Assimilate Compartmentation,* ed. J.L. Bonnemain. Nantes, France: Ouest, pp. 213-223.

Farrar, J.F. and S. Gunn (1996). Effects of temperature and atmospheric carbon dioxide on source-sink relations in the context of climate change. In *Photoassimilate Distribution in Plants and Crops,* eds. E. Zamski and A.A. Schaffer. New York: Marcel Dekker, Inc., pp. 389-406.

Farrar, J.F. and D.L. Jones (2000). The control of carbon acquisition by roots. *New Phytologist* 147: 43-53.

Farrar, J.F. and M.L. Williams (1991). The effects of increased atmospheric carbon dioxide and temperature on carbon partitioning, source-sink relations and respiration. *Plant, Cell and Environment* 14: 819-830.

Feder, W.A. (1968). Reduction in tobacco pollen germination and tube elongation, induced by low levels of ozone. *Science* 160: 1122.

Feder, W.A. and R. Shrier (1968). Combination of U.V.-B and ozone reduces pollen tube growth more than either stress alone. *Environmental and Experimental Botany* 30: 451-454.

Feder, W.A. and F. Sullivan (1969). Differential susceptibility of pollen grains to ozone injury. *Phytopathology* 59: 399.

Fehrmann, H., A. von Tiedemann, and P. Fabian (1986). Predisposition of wheat and barley to fungal leaf attack by preinoculative treatment with ozone and sulfur dioxide. *Journal of Plant Disease Protection* 93: 313-318.

Fenner, M. (1992). Environmental influences on seed size and composition. *Horticultural Reviews* 13: 183-213.

Ferris, R. and G. Taylor (1994). Increased root growth in elevated CO_2: A biophysical analysis of root cell elongation. *Journal of Experimental Botany* 45: 1603-1612.

Ferris, R., T.R. Wheeler, R.H. Ellis, and P. Hadley (1999). Seed yield after environmental stress in soybean grown under elevated CO_2. *Crop Science* 39: 710-718.

Field, C.B., J.T. Ball, and J.A. Berry (1989). Photosynthesis: Principles and field techniques. In *Plant Physiological Ecology: Field Methods and Instrumentation,* eds. R.W. Pearcy, J.R. Ehleringer, H.A. Mooney, and P.W. Rundel. London: Chapman and Hall, pp. 209-253.

Finn, G.A. and W.A. Brun (1982). Effect of atmospheric CO_2 enrichment on growth, nonstructural carbohydrate content, and root nodule activity in soybean. *Plant Physiology* 69: 327-331.

Finnan, J.M., A. Donnelly, J.I. Burke, and M.B. Jones (2002). The effects of elevated concentrations of carbon dioxide and ozone on potato (*Solanum tuberosum* L.) yield. *Agriculture, Ecosystems and Environment* 88: 11-22.

Finney, K.F. and H.C. Fryer (1958). Effect on loaf volume of high temperatures during the fruiting period of wheat. *Agronomy Journal* 50: 28-34.

Fischer, R.A. (1983). Wheat. In *Potential Productivity of Field Crops Under Different Environments,* eds. W.H. Smith and S.J. Banta. Los Baños, Laguna, Philippines: International Rice Research Institute, pp. 129-154.

Fischer, R.A. and R. Maurer O (1975). Crop temperature modification and yield potential in a dwarf spring wheat. *Crop Science* 16: 855-859.

Fischer, R.A. and J.T. Wood (1979). Drought resistance in spring wheat cultivars. III. Yield associated with morphological traits. *Australian Journal of Agricultural Research* 30: 1001-1020.

Fitter, A.H. (1996). Characterizations and functions of root systems. In *Plant Roots: The Hidden Half,* eds. Y. Waisel, A. Eshel, and U. Kafkafi. New York: Marcel Dekker, pp. 1-20.

Fitter, A.H., A. Heinenmeyer, and P.L. Staddon (2000). The impact of elevated CO_2 and global climate change on arbuscular mycorrhizas: A mycocentric approach. *New Phytologist* 147: 179-187.

Fitter, A.H., G.K. Self, J. Wolfenden, M.M.I. van Vuuren, T.K. Brown, D. Bogie, and T.A. Mansfield (1996). Root production and mortality under elevated atmospheric carbon dioxide. *Plant and Soil* 187: 299-306.

Flagler, R.B., R.P. Patterson, A.S. Heagle, and W.W. Heck (1987). Ozone and soil moisture deficit effects on nitrogen metabolism of soybean. *Crop Science* 27: 1177-1184.

Flint, E.P. and D.T. Patterson (1983). Interference and temperature effects on growth in soybean *(Glycine max)* and associated C_3 and C_4 weeds. *Weed Science* 31: 193-199.

Flint, E.P., D.T. Patterson, and J.L. Beyers (1983). Interference and temperature effects on growth of cotton *(Gossypium hirsutum),* spurred anoda *(Anoda cristata),* and velvetleaf *(Abutilon theophrasti). Weed Science* 31: 892-898.

Folland, C.K., T.R. Karl, J.R. Christy, R.A. Clarke, G.V. Gruza, J. Jouzel, M.E. Mann, J. Oerlemans, M.J. Salinger, S.-W. Wang, et al. (2001). Observed climate variability and change. In *Climate Change 2001: The Scientific Basis,* eds. J.T. Houghton, Y. Ding, D.J. Griggs, M. Noguer, P.J. van der Linden, X. Dai, K. Maskell, and C.A. Johnson. Cambridge, UK: Cambridge University Press, pp. 99-181.

Fortin, M.C.A. and K.L. Poff (1990). Temperature sensing by primary roots of maize. *Plant Physiology* 94: 367-369.

Fosket, D.E. (1994). *Plant Growth and Development.* San Diego: Academic Press.

Foster, K.R., D.M. Reid, and R.P. Pharis (1992). Ethylene biosynthesis and ethephon metabolism and transport in barley. *Crop Science* 32: 1345-1352.

Francis, D. (1992). The cell cycle in plant development. *New Phytologist* 122: 1-22.

Francis, D. and P.W. Barlow (1988). Temperature and the cell cycle. In *Plants and Temperature,* Volume 42, eds. S.P. Long and F.I. Woodward. Cambridge, UK: Society for Experimental Biology, pp. 181-201.

Francois, L.E. (1987). Salinity effects on asparagus yield and vegetative growth. *Journal of the American Society of Horticultural Science* 112: 432-436.

Francois, L.E., T.J. Donovan, K. Lorenz, and E.V. Maas (1989). Salinity effects on rye grain yield, quality, vegetative growth, and emergence. *Agronomy Journal* 81: 707-712.

Francois, L.E., T. Donovan, and E.V. Maas (1984). Salinity effects on seed yield, growth, and germination of grain sorghum. *Agronomy Journal* 76: 741-744.

Francois, L.E., C.M. Grieve, E.V. Maas, and S.M. Lesch (1994). Time of salt stress affects growth and yield components of irrigated wheat. *Agronomy Journal* 86: 100-107.

Francois, L.E. and E.V. Maas (1999). Crop response and management of salt-affected soils. In *Handbook of Plant and Crop Stress,* ed. M. Pessarakli. New York: Marcel Dekker, pp. 169-202.

Friedman, M. (1997). Chemistry, biochemistry, and dietary role of potato polyphenols: A review. *Journal of Agricultural and Food Chemistry* 45: 1523-1540.

Friedman, R., A. Altman, and N. Levin (1989). The effect of salt stress on polyamine biosynthesis and content in mung bean plants and in halophytes. *Physiologia Plantarum* 76: 295-302.

Fuhrer, J. (1994). Effects of ozone on managed pasture: I. Effects of open-top chambers on microclimate, ozone flux, and plant growth. *Environmental Pollution* 86: 297-305.

Fuhrer, J., A. Grandjean Grimm, W. Tschannen, and H. Shariat-Madari (1992). The response of spring wheat (*Triticum aestivum* L.) to ozone at higher elevations. *New Phytologist* 121: 211-219.

Fuhrer, J., H. Shariat-Madari, R. Perler, W. Tschannen, and A. Grub (1994). Effects of ozone on managed pasture: II. Yield, species composition, canopy structure, and forage quality. *Environmental Pollution* 86: 307-314.

Gagliardi, D., C. Breton, A. Chaboud, P. Vergne, and C. Dumas (1995). Expression of heat shock factor and heat shock protein 70 genes during maize pollen development. *Plant Molecular Biology* 29: 841-856.

Galiba, G., G. Kocsy, R. Kaur-Sawhney, J. Sutka, and A. Galston (1993). Chromosomal localization of osmotic and salt stress-induced differential alterations in polyamine content in wheat. *Plant Science* 92: 203-211.

Gallardo, M., I. Sánchez-Calle, P. Muñoz De Rueda, and A.J. Matilla (1996). Alleviation of thermoinhibition in chickpea seeds by putrescine involves the ethylene pathway. *Australian Journal of Plant Physiology* 23: 479-487.

Galliger, G.C. (1938). Temperature effects upon the growth of excised root tips. *Plant Physiology* 13: 835.

Galston, A.W. and R.K. Sawhney (1990). Polyamines in plant physiology. *Plant Physiology* 94: 406-410.

Garcia, R.L., S.P. Long, G.W. Wall, C.P. Osborne, B.A. Kimball, G.Y. Nie, P.J. Pinter, R.L. Lamorte, and F. Wechsung (1998). Photosynthesis and conductance of spring-wheat leaves: Field response to continuous free-air CO_2 enrichment. *Plant, Cell and Environment* 21: 659-669.

Garcia-Huidobro, J., J.L. Monteith, and G.R. Squire (1982). Time, temperature and germination in pearl millet (*Pennisetum typhoides* S. and H.). I. Constant temperature. *Journal of Experimental Botany* 33: 288-296.

Gawronska, H., R.B. Dwelle, J.J. Pavek, and P. Rowe (1984). Partitioning of photoassimilates by four potato clones. *Crop Science* 24: 1031-1036.

Geiger, D.R., K.E. Koch, and W.-J. Shieh (1996). Effect of environmental factors on whole plant assimilate partitioning and associated gene expression. *Journal of Experimental Botany* 47: 1229-1238.

Geiger, D.R. and J.C. Servaites (1991). Carbon allocation in response to stress. In *Response of Plants to Multiple Stresses,* eds. H.A. Mooney, W.E. Winner, and E.J. Pell. New York: Academic Press, pp. 103-127.

Gelang, J., H. Pleijel, E. Sild, H. Danielsson, S. Younis, and G. Selldén (2000). Rate and duration of grain filling in relation to flag leaf senescence and grain yield in spring wheat *(Triticum aestivum)* exposed to different concentrations of ozone. *Physiologia Plantarum* 110: 366-375.

Gelang, J., G. Selldén, S. Younis, and H. Pleijel (2001). Effects of ozone on biomass, non-structural carbohydrates and nitrogen in spring wheat with artificially manipulated source/sink ratio. *Environmental and Experimental Botany* 46: 155-169.

Gesch, R.W., J.C.V. Vu, K.J. Boote, L.H. Allen, and G. Bowes (2002). Sucrose-phosphate synthase activity in mature rice leaves following changes in growth CO_2 is unrelated to sucrose pool size. *New Phytologist* 154: 77-84.

Ghassemi, F., A.J. Jakeman, and H.A. Nix (1995). *Salinization of Land and Water Resources.* Waalingford, Oxon, UK: CAB International.

Giaquinta, R.T. and D.R. Geiger (1973). Mechanism of inhibition of translocation by localized chilling. *Plant Physiology* 51: 372-377.

Gifford, R.M. (1995). Whole plant respiration and photosynthesis of wheat under increased CO_2 concentration and temperature: Long-term vs. short-term distinctions for modelling. *Global Change Biology* 1: 385-396.

Gifford, R.M. (1977). Growth pattern, carbon dioxide exchange and dry weight distribution in wheat growing under differing photosynthetic environments. *Australian Journal of Plant Physiology* 4: 99-110.

Gill, K.S. (1979). Effect of soil salinity on grain filling and grain development in barley. *Biologia Plantarum* 21: 241-244.

Giorgi, F., B. Hewitson, J. Christensen, M. Hulme, H. Von Storch, P. Whetton, R. Jones, L. Mearns, C. Fu, et al. (2001). Regional climate information : Evaluation and projections. In *Climate Change 2001: The Scientific Basis,* eds. J.T. Houghton, Y. Ding, D.J. Griggs, M. Noguer, P.J. van der Linden, X. Dai, K. Maskell, and C.A. Johnson. Cambridge, UK: Cambridge University Press, pp. 583-638.

Glick, R.E., C.D. Schlagnhaufer, R.N. Arteca, and E.J. Pell (1995). Ozone-induced ethylene emission accelerates the loss of ribulose-1,5-bisphosphate carboxylase/oxygenase and nuclear-encoded mRNAs in senescing potato leaves. *Plant Physiology* 109: 891-898.

Glinski, J. and J. Lipiec (1990). *Soil Conditions and Plant Roots*. Boca Raton, FL: CRC Press.

Goldberg, R.B., S.J. Barker, and L. Perez-Grau (1989). Regulation of gene expression during plant embryogenesis. *Cell* 56: 149-160.

Gomes Filho, E. and L. Sodek (1988). Effect of salinity on ribonuclease activity of *Vigna unguiculata* cotyledons during germination. *Journal of Plant Physiology* 132: 307-311.

Gong, M., X. Li, M. Dai, M. Tian, and Z.-G. Li (1997). Involvement of calcium and calmodulin in the acquisition of heat-shock induced thermotolerance in maize seedlings. *Journal of Plant Physiology* 150: 615-621.

Gordon, D.C., M.M.I. van Vuuren, B. Marshall, and D. Robinson (1995). Plant growth chambers for the simultaneous control of soil and air temperature, and of atmospheric carbon dioxide concentration. *Global Change Biology* 1: 455-464.

Gorissen, A., P.J. Kuikman, J.H. van Ginkel, H. van de Beek, and A.G. Jansen (1996). ESPAS: An advanced phytotron for measuring carbon dynamics in a whole plant-soil system. *Plant and Soil* 179: 81-87.

Gossett, D.G., E.P. Millhollon, and M.C. Lucas (1994). Antioxidant response to NaCl stress in salt-tolerant and salt-sensitive cultivars of cotton. *Crop Science* 34: 706-714.

Grace, J. (1988). Temperature as a determinant of plant productivity. In *Plants and Temperature,* Volume 42, eds. S.P. Long and F.I. Woodward. Cambridge, UK: Society for Experimental Biology, pp. 91-107.

Graham, I.A., C.J. Baker, and C.J. Leaver (1994). Analysis of the cucumber malate synthase gene promotor by transient expression and get retardation assays. *Plant Journal* 6: 893-902.

Graham, I.A., K.J. Denby, and C.J. Leaver (1994). Carbon catabolite repression regulates glyoxylate cycle gene expression in cucumber. *Plant Cell* 6: 761-772.

Graham, P.H. and C.P. Vance (2000). Nitrogen fixation in perspective: An overview of research and extension needs. *Field Crops Research* 65: 93-106.

Grantz, D.A. and J.F. Farrar (2000). Ozone inhibits phloem loading from a transport pool: Compartmental efflux analysis in pima cotton. *Australian Journal of Plant Physiology* 27: 859-868.

Grantz, D.A. and S. Yang (1996a). Effect of O_3 on hydraulic architecture in pima cotton. *Plant Physiology* 112: 1649-1657.

Grantz, D.A. and S. Yang (1996b). Mineral nutrition and ozone damage to pima cotton. In *Proceedings, 1996 Beltwide Cotton Conferences*. Memphis, TN: National Cotton Council of America, pp. 1203-1204.

Grantz, D.A. and S. Yang (2000). Ozone impacts on allometry and root hydraulic conductance are not mediated by source limitation nor developmental age. *Journal of Experimental Botany* 51: 919-927.

Grantz, D.A., X. Zhang, and T. Carlson (1999). Observations and model simulations link stomatal inhibition to impaired hydraulic conductance following ozone exposure in cotton. *Plant Cell and Environment* 22: 1201-1210.

Grattan, S.R. and C.M. Grieve (1994). Mineral nutrient acquisition and response by plants grown in saline enironments. In *Handbook of Plant and Crop Stress,* ed. M. Pessarakli. New York: Marcel Dekker, Inc., pp. 203-226.

Grattan, S.R. and E.V. Maas (1984). Interactive effects of salinity and substrate phosphate on soybean. *Agronomy Journal* 76: 668-676.

Great Britain Ministry of Agriculture, Fisheries and Food (1968). *A Century of Agricultural Statistics, Great Britain 1866-1966*. London: Her Majesty's Stationery Office.

Gregory, P.J. (1986). Response to temperature in a stand of pearl millet. *Journal of Experimental Botany* 37: 379-388.

Grieve, C.M., L.E. Francois, and E.V. Maas (1994). Salinity affects the timing of phasic development in spring wheat. *Crop Science* 34: 1544-1549.

Grieve, C.M., S.M. Lesch, E.V. Maas, and L.E. Francois (1993). Leaf and spikelet primordia initiation in salt-stressed wheat. *Crop Science* 33: 1286-1292.

Grieve, C.M. and E.V. Maas (1988). Differential effects of sodium/calcium ratio on sorghum genotypes. *Crop Science* 28: 659-665.

Grieve, C.M. and D.L. Suarez (1997). Purslane (*Portulaca oleracea* L.): A halophytic crop for drainage water reuse systems. *Plant and Soil* 192: 277-283.

Grif, V.G. (1981). The use of temperature coefficients for studying the mitotic cycle in plants. *Tsitologiya* 23: 166-173 (In Russian).

Grimmer, C., T. Bachfischer, and E. Komor (1999). Carbohydrate partitioning into starch in leaves of *Ricinus communis* L. grown under elevated CO_2 is controlled by sucrose. *Plant, Cell and Environment* 22: 1275-1280.

Grimmer, C. and E. Komor (1999). Assimilate export by leaves of *Ricinus communis* L. growing under normal and elevated carbon dioxide concentrations: The same rate during the day, a different rate at night. *Planta* 209: 275-281.

Grodzinski, B., J. Jiao, and E.D. Leonardos (1998). Estimating photosynthesis and concurrent export rates in C_3 and C_4 species at ambient and elevated CO_2. *Plant Physiology* 117: 207-215.

Grotenhuis, T.P. and B. Bugbee (1997). Super-optimal CO_2 reduces seed yield but not vegetative growth in wheat. *Crop Science* 37: 1215-1222.

Grunwald, C. and A.G. Endress (1988). Oil, fatty acid and protein content of seeds harvested from soybeans exposed to O_3 and/or SO_2. *Botanical Gazette* 149: 283-288.

Gueta-Dahan, Y., Z. Yaniv, B.A. Zilinskas, and G. Ben-Hayyim (1997). Salt and oxidative stress: Similar and specific responses and their relation to salt tolerance in *Citrus*. *Planta* 203: 460-469.

Guidi, L., G. Bongi, S. Ciompi, and G.F. Soldatini (1999). In *Vicia faba* leaves photoinhibition from ozone fumigation in light precedes a decrease in quantum yield of functional PS2 centres. *Journal of Plant Physiology* 154: 167-172.

Gunning, B.E.S. and M.W. Steer (1996). *Plant Cell Biology*. London: Jones and Bartlett.

Hadley, P., G.R. Batts, R.H. Ellis, J.I.L. Morison, S. Pearson, and T.R. Wheeler (1995). Temperature gradient chambers for research on global environment change. II. A twin-wall tunnel system for low-stature, field-grown crops using a split heat pump. *Plant, Cell and Environment* 18: 1055-1063.

Hafeez, F.Y., S. Asad, and K.A. Malik (1991). The effect of high temperature on *Vigna radiata* nodulation and growth with different bradyrhizobial strains. *Environmental and Experimental Botany* 31: 285-294.

Hall, A.E. (2001). *Crop Responses to Environment.* Boca Raton, FL: CRC Press.

Hampson, C.R. and G.M. Simpson (1990). Effects of temperature, salt, and osmotic potential on early growth of wheat *(Triticum aestivum):* I. Germination. *Canadian Journal of Botany* 68: 524-528.

Harmens, H., C.M. Stirling, C. Marshall, and J.F. Farrar (2000). Does down-regulation of photosynthetic capacity by elevated CO_2 depend on N supply in *Dactylis glomerata? Physiologia Plantarum* 108: 43-50.

Harrison, B.H. and W.A. Feder (1974). Ultrastructural changes in pollen exposed to ozone. *Phytopathological Notes* 64: 257-258.

Hartmann, T. (1996). Diversity and variability of plant secondary metabolism: A mechanistic view. *Entomologia Experimentalis et Applicata* 80: 177-188.

Harwood, J.L. (1997). Plant lipid metabolism. In *Plant Biochemistry,* eds. P.M. Dey and J.B. Harborne. San Diego: Academic Press, pp. 237-272.

Hasegawa, P.M., R.A. Bressan, and J.M. Pardo (2000). The dawn of plant salt tolerance genetics. *Trends in Plant Science* 5: 317-319.

Hasegawa, P.M., R.A. Bressan, J.K. Zhu, and H.J. Bohnert (2000). Plant cellular and molecular responses to high salinity. *Annual Review of Plant Physiology and Molecular Biology* 51: 463-499.

Hashimoto, T. and Y. Yamada (1994). Alkaloid biogenesis: Molecular aspects. *Annual Review of Plant Physiology and Molecular Biology* 45: 257-285.

Hausladen, A. and R.G. Alscher (1993). Glutathione. In *Antioxidants in Higher Plants,* eds. R.G. Alscher and J.L. Hess. Boca Raton, FL: CRC Press, pp. 1-30.

Hawkins, H.J. and O.A.M. Lewis (1993a). Combination effect of NaCl salinity, nitrogen form and calcium concentration on the growth, ionic content and gaseous exchange properties of *Triticum aestivum* L. cv. Gamtoos. *New Phytologist* 124: 161-170.

Hawkins, H.J. and O.A.M. Lewis (1993b). Effect of NaCl salinity, nitrogen form, calcium and potassium concentrations on nitrogen uptake and kinetics in *Triticum aestivum* L. cv. Gamtoos. *New Phytologist* 124: 171-177.

Hay, R.K.M. (1999). Physiological control of growth and yield in wheat: Analysis and synthesis. In *Crop Yield: Physiology and Processes,* eds. D.L. Smith and C. Hamel. Berlin: Springer, pp. 1-38.

Heagle, A.S. (1979). Effects of growth media, fertilizer rates and hour and season of exposure on sensitivity of four soybean cultivars to ozone. *Environmental Pollution* 18: 313-322.

Heagle, A.S. (1989). Ozone and crop yield. *Annual Review of Phytopathology* 27: 397-423.

Heagle, A.S., D.E. Body, and W.W. Heck (1973). An open-top field chamber to assess the impact of air pollution on plants. *Journal of Environmental Quality* 2: 365-368.

Heagle, A.S., W.W. Heck, V.M. Lesser, J.O. Rawlings, and F.L. Mowry (1986). Injury and yield response of cotton to chronic doses of ozone and sulfur dioxide. *Journal of Environmental Quality* 15: 375-382.

Heagle, A.S. and L.W. Key (1973). Effect of *Puccinia graminis* f. sp. *tritici* on ozone injury in wheat. *Phytopathology* 63: 609-613.

Heagle, A.S., J.E. Miller, and F.L. Booker (1998). Influence of ozone stress on soybean response to carbon dioxide enrichment: I. Foliar properties. *Crop Science* 38: 113-121.

Heagle, A.S., J.E. Miller, and W.A. Pirsley (2000). Growth and yield responses of winter wheat to mixtures of ozone and carbon dioxide. *Crop Science* 40: 1656-1664.

Heckathorn, S.A., G.J. Poeller, J.S. Coleman, and R.L. Hallberg (1996). Nitrogen availability and vegetative development influence the response of ribulose-1,5-bisphosphate carboxylase/oxygenase, phosphoenolpyruvate carboxylase, and heat-shock protein content to heat stress in *Zea mays* L. *International Journal of Plant Sciences* 157: 546-553.

Heim, U., H. Weber, H. Bäumlein, and U. Wobus (1993). A sucrose-synthase gene of *Vicia faba* L.: Expression pattern in developing seeds in relation to starch synthesis and metabolic regulation. *Planta* 191: 394-401.

Heitholt, J.J. (1999). Cotton: Factors associated with assimilate capacity, flower production, boll set, and yield. In *Crop Yield: Physiology and Processes,* eds. D.L. Smith and C. Hamel. Berlin: Springer, pp. 235-270.

Held, A.A., H.A. Mooney, and J.N. Gorham (1991). Acclimation to ozone stress in radish: Leaf demography and photosynthesis. *New Phytologist* 118: 417-423.

Hellmann, H., L. Barker, D. Funck, and W.B. Frommer (2000). The regulation of assimilate allocation and transport. *Australian Journal of Plant Physiology* 27: 583-594.

Hellmuth, E.O. (1971). The effect of varying air-CO_2 level, leaf temperature, and illuminance on the CO_2 exchange of the dwarf pea, *Pisum sativum* L. var Meteor. *Photosynthetica* 5: 190-194.

Hendrey, G.R. and B.A. Kimball (1994). The FACE program. *Agricultural and Forest Meteorology* 70: 3-14.

Hendriks, J.H.M., A. Kolbe, Y. Gibon, M. Stitt, and P. Geigenberger (2003). ADP-glucose pyrophosphorylase is activated by posttranslational redox-modification in response to light in leaves in response to light and to sugars in leaves of *Arabidopsis* and other plant species. *Plant Physiology* 133: 838-849.

Hendrix, D.L. and R.I. Grange (1991). Carbon partitioning and export from mature cotton leaves. *Plant Physiology* 95: 228-233.

Hendrix, J.E. (1995). Assimilate transport and partitioning. In *Handbook of Plant and Crop Physiology,* ed. M. Pessarakli. New York: Marcel Dekker Inc., pp. 357-385.

Hernández, J.A., A. Jiménez, P. Mullineaux, and F. Sevilla (2000). Tolerance of pea (*Pisum sativum* L.) to long-term salt stress is associated with induction of antioxidant defences. *Plant, Cell and Environment* 23: 853-862.

Herold, A. (1980). Regulation of photosynthesis by sink activity: The missing link. *New Phytologist* 86: 131-144.

Hertstein, U., J. Colls, F. Ewert, and M. van Oijen (1999). Climatic conditions and concentrations of carbon dioxide and air pollutants during "ESPACE-wheat" experiments. *European Journal of Agronomy* 10: 163-169.

Hesse, H., U. Sonnewald, and L. Willmitzer (1995). Cloning and expression analysis of sucrose-phosphate-synthase from sugar beet (*Beta vulgaris* L.). *Molecular Genetics* 247: 515-520.

Hibberd, J.M., R. Whitbread, and J.F. Farrar (1996). The effect of elevated concentrations of CO_2 on infection of barley by *Erysiphe graminis. Physiological & Molecular Plant Pathology* 48: 37-53.

Hileman, D.R., G. Huluka, P.K. Kenjige, N. Sinha, N. Bhattacharya, P.K. Biswas, K.F. Lewin, J. Nagy, and G.R. Hendrey (1994). Canopy photosynthesis and transpiration of field grown cotton exposed to free-air CO_2 enrichment (FACE) and differential irrigation. *Agricultural and Forest Meteorology* 70: 189-207.

Ho, L.C. and J.H.M. Thornley (1978). Energy requirements for assimilate translocation from mature tomato leaves. *Annals of Botany* 42: 481-483.

Hodge, A., E. Paterson, S.J. Grayston, C.D. Campbell, B.G. Ord, and K. Killham (1998). Characterization and microbial utilization of exudate material from the rhizosphere of *Lolium Perenne* grown under CO_2 enrichment. *Soil Biology and Biochemistry* 30: 1033-1043.

Hofer, R.-M. (1996). Root hairs. In *Plant Roots: The Hidden Half*, eds. Y. Waisel, A. Eshel, and U. Kafkafi. New York: Marcel Dekker, pp. 111-126.

Hofstra, D. and C.D. Nelson (1969). The translocation of photosynthetically assimilated ^{14}C in corn. *Canadian Journal of Botany* 47: 1435-1442.

Hogsett, W.E., D. Olszyk, D.P. Ormrod, G.E. Taylor Jr., and D.T. Tingey (1987). *Air Pollution Exposure Systems and Experimental Protocols:* Volume 2: *Description of Facilities*. EPA 600/3-87/037b. Corvallis, OR: Environmental Research Laboratory, Office of Research and Development, U.S. Environmental Protection Agency.

Holbrook, N.M. and M.A. Zwieniecki (2003). Water gate. *Nature* 425: 361.

Horie, T., J.T. Baker, H. Nakagawa, T. Matsui, and H.Y. Kim (2000). Crop ecosystem responses to climatic change: Rice. In *Climate Change and Global Crop Productivity,* eds. K.R. Reddy and H.F. Hodges. Wallingford, UK: CABI Publishing, pp. 81-106.

Horie, T., H. Nakagawa, J. Nakano, K. Hamotani, and H.Y. Kim (1995). Temperature gradient chambers for research on global environment change. III. A system designed for rice in Kyoto, Japan. *Plant, Cell and Environment* 18: 1064-1069.

Hoshida, H., Y. Tanaka, T. Hibino, U. Hayashi, A. Tanaka, T. Takabe, and T. Takabe (2000). Enhanced tolerance to salt stress in transgenic rice that overexpresses chloroplast glutamine synthetase. *Plant Molecular Biology* 43: 103-111.

Hough, A.M. and R.G. Derwent (1990). Changes in the global concentration of tropospheric ozone due to human activities. *Nature* 344: 645-648.

Houghton, R.A., J.E. Hobbie, J.M. Melillo, B. Moore, B.J. Peterson, G.R. Shaver, and G.M. Woodwell (1983). Changes in the carbon content of terrestrial biota and soils between 1860 and 1980: A net release of CO_2 to the atmosphere. *Ecological Monographs* 53: 235-262.

Houghton, S.K. and T.H. Thomas (1996). Effects of elevated carbon-dioxide concentration and temperature on the growth and competition between sugar beet *(Beta vulgaris)* and fat-hen *(Chenopodium album). Aspects of Applied Biology* 45: 197-204.

Howell, R.K. and J.H. Graham (1977). Interaction of ozone and bacterial leaf spot of alfalfa. *Plant Disease Reports* 61: 565-567.

Hsiao, T.C. and R.B. Jackson (1999). Interactive effects of water stress and elevated CO_2 on growth, photosynthesis, and water use efficiency. In *Carbon Dioxide and Environmental Stress,* eds. Y. Luo and H.A. Mooney. San Diego: Academic Press, pp. 3-31.

Hu, Y., H. Schnyder, and U. Schmidhalter (2000). Carbohydrate deposition and partitioning in elongating leaves of wheat under saline conditions. *Australian Journal of Plant Physiology* 27: 363-370.

Huang, B., H.M. Taylor, and B.L. McMichael (1991). Growth and development of seminal roots of wheat seedlings as affected by temperature. *Environmental and Experimental Botany* 31: 471-477.

Huang, B. and Q. Xu (2000). Root growth and nutrient element status of creeping bentgrass cultivars difference in heat tolerance as influenced by supraoptimal shoot and root temperatures. *Journal of Plant Nutrition* 23: 979-990.

Hui, D., Y. Luo, W. Cheng, J.S. Coleman, D.W. Johnson, and D.A. Sims (2001). Canopy radiation- and water-use efficiencies as affected by elevated [CO_2]. *Global Change Biology* 7: 75-91.

Hunsaker, D.J., B.A. Kimball, P.J. Pinter Jr., G.W. Wall, R.L. LaMorte, F.J. Adamson, S.W. Leavitt, T.L. Thompson, A.D. Matthias, and T.J. Brooks (2000). CO_2 enrichment and soil nitrogen effects on wheat evapotranspiration and water use efficiency. *Agricultural and Forest Meteorology* 104: 85-105.

Hur, J.-S., P.G. Kim, S.-C. Yun, and E.W. Park (2000). Indicative responses of rice plant to atmospheric ozone. *Plant Pathology Journal* 16: 130-136.

Hurburgh, C.R. (2001). Quality of the 2001 soybean crop from the United States. <www.exnet.iastate.edu/Pages/grain/test/soybean/01sbqual.pdf>, accessed May 22, 2002.

Huxman, K.A., S.D. Smith, and D.S. Neuman (1999). Root hydraulic conductivity of *Larrea tridentata* and *Helianthus annuus* under elevated CO_2. *Plant, Cell and Environment* 22: 325-330.

Hymus, G.J., N.R. Baker, and S.P. Long (2001). Growth in elevated CO_2 can both increase and decrease photochemistry and photoinhibition of photosynthesis in a predictable manner. *Dactylis glomerata* growing in two levels of nitrogen nutrition. *Plant Physiology* 127:1204-1211.

Irvine, J.E. (1983). Sugarcane. In *Potential Productivity of Field Crops Under Different Environments*, eds. W.H. Smith and S.J. Banta. Los Baños, Laguna, Philippines: International Rice Research Institute, pp. 361-381.

Ishiguro, S. and K. Nakamura (1994). Characterization of a cDNA encoding a novel DNA binding protein, SPF1, that recognizes SP8 sequences in the 5' upstream regions of genes coding for sporamin and beta-amylase from sweet potato. *Molecular Genetics* 244: 563-571.

Ismail, A.M. and A.E. Hall (1999). Reproductive-stage heat tolerance, leaf membrane thermostability and plant morphology in cowpea. *Crop Science* 39: 1762-1768.

Israel, D.W., T.W. Rufty Jr., and J.D. Cure (1990). Nitrogen and phosphorus nutritional interactions in a CO_2 enriched environment. *Journal of Plant Nutrition* 13: 1419-1433.

Jablonski, L.M., X. Wang, and P.S. Curtis (2002). Plant reproduction under elevated CO_2 conditions: A meta-analysis of reports on 79 crop and wild species. *New Phytologist* 156: 9-26.

Jackson, D., A.S. Heagle, and R.V.W. Eckel (1999). Ovipositional response of tobacco hornworm moths (Lepidoptera: Sphingidae) to tobacco plants grown under elevated levels of ozone. *Physiological and Chemical Ecology* 28: 566-571.

Jackson, R.B. and H.L. Reynolds (1996). Nitrate and ammonium uptake for single- and mixed-species communities grown at elevated CO_2. *Oecologia* 105: 74-80.

Jackson, S.D., U. Sonnewald, and L. Willmitzer (1993). Cloning and expression analysis of ß-isopropylmalate dehydrogenase from potato. *Molecular Genetics* 236: 309-314.

Jacob, D.J., J.A. Logan, and P.P. Murti (1999). Effect of rising Asian emissions on surface ozone in the United States. Geophysical Research Letters 26: 2175-2178.

Jacobs, T. (1997). Why do plant cells divide? *Plant Cell* 9: 1021-1029.

Jacobsen, T. and R.M. Adams (1958). Salt and silt in ancient Mesopotamian agriculture. *Science* 128: 1251-1258.

Jacoby, B. (1995). Nutrient uptake by plants. In *Handbook of Plant and Crop Physiology,* ed. M. Pessarakli. Marcel Dekker Inc., New York, pp. 1-22.

Jäger, H.J., U. Hertstein, and A. Fangmeier (1999). The European Stress Physiology and Climate Experiment—Project 1: Wheat (ESPACE-wheat): Introduction, aims and methodology. *European Journal of Agronomy* 10: 155-162.

Jahnke, S. (2001). Atmospheric CO_2 concentration does not directly affect leaf respiration in bean or poplar. *Plant, Cell and Environment* 24: 1139-1151.

Jahnke, S. and M. Krewitt (2002). Atmospheric CO_2 concentration may directly affect leaf respiration measurement in tobacco, but not respiration itself. *Plant, Cell and Environment* 25: 641-651.

Jefferson, P.G., D.A. Johnson, and K.H. Asay (1989). Epicuticular wax production, water status and leaf temperature in *Triticeae* range grasses of contrasting visible glaucousness. *Canadian Journal of Plant Sciences* 69: 513-519.

Jeschke, W.D., O. Wolf, and W. Hartung (1992). Effect of NaCl salinity on flows and partitioning of C, N, and mineral ions in whole plants of white lupin, *Lupinus albus* L. *Journal of Experimental Botany* 43: 777-788.

Jitla, D.S., G.S. Rogers, S.P. Seneweera, A.S. Basra, R.J. Oldfield, and J.P. Conroy (1997). Accelerated early growth of rice at elevated CO_2. *Plant Physiology* 115: 15-22.

Johansson, I., M. Karlsson, U. Johanson, C. Larsson, and P. Kjellbom (2000). The role of aquaporins in cellular and whole plant water balance. *Biochimica et Biophysica Acta* 1465: 324-342.

Johnson, B.G., B.A. Hale, and D.P. Ormrod (1996) Carbon dioxide and ozone effects on growth of a legume-grass mixture. *Journal of Environmental Quality* 25: 908-916.

Johnson, I.R. and J.H.M. Thornley (1985). Temperature dependence of plant and crop processes. *Annals of Botany* 55: 1-24.

Johnson, R. and C.A. Ryan (1990). Wound-inducible potato inhibitor II genes: Enhancement of expression by sucrose. *Plant Molecular Biology* 14: 527-536.

Joly, R.J. (1989). Effects of sodium chloride on the hydraulic conductivity of soybean root systems. *Plant Physiology* 91: 1262-1265.

Jones, H.G. (1985). Adaptive significance of leaf development and structural responses to environment. In *Control of Leaf Growth,* eds. N.R. Baker, W.J. Davies, and C.K. Ong. Cambridge, UK: Cambridge University Press, pp. 155-173.

Jones, H.G. (1992). *Plants and Microclimate,* Second Edition. Cambridge, UK: Cambridge University Press.

Jones, P.D., M. New, D.E. Parker, S. Martin, and I.G. Rigor (1999). Surface air temperature and its change over the past 150 years. *Reviews of Geophysics* 37: 173-199.

Jones, P.D., D.E. Parker, T.J. Osborn, and K.R. Briffa (2001). Global and hemispheric temperature anomalies—land and marine instrumental records. In *Trends: A Compendium of Data on Global Change.* Oak Ridge, TN: Carbon Dioxide Information Analysis Center, Oak Ridge National Laboratory, U.S. Department of Energy. <cdiac.esd.ornl.gov>, accessed February 7, 2003.

Jongen, M., M.B. Jones, T. Hebeisin, H. Blum, and G. Hendrey (1995). The effects of elevated CO_2 concentration on the root growth of *Lolium perenne* and *Trifolium repens* grown in a FACE system. *Global Change Biology* 1: 361-371.

Jordan, W.R., P.J. Shouse, A. Blum, F.R. Miller, and R.L. Monk (1984). Environmental physiology of sorghum. II. Epicuticular wax load and cuticular transpiration. *Crop Science* 24: 1168-1173.

Jouzel, J., C. Lorius, J.R. Petit, C. Genthon, N.I. Barkov, V.M. Kotlyakov, and V.M. Petrov (1987). Vostok ice core: A continuous isotopic temperature record over the last climatic cycle (160,000 years). *Nature* 329: 403-408.

Jwa, N.-S. and L.L. Walling (2002). Influence of elevated CO_2 concentration on disease development in tomato. *New Phytologist* 149: 509-518.

Kafkafi, U. (1990). Root temperature, concentration and the ratio NO_3^-/NH_4^+ effect on plant development. *Journal of Plant Nutrition* 13: 1291-1306.

Kafkafi, U. and N. Bernstein (1996). Root growth under salinity stress. In *Plant Roots: The Hidden Half,* eds. Y. Waisel, A. Eshel, and U. Kafkafi. New York: Marcel Dekker, pp. 435-452.

Kakkar, R.K., S. Bhaduri, V.K. Rai, and S. Kumar (2000). Amelioration of NaCl stress by arginine in rice seedlings: Changes in endogenous polyamines. *Biologia Plantarum* 43: 419-422.

Kakkar, R.K., P.K. Nagar, P.S. Ahuja, and V.K. Rai (2000). Polyamines and plant morphogenesis. *Biologia Plantarum* 43: 1-11.

Kakkar, R.K. and V.K. Rai (1997). Polyamines under salt stress. In *Strategies for Improving Salt Tolerance in Higher Plants*, eds. P.K. Jaiwal, R.P. Singh, and A. Gulati. Enfield, NH: Science Publishers, Inc., pp. 191-204.

Kang, S., F. Zhang, X. Hu, and J. Zhang (2002). Benefits of CO_2 enrichment on crop plants are modified by soil water status. *Plant and Soil* 238: 69-77.

Kangasjärvi, J., J. Talvinen, M. Utriainen, and R. Karjalainen (1994). Plant defence systems induced by ozone. *Plant, Cell and Environment* 17: 783-794.

Kar, S., S.B. Varade, T.K. Subramanyam, and P.B. Ghildyal (1976). Soil physical conditions affecting rice root growth: Bulk density and submerges soil temperature regime effects. *Agronomy Journal* 68: 23-26.

Karpinski, S., H. Reynolds, B. Karpinska, G. Wingsle, G. Creissen, and P. Mullineaux (1999). Systemic signaling and acclimation in response to excess excitation energy in *Arabidopsis*. *Science* 284: 654-657.

Karrer, E.E. and R.L. Rodriguez (1992). Metabolic regulation of rice α-amylase and sucrose synthase genes in plants. *Plant Journal* 2: 517-523.

Kaspar, T.C. and W.L. Bland (1992). Soil temperature and root growth. *Soil Science* 154: 290-299.

Kaspar, T.C., D.G. Woolley, and H.M. Taylor (1981). Temperature effect on the inclination of lateral roots of soybeans. *Agronomy Journal* 73: 383-385.

Katerji, N., J.W. van Hoorn, A. Hamdy, M. Mastrorilli, and E.M. Karzel (1997). Osmotic adjustment of sugar beets in response to soil salinity and its influence on stomatal conductance, growth and yield. *Agricultural Water Management* 34: 57-69.

Kats, G., P.J. Dawson, A. Bytnerowicz, J.W. Wolf, C.R. Thompson, and D.M. Olszyk (1985). Effects of ozone or sulfur dioxide on growth and yield of rice. *Agriculture, Ecosystems and Environment* 14: 103-117.

Katz, R.W. (1977). Assessing the impact of climatic change on food production. *Climatic Change* 1: 85-96.

Kearns, E.V. and S.M. Assmann (1993). The guard cell-environment connection. *Plant Physiology* 102: 711-715.

Keeling, C.D. and T.P. Whorf (2001). Atmospheric CO_2 records from sites in the SIO air sampling network. In *Trends: A Compendium of Data on Global Change*. Oak Ridge, TN: Carbon Dioxide Information Analysis Center, Oak Ridge National Laboratory, U.S. Department of Energy. <cdiac.esd.ornl.gov>.

Keigley, P.J. and R.E. Mullen (1986). Changes in soybean seed quality from high temperature during seed fill and maturation. *Crop Science* 26: 1212-1216.

Kelman, W.M. and C.O. Qualset (1993). Responses of recombinant inbred lines of wheat to saline irrigation: Milling and baking qualities. *Crop Science* 33: 1223-1228.

Kendall, A.C., J.C. Turner, and S.M. Thomas (1985). Effects of CO_2 enrichment at different irradiances on growth and yield of wheat. *Journal of Experimental Botany* 36: 252-260.

Kende, H. and J.A.D. Zeevaart (1997). The five "classical" plant hormones. *Plant Cell* 9: 1197-1210.

Kerk, N. (1998). The root meristem and its relationship to root system architecture. In *Root Demographics and Their Efficiencies in Sustainable Agriculture, Grasslands and Forest Ecosystems*, ed. J.E. Box. Dordrecht, the Netherlands: Kluwer Academic, pp. 509-521.

Kerkeb, L., J.P. Donaire, K. Venema, and M.P. Rodriguez-Rosales (2001). Tolerance to NaCl induces changes in plasma membrane lipid composition, fluidity and H^+-ATPase activity of tomato calli. *Physiologia Plantarum* 113: 217-224.

Kerstiens, G. and K.J. Lendzian (1989). Interactions between ozone and plant cuticles. I. Ozone deposition and permeability. *New Phytologist* 112: 13-19.

Key, J.L., C.Y. Lin, and Y.M. Chen (1981). Heat shock proteins of higher plants. *Proceedings of the National Academy of Sciences, USA* 78: 3526-3530.

Khalil, M.A., F. Amer, and M.M. Elgabaly (1967). A salinity-fertility interaction study on corn and cotton. *Soil Science Society of America Proceedings* 31: 683-686.

Khan, M.G., M. Silberbush, and S.H. Lips (1994). Physiological studies on salinity and nitrogen interaction in alfalfa. II. Photosynthesis and transpiration. *Journal of Plant Nutrition* 17: 669-682.

Khatun, S. and T.J. Flowers (1995). Effects of salinity on seed set in rice. *Plant, Cell and Environment* 18: 61-67.

Khatun, S., C.A. Rizzo, and T.J. Flowers (1995). Genotypic variation in the effect of salinity on fertility in rice. *Plant and Soil* 173: 239-250.

Kibite, S. and L.E. Evans (1984). Causes of negative correlations between grain yield and grain protein concentration in common wheat. *Euphytica* 33: 801-810.

Kiegle, E., C. Moore, J. Haseloff, M.A. Tester, and M.R. Knight (2000). Cell-type-specific calcium responses to drought, salt and cold in the *Arabidopsis* root. *The Plant Journal* 23: 267-278.

Kim, H.Y., M. Lieffering, S. Miura, K. Kobayashi, and M. Okada (2001). Growth and nitrogen uptake of CO_2-enriched rice under field conditions. *New Phytologist* 150: 223-229.

Kim, S.Y., G.D. May, and W.D. Park (1994). Nuclear-protein factors binding to a class-I patatin promoter region are tuber-specific and sucrose-inducible. *Plant Molecular Biology* 26: 603-615.

Kimball, B.A. (1983). Carbon dioxide and agricultural yield: An assemblage and analysis of 430 prior observations. *Agronomy Journal* 75: 779-788.

Kimball, B.A., K. Kobayashi, and M. Bindi (2002). Responses of agricultural crops to free-air CO_2 enrichment. *Advances in Agronomy* 77: 293-368.

Kimball, B.A., C.F. Morris, P.J. Pinter, G.W. Wall, D.J. Hunsaker, F.J. Adamsen, R.L. LaMorte, S.W. Leavitt, T.L. Thompson, A.D. Matthias, and T.J. Brooks (2001). Elevated CO_2, drought and soil nitrogen effects on wheat grain quality. *New Phytologist* 150: 295-303.

Kimball, B.A., P.J. Pinter Jr., R.L. Garcia, R.L. LaMorte, G.W. Wall, D.J. Hunsaker, G. Wechsung, F. Wechsung, and T. Kartschall (1995). Productivity and water use of wheat under free-air CO_2 enrichment. *Global Change Biology* 1: 429-442.

Kimball, B.A., P.J. Pinter Jr., G.W. Wall, R.L. Garcia, R.L. LaMorte, P.M.C. Jak, K.F.A. Fruman, and H.F. Vugts (1997). Comparisons of responses of vegetation to elevated carbon dioxide in free-air and open-top chamber facilities. In *Advances in Carbon Dioxide Effects Research,* eds. L.H. Allen Jr., M.B. Kirkham, D.M. Olszyk, and C.E. Whitman. Madison, WI: American Society of Agronomy, pp. 113-130.

King, K.M. and D.H. Greer (1986). Effects of carbon dioxide enrichment and soil water on maize. *Agronomy Journal* 78: 515-521.

Kinsman, E.A., C. Lewis, M.S. Davies, J.E. Young, D. Francis, B. Vilhar, and H.J. Ougham (1997). Elevated CO_2 stimulates cells to divide in grass meristems: A differential effect in two natural populations of *Dactylis glomerata. Plant, Cell and Environment* 20: 1309-1316.

Klock, K.A., W.R. Graves, and H.G. Taber (1996). Growth and phosphorus, zinc, and manganese content of tomato, muskmelon, and honey locust at high root-zone temperatures. *Journal of Plant Nutrition* 19: 795-806.

Klueva, N.Y., E. Maestri, N. Marmiroli, and H.T. Nguyen (2001). Mechanisms of thermotolerance in crops. In *Crop Responses and Adaptations to Temperature Stress,* ed. A.S. Basra. Binghamton, NY: Food Products Press, pp. 177-218.

Knight, H. and M.R. Knight (2001). Abiotic stress signaling pathways: Specificity and cross-talk. *Trends in Plant Sciences* 6: 262-267.

Knight, J.S. and J.C. Gray (1994). Expression of genes encoding the tobacco chloroplast phosphate translocator is not light regulated and is repressed by sucrose. *Molecular Genetics* 242: 586-594.

Kobayashi, K., M. Lieffering, and H.Y. Kim (2001). Growth and yield of paddy rice under free-air CO_2 enrichment. In *Structure and Function in Agroecosystem Design and Management,* eds. M. Shiyomi and H. Koizumi. Boca Raton, FL: CRC Press, pp. 371-395.

Koch, K.E. (1996). Carbohydrate-modulated gene expression in plants. *Annual Review of Plant Physiology and Plant Molecular Biology* 47: 509-540.

Koch, K.E., K.D. Nolte, E.R. Duke, D.R. McCarty, and W.T. Avigne (1992). Sugar levels modulate differential expression of maize sucrose synthase genes. *Plant Cell* 4: 59-69.

Koch, K.E., Xu, J., Duke, E.R., McCarty, D.R., Yuan, C.X., Tan, B.C., and Avigne, W.T. (1995). Sucrose provides a long distance signal for coarse control of genes affecting its metabolism. In *Sucrose Metabolism, Biochemistry, and Molecular Biology,* eds. H.G. Pontis, G. Salerno, and E. Echeverria. American Society of Plant Physiologists, Maryland, pp. 266-277.

Kohut, R.J., R.G. Amundson, and J.A. Laurence (1986). Evaluation of growth and yield of soybean exposed to ozone in the field. *Environmental Pollution* 41: 219-234.

Komor, E. (2000). Source physiology and assimilate transport: The interaction of sucrose metabolism, starch storage and phloem export in source leaves and the effects on sugar status in phloem. *Australian Journal of Plant Physiology* 27: 497-505.

Körner, Ch., S. Pelaez-Riedl, and A.J.E. van Bel (1995). CO_2 responsiveness of plants: A possible link to phloem loading. *Plant, Cell and Environment* 18: 595-600.

Kossmann, J., R.G.F. Visser, B.T. Müller-Röber, L. Willmitzer, and U. Sonnewald (1991). Cloning and expression analysis of a potato cDNA that encodes branching enzyme: Evidence for coexpression of starch biosynthetic genes. *Molecular Genetics* 230: 39-44.

Kouchi, H., M. Sekine, and S. Hata (1995). Distinct classes of mitotic cyclins are differentially expressed in the soybean shoot apex during the cell cycle. *Plant Cell* 7: 1143-1155.

Kramer, P.J., H. Hellmers, and R.J. Downs (1970). SEPEL: New phytotrons for environmental research. *BioScience* 20: 1201-1208.

Krapp, A., B. Hofmann, C. Schäfer, and M. Stitt (1993). Regulation of the expression of rbcS and other photosynthetic genes by carbohydrates: A mechanism for the "sink regulation" of photosynthesis? *Plant Journal* 3: 817-828.

Krapp, A. and M. Stitt (1994). Influence of high-carbohydrate content on activity of pastidic and cytosolic isoenzyme pairs in photosynthetic tissues. *Plant, Cell and Environment* 17: 861-866.

Kremer, M. (1993). Population growth and technological change: One million B.C. to 1990. *Quarterly Journal of Economics* 108: 681-716.

Kress, L.W. and J.E. Miller (1983). Impact of ozone on soybean yield. *Journal of Environmental Quality* 12: 276-281.

Krupa, S., F.L. Booker, K.O. Burkey, B.I. Chevone, M.T. McGrath, A.H. Chappelka, E.J. Pell, C.P. Anderson, and B.A. Zilinskas (2000). Ambient ozone and plant health. *Plant Disease* 85: 1-12.

Krupa, S.V. and H.-J. Jäger (1996). Adverse effects of elevated levels of ultraviolet (UV)-B radiation and ozone (O_3) on crop growth and productivity. In *Global Climate Change and Agricultural Production,* eds. F. Bazzaz and W. Sombroek. Rome: FAO, pp. 141-169.

Kumar, A., T. Altabella, M.A. Taylor, and A.F. Tiburcio (1997). Recent advances in polyamine research. *Trends in Plant Science* 2: 124-130.

Kurth, E., G.R. Cramer, A. Läuchli, and E. Epstein (1986). Effects of NaCl and $CaCl_2$ on cell enlargement and cell production in cotton roots. *Plant Physiology* 82: 1102-1106.

Labanauskas, C.K., L.H. Stolzy, and M.F. Handy (1981). Protein and free amino acids in wheat grain as affected by soil types and salinity levels in irrigation water. *Plant and Soil* 59: 299-316.

Lalonde, S., E. Boles, H. Hellmann, L. Barker, J.W. Patrick, W.B. Frommer, and M. Ward (1999). The dual function of sugar carriers: Transport and sugar sensing. *Plant Cell* 11: 707-726.

Lam, H.-M., S.S.-Y. Peng, and G.M. Coruzzi (1994). Metabolic regulation of the gene encoding glutamine-dependent asparagine synthetase in *Arabidopsis thaliana. Plant Physiology* 106: 1347-1357.

Lambers, H., F.S. Chapin, and T.L. Pons (1998). *Plant Physiological Ecology*. New York: Springer-Verlag.

Lambers, H., I. Stulen, and A. van der Werf (1996). Carbon use in root respiration as affected by elevated CO_2. *Plant and Soil* 187: 251-263.

Landsberg, J. and M.S. Smith (1992). A functional scheme for predicting the outbreak potential of herbivorous insects under global atmospheric change. *Australian Journal of Botany* 40: 565-577.

Large, E.C. (1954). Growth stages in cereals. Illustrations of the Feekes' scale. *Plant Pathology* 3: 128-129.

Laurence, J.A. and F.A. Wood (1978). Effects of ozone on infection of soybean by *Pseudomonas glycinea. Phytopathology* 68: 441-445.

Law, D.R. and S.J. Crafts-Brandner (1999). Inhibition and acclimation of photosynthesis to heat stress is closely correlated with activation of ribulose-1,5-bisphosphate carboxylase/oxygenase. *Plant Physiology* 120: 173-181.

Law, D.R. and S.J. Crafts-Brandner (2001). High temperature stress increases the expression of wheat leaf ribulose-1,5-bisphosphate carboxylase/oxygenase activase protein. *Archives of Biochemistry and Biophysics* 386: 261-267.

Lawlor, D.W. and R.A.C. Mitchell (1991). The effects of increasing CO_2 on crop photosynthesis and productivity: A review of field studies. *Plant, Cell and Environment* 14: 807-818.

Lawlor, D.W. and R.A.C. Mitchell (2000). Crop ecosystem responses to climate change: Wheat. In *Climate Change and Global Crop Productivity,* eds. K.R. Reddy and H.F. Hodges. Wallingford, UK: CABI Publishing, pp. 57-80.

Lawlor, D.W., R.A.C. Mitchell, J. Franklin, V.J. Mitchell, S.P. Driscoll, and E. Delgado (1993). Facility for studying the effects of elevated carbon dioxide concentration and increased temperature on crops. *Plant, Cell and Environment* 16: 603-608.

Lazof, D.B. and N. Bernstein (1999). The NaCl induced inhibition of shoot growth: The case for disturbed nutrition with special consideration of calcium. *Advances in Botanical Research* 29: 115-189.

Lee, E.H., D.T. Tingey, and W.E. Hogsett (1988). Evaluation of ozone exposure indices in exposure-response modelling. *Environmental Pollution* 53: 43-62.

Lee, G.J. and E. Vierling (2000). A small heat shock protein cooperates with heat shock protein 70 systems to reactivate a heat-denatured protein. *Plant Physiology* 122: 189-197.

Leone, I.A. and E. Brennan (1970). Ozone toxicity in tomato as modified by phosphorus nutrition. *Phytopathology* 60: 1521-1524.

Leone, I.A., E. Brennan, and R.H. Daines (1966). Effect of nitrogen nutrition on response of tobacco to ozone in the atmosphere. *Air Pollution Control Association Journal* 16: 191-196.

Levine, A., R. Tenhaken, R. Dixon, and C. Lamb (1994). H_2O_2 from the oxidative burst orchestrates the plant hypersensitive disease resistance response. *Cell* 79: 583-593.

Lewis, O.A.M., E.O. Leidi, and S.H. Lips (1989). Effect of nitrogen source on growth response to salinity stress in maize and wheat. *New Phytologist* 111: 155-160.

Li, A.-G., Y.-S. Hou, G.W. Wall, A. Trent, B.A. Kimball, and P.J. Pinter, Jr. (2000). Free-air CO_2 enrichment and drought stress effects on grain filling rate and duration in spring wheat. *Crop Science* 40: 1263-1270.

Li, A., Y. Hou, and A. Trent (2001). Effects of elevated atmospheric CO_2 and drought stress on individual grain filling rates and durations of the main stem in spring wheat. *Agricultural and Forest Meteorology* 106: 289-301.

Li, Y., Y.Q. Zu, J.J. Chen, and H.Y. Chen (2002). Intraspecific responses in crop growth and yield of 20 soybean cultivars to enhanced ultraviolet-B radiation under field conditions. *Field Crops Research* 78: 1-8.

Lian, S. and A. Tanaka (1967). Behaviour of photosynthetic products associated with growth and grain production in the rice plant. *Plant and Soil* 26: 333-347.

Lilley, J.M., T.P. Bolger, M.B. Peoples, and R.M. Gifford (2001). Nutritive value and the nitrogen dynamics of *Trifolium subterraneum* and *Phalaris aquatica* under warmer, high CO_2 conditions. *New Phytologist* 150: 385-395.

Lin, C.C. and C.H. Kao (1996). Disturbed ammonium assimilation is associated with growth inhibition of roots in rice seedlings caused by NaCl. *Plant Growth Regulators* 18: 233-238.

Liu, K., H. Fu, Q. Bei, and S. Luan (2000). Inward potassium channel in guard cells as a target for polyamine regulation of stomatal movement. *Plant Physiology* 124: 1315-1325.

Lockhart, J.A. (1965). An analysis of irreversible plant cell elongation. *Journal of Theoretical Biology* 8: 264-275.

Long, S.P. (1991). Modification of the response of photosynthetic productivity to rising temperature by atmospheric CO_2 concentrations: Has its importance been underestimated? *Plant, Cell and Environment* 14: 729-739.

Long, S.P. and S.L. Naidu (2002). Effects of oxidants at the biochemical, cell and physiological levels, with particular reference to ozone. In *Air Pollution and Plant Life,* eds. J.N.B. Bell and M. Treshow. West Sussex, UK: John Wiley and Sons Ltd., pp. 69-88.

Loomis, R.S. and D.J. Connor (1992). *Crop Ecology: Productivity and Management in Agricultural Systems.* Cambridge, UK: Cambridge University Press.

Loreto, F., G. Di Marco, D. Tricoli, and T.D. Sharkey (1994). Measurements of mesophyll conductance, photosynthetic electron transport and alternative electron sinks of field grown wheat leaves. *Photosynthesis Research* 41: 397-403.

Low, P.S. and J.R. Merida (1996). The oxidative burst in plant defense: Function and signal transduction. *Physiologia Plantarum* 96: 533-542.

Lu, Z., J.W. Radin, E.L. Turcotte, R. Percy, and E. Zeiger (1994). High yields in advanced lines of Pima cotton are associated with higher stomatal conductance, reduced leaf area and lower leaf temperature. *Physiologia Plantarum* 92: 266-272.

Lunin, J. and M.H. Gallatin (1965). Salinity-fertility interactions in relation to the growth and composition of beans: II. Varying levels of N and P. *Agronomy Journal* 57: 342-345.

Maas, E.V. and C.M. Grieve (1987). Sodium-induced calcium deficiency in salt-stressed corn. *Plant, Cell and Environment* 10: 559-564.

Maas, E.V. and C.M. Grieve (1990). Spike and leaf development in salt-stressed wheat. *Crop Science* 30: 1309-1313.

Maas, E.V., G.J. Hoffman, G.D. Chaba, J.A. Poss, and M.C. Shannon (1983). Salt sensitivity of corn at various growth stages. *Irrigation Science* 4: 45-57.

Maas, E.V. and J.A. Poss (1989). Salt sensitivity of wheat at various growth stages. *Irrigation Science* 10: 29-40.

MacDowall, F.D.H. (1965). Predisposition of tobacco to ozone damage. *Canadian Journal of Plant Sciences* 45: 1-12.

MacDuff, J.H. and S.B. Jackson (1991). Growth and preference for ammonium or nitrate uptake by barley in relation to root temperature. *Journal of Experimental Botany* 42: 521-530.

MacDuff, J.H., S.C. Jarvis, and J.E. Cockburn (1994). Acclimation of NO_3^- fluxes to low temperature by *Brassica napus* in relation to NO_3^- supply. *Journal of Experimental Botany* 45: 1045-1056.

Machado, S. and G.M. Paulsen (2001). Combined effects of drought and high temperature on water relations of wheat and sorghum. *Plant and Soil* 233: 179-187.

Madson, E. (1968). Effect of CO_2 concentration on the accumulation of starch and sugar in tomato leaves. *Physiologia Plantarum* 21: 168-175.

Maggio, A. and R.J. Joly (1995). Effects of mercuric chloride on the hydraulic conductivity of tomato root systems. Evidence for a channel-mediated pathway. *Plant Physiology* 109: 331-335.

Magnuson, J.J., D.M. Robertson, B.J. Benson, R.H. Wynne, D.M. Livingston, T. Arai, R.A. Assel, R.G. Barry, V. Card, E. Kuusisto, et al. (2000). Historical trends in lake and river ice cover in the Northern Hemisphere. *Science* 289: 1743-1746.

Mahon, J.D. (1979). Environmental and genotypic effects on the respiration associated with symbiotic nitrogen fixation in peas. *Plant Physiology* 63: 892-897.

Malmström, C.M and C.B. Field (1997). Virus-induced differences in the response of oat plants to elevated carbon dioxide. *Plant, Cell and Environment* 20: 178-188.

Manderscheid, R. and H.J. Weigel (1995). Do increasing atmospheric CO_2 concentrations contribute to yield increases of German crops? *Journal of Agronomy and Crop Science* 175: 73-82.

Manning, W.J., W.A. Feder, I. Perkins, and M. Glickman (1969). Ozone injury and infection of potato leaves by *Botrytis cinerea. Plant Disease Reports* 53: 691-693.

Manning, W.J. and A. von Tiedemann (1995). Climate change: Potential effects of increased atmospheric carbon dioxide (CO_2), ozone (O_3), and ultraviolet-B radiation on plant diseases. *Environmental Pollution* 88: 219-245.

Mansour, M.M.F. and M.M. Al-Mutawa (1999). Stabilization of plasma membrane by polyamines against salt stress. *Cytobios* 100: 7-17.

Marcelis, L.F.M. and L.R. Baan Hofman-Eijer (1995). Growth analysis of sweet pepper fruits (*Capsicum annuum* L.). *Acta Horticulturae* 412: 470-478.

Marenco, A., H. Gouget, P. Nédélec, J.-P. Pagés, and F. Karcher (1994). Evidence of a long-term increase in tropospheric ozone from Pic du Midi data series: Consequences: Positive radiative forcing. *Journal of Geophysical Research* 99: 16,617-16,632.

Marland, G., T.A. Boden, and R.J. Andres (2000). Global, regional, and national fossil fuel CO_2 emissions. In *Trends: A Compendium of Data on Global Change.* Oak Ridge, TN: Carbon Dioxide Information Analysis Center, Oak Ridge National Laboratory, U.S. Department of Energy. <cdiac.esd.ornl.gov>, accessed March 8, 2002.

Marschner, H. (1995). *Mineral Nutrition of Higher Plants.* San Diego, CA: Academic Press.

Marten, G.G. (1988). Productivity, stability, sustainability, equitability and autonomy as properties for agroecosystem assessment. *Agricultural Systems* 26: 291-316.

Martinez, V., N. Bernstein, and A. Läuchli (1996). Salt-induced inhibition of phosphorus transport in lettuce plants. *Physiologia Plantarum* 97: 118-122.

Matsui, T., O.S. Namuco, L.H. Ziska, and T. Horie (1997). Effects of high temperature and CO_2 concentration on spikelet sterility in indica rice. *Field Crops Research* 51: 213-219.

Mauney, J.R., B.A. Kimball, P.J. Pinter Jr., R.L. LaMorte, K.F. Lewin, J. Nagy, and G.R. Hendrey (1994). Growth and yield of cotton in response to a free-air carbon dioxide enrichment (FACE) environment. *Agricultural and Forest Meteorology* 70: 49-67.

Mauney, J.R., K.F. Lewin, G.R. Hendrey, and B.A. Kimball (1992). Growth and yield of cotton exposed to free-air CO_2 enrichment (FACE). *Critical Reviews in Plant Sciences* 11: 213-222.

Mayeux, H.S., H.B. Johnson, H.W. Polley, M.J. Dumesnil, and G.A. Spanel (1993). A controlled environment chamber for growing plants across a subambient CO_2 gradient. *Functional Ecology* 7: 125-133.

Mazza, C.A., D. Battista, A.M. Zima, M. Szwarcberg-Bracchitta, C.V. Giordano, A. Acevedo, A.L. Scopel, and C.L. Ballare (1999). The effects of solar ultraviolet-B radiation on the growth and yield of barley are accompanied by increased DNA damage and antioxidant responses. *Plant, Cell and Environment* 22: 61-70.

McAinsh, M.R., N. Evans, L.T. Montgomery, and K.A. North (2002). Stomatal responses to gaseous pollutants. *New Phytologist* 153: 441-448.

McCree, K.J. (1981). Photosynthetically active radiation. In *Encyclopedia of Plant Physiology,* Volume 12A, *Physiological Plant Ecology I: Responses to the Physical Environment,* eds. O.L. Lange, P.S. Nobel, C.B. Osmond, and H. Ziegler. Berlin: Springer-Verlag, pp. 41-55.

McCree, K.J. (1986). Whole-plant carbon balance during osmotic adjustment to drought and salinity. *Australian Journal of Plant Physiology* 13: 33-43.

McKee, I.F., J.F. Bullimore, and S.P. Long (1997). Will elevated CO_2 concentrations protect the yield of wheat from O_3 damage? *Plant, Cell and Environment* 20: 77-84.

McKee, I.F. and S.P. Long (2001). Plant growth regulators control ozone damage to wheat yield. *New Phytologist* 152: 41-51.

McLeod, A.R. (1993). Open-air exposure systems for air pollutant studies: Their potential and limitations. In *Design and Execution of Experiments on CO_2 Enrichment,* eds. E.-D. Schulze and H.A. Mooney. Brussels, Luxembourg: Office for Official Publications of the European Communities, pp. 353-365.

McLeod, A.R., J.E. Fackrell, and K. Alexander (1985). Open-air fumigation of field crops: Criteria and design for a new experimental system. *Atmospheric Environment* 19: 1639-1649.

McLeod, A.R. and S.P. Long (1999). Free-air carbon dioxide enrichment (FACE) in global change research: A review. *Advances in Ecological Research* 28: 1-56.

McMichael, B.L. and J.J. Burke (1996). Temperature effects on root growth. In *Plant Roots: The Hidden Half,* Third Edition, eds. Y. Waisel, A. Eshel, and U. Kafkafi. New York: Marcel Dekker, pp. 717-728.

McQueen, S.J. and F. Rochange (1999). Expansins in plant growth and development: An update on an emerging topic. *Plant Biology* 1: 19-25.

Mehlhorn, H., J.M. O'Shea, and A.R. Wellburn (1991). Atmospheric ozone interacts with stress ethylene formation by plants to cause visible plant injury. *Journal of Experimental Botany* 42: 17-24.

Meneguzzo, S., F. Navari-Izzo, and R. Izzo (1999). Antioxidative responses of shoots and roots of wheat to increasing NaCl concentrations. *Journal of Plant Physiology* 155: 274-280.

Meneguzzo, S., F. Navari-Izzo, and R. Izzo (2000). NaCl effects on water relations and accumulation of mineral nutrients in shoots, roots and cell sap of wheat seedlings. *Journal of Plant Physiology* 156: 711-716.

Menser, H.A. and G.H. Hodges (1967). Nitrogen nutrition and susceptibility of tobacco leaves to ozone. *Tobacco Science* 11: 151-154.

Meyer, U., B. Köllner, J. Willenbrink, and G.H.M. Krause (1997). Physiological changes on agricultural crops induced by different ambient ozone exposure regimes. *New Phytologist* 136: 645-652.

Meyer, U., B. Köllner, J. Willenbrink, and G.H.M. Krause (2000). Effects of different ozone exposure regimes on photosynthesis, assimilates and thousand grain weight in spring wheat. *Agriculture, Ecosystems and Environment* 78: 49-55.

Michiels, J., C. Verreth, and J. Vanderleyden (1994). Effects of temperature stress on bean-nodulating *Rhizobium* strains. *Applied and Environmental Microbiology* 60: 1206-1212.

Middleton, J.T., E.F. Darley, and R.F. Brewer (1958). Damage to vegetation from polluted atmospheres. *Journal of Air Pollution Control Association* 8: 9-15.

Miflin, B. (2000). Crop improvement in the 21st century. *Journal of Experimental Botany* 51: 1-8.

Miglietta, F., V. Magliulo, M. Bind, L. Cerio, F.P. Vaccari, V. Loduca, and A. Peressotti (1998). Free air CO_2 enrichment of potato (*Solanum tuberosum* L.): Development, growth and yield. *Global Change Biology* 4: 163-172.

Miller, G.T. (2002). *Sustaining the Earth.* Belmont, CA: Wadsworth/Thomson Learning.

Ming-hong, J., F. Zong-wei, and Z. Fu-zhu (2001). Impacts of ozone on the biomass and yield of rice in open-top chambers. *Journal of Environmental Sciences* 13: 233-236.

Minkov, I.N., G.T. Jahoubjan, I.D. Denev, and V.T. Toneva (1999). Photooxidative stress in higher plants. In *Handbook of Plant and Crop Stress,* ed. M. Pessarakli. New York: Marcel Dekker, pp. 499-525.

Mironov, V., L. De Veylder, M. Van Montagu, and D. Inzé (1999). Cyclin-dependent kinases and cell division in plants: The nexus. *Plant Cell* 11: 509-521.

Mitchell, R.A.C., V.J. Mitchell, S.P. Driscoll, J. Franklin, and D.W. Lawlor (1993). Effects of increased CO_2 concentration and temperature on growth and yield of winter wheat at two levels of nitrogen application. *Plant, Cell and Environment* 16: 521-529.

Mitchell, R.A.C., V.J. Mitchell, and D.W. Lawlor (2001). Response of wheat canopy CO_2 and water gas-exchange to soil water content under ambient and elevated CO_2. *Global Change Biology* 7: 599-611.

Moldau, H., I. Bichele, and K. Hüve (1998). Dark-induced ascorbate deficiency in leaf cell walls increases plasmalemma injury under ozone. *Planta* 207: 60-66.

Molina, M. and F.S. Rowland (1974). Stratospheric sink for chlorofluoromethanes: Chlorine atom catalyzed destruction of ozone. *Nature* 249: 810-812.

Monteith, J.L., G. Szeicz, and K. Yabuki (1964). Crop photosynthesis and the flux of carbon dioxide below the canopy. *Journal of Applied Ecology* 1: 321-337.

Monz, C.A., H.W. Hunt, F.B. Reeves, and E.T. Elliott (1994). The response of mycorrhizal colonization to elevated CO_2 and climate change in *Pascopyrum smithii* and *Bouteloua gracilis*. *Plant and Soil* 165: 75-80.

Mooney, H.A. and W.E. Winner (1991). Partitioning responses of plants to stress. In *Response of Plants to Multiple Stresses,* eds. H.A. Mooney, W.E. Winner, and E.J. Pell. New York: Academic Press, pp. 129-141.

Moorby, H. and P.H. Nye (1984). The effect of temperature variation over the root system on root extension and phosphate uptake by rape. *Plant and Soil* 78: 283-293.

Moore, B.D., S.-H. Cheng, D. Sims, and J.R. Seemann (1999). The biochemical and molecular basis for photosynthetic acclimation to elevated atmospheric CO_2. *Plant, Cell and Environment* 22: 567-582.

Moot, D.J., A.L. Henderson, J.R. Porter, and M.A. Semenov (1996). Temperature, CO_2 and the growth and development of wheat: Changes in the mean and variability of growing conditions. *Climatic Change* 33: 351-368.

Morecroft, M.D., C.E. Bealey, O. Howells, S. Rennie, and I.P. Woiwod (2002). Effects of drought on contrasting insect and plant species in the UK in the mid-1990s. *Global Ecology and Biogeography* 11: 7-22.

Morgan, P.B., E.A. Ainsworth, and S.P. Long (2003). How does elevated ozone impact soybean? A meta-analysis of photosynthesis, growth and yield. *Plant, Cell and Environment* 26: 1317-1328.

Morgan, P.B., C.J. Bernacchi, D.R. Ort, and S.P. Long (2004). An in vivo analysis of the effect of season-long open-air elevation of ozone to anticipated 2050 levels on photosynthesis in soybean. *Plant Physiology* 135: 2348-2357.

Morison, J.I.L. (1998). Stomatal responses to increased CO_2 concentration. *Journal of Experimental Botany* 49: 443-452.

Morison, J.I.L. and Lawlor (1999). Interactions between increasing CO_2 concentration and temperature on plant growth. *Plant, Cell and Environment* 22: 659-682.

Morowitz, H.J. (1968). *Energy Flow in Biology*. New York and London: Academic Press.

Morse, R.N. and L.T. Evans (1962). Design and development of CERES: An Australian phytotron. *Journal of Agricultural Engineering Research* 7: 128-140.

Mortensen, L. and K.C. Engvild (1995). Effects of ozone on ^{14}C translocation velocity and growth of spring wheat (*Triticum aestivum* L.) exposed in open-top chambers. *Environmental Pollution* 87: 135-140.

Mosher, P.N. and M.H. Miller (1972). Influence of soil temperature on the geotropic response of corn roots (*Zea mays* L.). *Agronomy Journal* 64: 469-452.

Moss, D.N., R.B. Musgrave, and E.R. Lemon (1961). Photosynthesis under field conditions. III. Some effects of light, carbon dioxide, temperature, and soil moisture on photosynthesis, respiration, and transpiration of corn. *Crop Science* 1: 83-87.

Mulchi, C.L., E. Lee, K. Tuthill, and E.V. Olinick (1988). Influence of ozone stress on growth processes, yields and grain quality characteristics among soybean cultivars. *Environmental Pollution* 53: 151-169.

Mulchi, C.L., D.J. Sammons, and P.S. Baenziger (1986). Yield and grain quality response of soft red winter wheat exposed to ozone during anthesis. *Agronomy Journal* 78: 593-600.

Mulchi, C.L., L. Slaughter, M. Saleem, E.H. Lee, R. Pausch, and R. Rowland (1992). Growth and physiological characteristics of soybean in open-top-chambers in response to ozone and increased atmospheric CO_2. *Agriculture, Ecosystems and Environment* 38: 107-118.

Mulholland, B.J., J. Craigon, C.R. Black, J.J. Colls, J. Atherton, and G. Landon (1997a). Effects of elevated carbon dioxide and ozone on the growth and yield of spring wheat (*Triticum aestivum* L.). *Journal of Experimental Botany* 48: 113-122.

Mulholland, B.J., J. Craigon, C.R. Black, J.J. Colls, J. Atherton, and G. Landon (1997b). Impact of elevated atmospheric CO_2 and O_3 on gas exchange and chlorophyll content of spring wheat (*Triticum aestivum* L.). *Journal of Experimental Botany* 48: 1853-1863.

Mulholland, B.J., J. Craigon, C.R. Black, J.J. Colls, J. Atherton, and G. Landon (1998a). Effects of elevated CO_2 and O_3 on the rate and duration of grain growth and harvest index in spring wheat (*Triticum aestivum* L.). *Global Change Biology* 4: 627-635.

Mulholland, B.J., J. Craigon, C.R. Black, J.J Colls, J. Atherton, and G. Landon (1998b). Growth, light interception and yield responses of spring wheat (*Triticum aestivum* L.) grown under elevated CO_2 and O_3 in open-top chambers. *Global Change Biology* 4: 121-130.

Müller-Röber, B.T., W. Sonnewald, and L. Willmitzer (1992). Inhibition of the ADP-glucose pyrophosphorylase in transgenic potatoes leads to sugar storing tubers and influences tuber formation and expression of tuber storage protein genes. *EMBO Journal* 11: 1229-1238.

Munns, R. (1993). Physiological processes limiting plant growth in saline soils: Some dogmas and hypotheses. *Plant, Cell and Environment* 16: 15-24.

Munns, R. (2002). Comparative physiology of salt and water stress. *Plant, Cell and Environment* 25: 239-250.

Munns, R., G.R. Cramer, and M.C. Ball (1999). Interactions between rising CO_2, soil salinity, and plant growth. In *Carbon Dioxide and Environmental Stress*, eds. Y. Luo and H.A. Mooney. San Diego: Academic Press, pp. 139-168.

Munns, R., J.B. Passioura, J. Guo, O. Chazen, and G.R. Cramer (2000). Water relations and leaf expansion: Importance of time scale. *Journal of Experimental Botany* 51: 1495-1504.

Munns, R. and A. Termaat (1986). Whole-plant responses to salinity. *Australian Journal of Plant Physiology* 13: 143-160.

Murakami, Y., M. Tsuyama, Y. Kobayashi, H. Kodama, and K. Iba (2000). Trienoic fatty acids and plant tolerance of high temperatures. *Science* 287: 476-479.

Murkowski, A. (2001). Heat stress and spermidine: Effect on chlorophyll fluorescence in tomato plants. *Biologia Plantarum* 44: 53-57.

Murray, D.R. (1995). Plant responses to carbon dioxide. *American Journal of Botany* 82: 690-697.

Murtadha, H.M., J.W. Maranville, R.B. Clark, and M.D. Clegg (1989). Effects of temperature and relative humidity on growth and calcium uptake, translocation, and accumulation in sorghum. *Journal of Plant Nutrition* 12: 535-545.

Musgrave, R.B. and D.N. Moss (1961). Photosynthesis under field conditions. I. A portable, closed system for determining net assimilation and respiration of corn. *Crop Science* 1: 37-41.

Nakagawa, H., T. Horie, H.Y. Kim, H. Ohnishi, and K. Homma (1997). Rice responses to elevated CO_2 concentrations and high temperatures. *Journal of Agricultural Meteorology* 52: 797-800.

Nakamoto, H. and T. Hiyama (1999). Heat-shock proteins and temperature stress. In *Handbook of Plant and Crop Stress,* ed. M. Pessarakli. New York: Marcel Dekker, pp. 399-416.

Nakamura, K., M. Ohto, N. Yoshida, and K. Nakamura (1991). Sucrose-induced accumulation of ß-amylase occurs concomitant with the accumulation of starch and sporamin in leaf-petiole cuttings of sweet potato. *Plant Physiology* 96: 902-909.

National Assessment Synthesis Team (2001). *Climate Change Impacts on the United States: The Potential Consequences of Climate Variability and Change.* Report for the U.S. Global Change Research Program. Cambridge, UK: Cambridge University Press.

Nátr, L., S. Driscoll, and D.W. Lawlor (1996). The effect of increased CO_2 concentrations on the dark respiration rate of etiolated wheat seedlings. *Cereal Research Communications* 24: 53-59.

Navari-Izzo, F. and N. Rascio (1999). Plant response to water-deficit conditions. In *Handbook of Plant and Crop Stress,* ed. M. Pessarakli. New York: Marcel Dekker, pp. 231-270.

Neumann, P. (1997). Salinity resistance and plant growth revisited. *Plant, Cell and Environment* 20: 1193-1198.

Neumann, P.M., H. Azaizeh, and D. Leon (1994). Hardening of root cell walls: A growth inhibitory response to salinity stress. *Plant, Cell and Environment* 17: 303-309.

Newbery, R.M., J. Wolfenden, T.A. Mansfield, and A.F. Harrison (1995). Nitrogen, phosphorus and potassium uptake and demand in *Agrostis capillaris:* The influence of elevated CO_2 and nutrient supply. *New Phytologist* 130: 565-574.

Nicol, F. and H. Hofte (1998). Plant cell wall expansion: Scaling the wall. *Current Opinions in Plant Biology* 1: 12-17.

Nieden, U.Z., D. Neumann, A. Bucka, and L. Nover (1995). Tissue-specific localization of heat-stress proteins during embryo development. *Planta* 196: 530-538.

Nielsen, K.F. and E.C. Humphries (1966). Effects of root temperature on plant growth. *Soils and Fertility* 29: 1-7.

Nilsen, E.T. and D.M. Orcutt (1996). *Physiology of Plants Under Stress: Abiotic Factors*. New York: John Wiley and Sons.

Nissenbaum, A. (1994). Sodom, Gomorrah and the other lost cities of the plain: A climatic perspective. *Climatic Change* 26: 435-446.

Nobel, P.S. (1999). *Physicochemical and Environmental Plant Physiology*. San Diego: Academic Press.

Nobel, P.S. (2002). *Physicochemical and Environmental Plant Physiology*. San Diego, CA: Academic Press.

Offler, C.E., M.R. Thorpe, and J.W. Patrick (2000). Assimilate transport and partitioning. Integration of structure, physiology and molecular biology. *Australian Journal of Plant Physiology* 27: 473-476.

Ohashi, Y. and Y. Ohshima (1992). Stress-induced expression of genes for pathogenesis-related proteins in plants. *Plant and Cell Physiology* 33: 819-826.

Ohlson, R. and R. Sepp (1975). Rapeseed and other crucifers. In *Food Protein Sources,* ed. N.W. Pirie. Cambridge, UK: Cambridge University Press, pp. 65-78.

Ojanperä, K., E. Pätsikkä, and T. Yläranta (1998). Effects of low ozone exposure of spring wheat on net CO_2 uptake, rubisco, leaf senescence and grain filling. *New Phytologist* 138: 451-460.

Okada, M., M. Lieffering, H. Nakamura, M. Yoshimoto, H.Y. Kim, and K. Kobayashi (2001). Free-air CO_2 enrichment (FACE) using pure CO_2 injection: System description. *New Phytologist* 150: 251-260.

Ollerenshaw, J.H. and T. Lyons (1999). Impacts of ozone on the growth and yield of field-grown winter wheat. *Environmental Pollution* 106: 67-72.

Ollerenshaw, J.H., T. Lyons, and J.D. Barnes (1999). Impacts of ozone on the growth and yield of field-grown winter oilseed rape. *Environmental Pollution* 104: 53-59.

Olmos, E., J.A. Hernandez, F. Sevilla, and E. Hellin (1994). Induction of several antioxidant enzymes in the selection of a salt-tolerant cell line of *Pisum sativum*. *Journal of Plant Physiology* 144: 594-598.

Olszyk, D.M. and C.W. Wise (1997). Interactive effects of elevated CO_2 and O_3 on rice and flacca tomato. *Agriculture, Ecosystems and Environment* 66: 1-10.

Ommen, O., A. Donnelly, S. Vanhoutvin, M. Van Oijen, and R. Manderscheid (1999). Chlorophyll content of spring wheat flag leaves grown under elevated CO_2 concentrations and other environmental stresses within the "ESPACE-wheat" project. *European Journal of Agronomy* 10: 197-203.

Onderdonk, J.J. and J.W. Ketcheson (1973). Effect of soil temperature on direction of corn root growth. *Plant and Soil* 39: 177-186.

Ong, C.K. (1983a). Response to temperature in a stand of pearl millet (*Pennisetum typhoides* S. and H.). I. Vegetative development. *Journal of Experimental Botany* 34: 322-336.

Ong, C.K. (1983b). Response to temperature in a stand of pearl millet (*Pennisetum typhoides,* S. and H.). II. Reproductive development. *Journal of Experimental Botany* 34: 337-348.

Oosterhuis, D.M., R.E. Hampton, and S.D. Wullschleger (1991). Water deficit effects on the cotton leaf cuticle and the efficiency of defoliants. *Journal of Production Agriculture* 4: 260-265.

Orcutt, D.M. and Nilsen, E.T. (2000). *Physiology of Plants Under Stress: Soil and Biotic Factors.* New York: John Wiley and Sons.

Ormrod, D.P., N.O. Adedipe, and G. Hofstra (1973). Ozone effects on growth of radish plants as influenced by nitrogen and phosphorus nutrition and by temperature. *Plant and Soil* 39: 437-439.

Osborne, C.P., L. LaRoche, R.L. Garcia, B.A. Kimball, G.W. Wall, P.J. Pinter, R.L. LaMorte, G.R. Hendrey, and S.P. Long (1998). Does leaf position within a canopy affect acclimation of photosynthesis to elevated CO_2? *Plant Physiology* 117:1037-1045.

Ottman, M.J., B.A. Kimball, P.J. Pinter, G.W. Wall, R.L. Vanderlip, S.W. Leavitt, R.L. LaMorte, A.D. Matthias, and T.J. Brooks (2001). Elevated CO_2 increases sorghum biomass under drought conditions. *New Phytologist* 150: 261-273.

Ougham, H.J. and C.J. Howarth (1988). Temperature shock proteins in plants. In *Plants and Temperature*, eds. S.P. Long and F.I. Woodward. Cambridge, UK: Society for Experimental Biology, pp. 181-201.

Pahlavanian, A.M. and W.K. Silk (1988). Effect of temperature on spatial and temporal aspects of growth in the primary maize root. *Plant Physiology* 87: 529-532.

Palutikof, J. (2002). Global temperature record. Climate Research Unit. <www.cru.uea.ac.uk/cru/info/warming>, accessed March 5, 2002.

Pankhurst, C.E., S. Yu, B.G. Hawke, and B.D. Harch (2001). Capacity of fatty acid profiles and substrate utilization patterns to describe differences in soil microbial communities associated with increased salinity or alkalinity at three locations in South Australia. *Biology and Fertility of Soils* 33: 204-217.

Papadopoulos, I. and V.V. Rendig (1983). Interactive effects of salinity and nitrogen on growth and yield of tomato plants. *Plant and Soil* 73: 47-57.

Papadopoulos, Y.A., R.J. Gordon, K.B. McRae, R.S. Bush, G. Belanger, E.A. Butler, S.A.E. Fillmore, and M. Morrison (1999). Current and elevated levels of UV-B radiation have few impacts on yields of perennial forage crops. *Global Change Biology* 5: 847-856.

Passioura, J.B. and R. Munns (2000). Rapid environmental changes that affect water status induce transient surges or pauses in leaf expansion rates. *Australian Journal of Plant Physiology* 27: 941-948.

Patterson, D.T. (1993). Implications of global climate change for impact of weeds, insects, and plant diseases. In *International Crop Science I,* eds. D.R. Buxton,

R. Shibles, R.A. Forsberg, B.L. Blad, K.H. Asay, G.M. Paulsen, and R.F. Wilson. Madison, WI: Crop Science Society of America, pp. 273-280.

Patterson, D.T. (1995). Weeds in a changing climate. *Weed Science* 43: 685-700.

Patterson, D.T., E.P. Flint, and J.L. Beyers (1984). Effects of CO_2 enrichment on competition between a C_4 weed and a C_3 crop. *Weed Science* 32: 101-105.

Patterson, D.T., M.T. Highsmith, and E.P. Flint (1988). Effects of temperature and CO_2 concentration on the growth of cotton *(Gossypium hirsutum),* spurred anoda *(Anoda cristata),* and velvetleaf *(Abutilon theophrasti). Weed Science* 36: 751-757.

Patterson, D.T., A.E. Russell, D.A. Mortensen, R.D. Coffin, and E.P. Flint (1986). Effects of temperature and photperiod on Texas panicum *(Panicum texanum)* and wild proso millet *(Panicum miliaceum). Weed Science* 34: 876-882.

Paul, M.J. and C.H. Foyer (2001). Sink regulation of photosynthesis. *Journal of Experimental Botany* 52: 1383-1400.

Paul, M.J. and T.K. Pellny (2003). Carbon metabolite feedback regulation of leaf photosynthesis and development. *Journal of Experimental Botany* 54: 539-547.

Paul, N.D. (2000). Stratospheric ozone depletion, UV-B radiation and crop disease. *Environmental Pollution* 108: 343-355.

Pearcy, R.W., E.-D. Schulze, and R. Zimmermann (1989). Measurement of transpiration and leaf conductance. In *Plant Physiological Ecology: Field Methods and Instrumentation,* eds. R.W. Pearcy, J.R. Ehleringer, H.A. Mooney, and P.W. Rundel. London: Chapman and Hall, pp. 137-160.

Pearson, G.A. and L. Bernstein (1959). Salinity effects at several growth stages of rice. *Agronomy Journal* 51: 654-657.

Pell, E.J., C.D. Schlagnhaufer, and R.N. Arteca (1997). Ozone-induced oxidative stress: Mechanisms of action and reaction. *Physiologia Plantarum* 100: 264-273.

Pell, E.J., W.E. Winner, C. Vinten-Johansen, and H.A. Mooney (1990). Response of radish to multiple stresses. I. Physiological and growth responses to changes in ozone and nitrogen. *New Phytologist* 115: 439-446.

Peltonen-Sainio, P. (1999). Growth and development of oat with special reference to source-sink interaction and productivity. In *Crop Yield: Physiology and Processes,* eds. D.L. Smith and C. Hamel. Berlin: Springer-Verlag, pp. 39-66.

Peñarrubia, L. and J. Moreno (1999). Molecular mechanisms of plant response to elevated levels of tropospheric ozone. In *Handbook of Plant and Crop Stress,* ed. M. Pessarakli. New York: Marcel Dekker, pp. 769-793.

Penning de Vries, F.W.T. (1975). The cost of maintenance processes in plant cells. *Annals of Botany* 39: 77-92.

Penning de Vries, F.W.T., A.H.M. Brunsting, and H.H. van Laar (1974). Products, requirements and efficiency of biosynthesis: A quantitative approach. *Journal of Theoretical Biology* 45: 339-377.

Penning de Vries, F.W.T., H.H. van Laar, and M.C.M. Chardon (1983). Bioenergetics of growth of seeds, fruits, and storage organs. In *Potential Productivity of Field Crops Under Different Environments,* eds. W.H. Smith and S.J.

Banata. Los Banos, Laguna, Philippines: International Rice Research Institute, pp. 37-59.

Percival, J. (1934). *Wheat in Great Britain.* Leighton, Shinfield, Reading, UK: Published by the author.

Petit, J.R., D. Raynaud, C. Lorius, J. Jouzel, G. Delaygue, N.I. Barkov, and V.M. Kotlyakov (2000). Historical isotopic temperature record from the Vostok ice core. In *Trends: A Compendium of Data on Global Change.* Oak Ridge, TN: Carbon Dioxide Information Analysis Center, Oak Ridge National Laboratory, U.S. Department of Energy. <cdiac.esd.ornl.gov>.

Petruzzelli, M.T., M.T. Melillo, T.G. Zache, and G. Taranto (1991). Physiological and ultrastructural changes in isolated wheat embryos during salt and osmotic shock. *Annals of Botany* 69: 25-31.

Pfleeger, T.G., M.A. da Luz, and C.C. Mundt (1999). Lack of synergistic interaction between ozone and wheat leaf rust in wheat swards. *Environmental and Experimental Botany* 41: 195-207.

Phene, C.J., D.N. Baker, J.R. Lambert, J.E. Parsons, and J.M. McKinion (1978). SPAR: A soil-plant-atmosphere-research system. *Transactions of the American Society of Agricultural Engineering* 21: 924-930.

Phillips, D.A., K.D. Newell, S.A. Hassell, and C.E. Felling (1976). The effect of CO_2 enrichment on root nodule development and symbiotic N_2 reduction in *Pisum sativum* L. *American Journal of Botany* 63: 356-362.

Pike, C.S. and J.A. Berry (1980). Membrane phospholipid phase separations in plants adapted to or acclimated to different thermal regimes. *Plant Physiology* 66: 238-241.

Pinter, P.J., Jr., B.A. Kimball, R.L. Garcia, G.W. Wall, D.J. Hunsaker, and R.L. LaMorte (1996). Free-air CO_2 enrichment: Responses of cotton and wheat crops. In *Carbon Dioxide and Terrestrial Ecosystems,* eds. G.W. Koch and H.A. Mooney. San Diego, CA: Academic Press, pp. 215-248.

Pinter, P.J., Jr., B.A. Kimball, G.W. Wall, R.L. LaMorte, D.J. Hunsaker, F.J. Adamsen, K.F.A. Frumau, H.F. Vugts, G.R. Hendrey, K.F. Lewin, et al. (2000). Free-air CO_2 enrichment (FACE): Blower effects on wheat canopy microclimate and plant development. *Agricultural and Forest Meteorology* 103: 319-333.

Platt, B.S. (1962). *Tables of Representative Values of Foods Commonly Used in Tropical Countries.* London: HMSO.

Plaut, Z., C.M. Grieve, and E.V. Maas (1990). Salinity effects on CO_2 assimilation and diffusive conductance of cowpea leaves. *Physiologia Plantarum* 79: 31-38.

Plazek, A., M. Rapacz, and A. Skoczowski (2000). Effects of ozone fumigation on photosynthesis and membrane permeability in leaves of spring barley, meadow fesque, and winter rape. *Photosynthetica* 38: 409-413.

Pleijel, H., H. Danielsson, J. Gelang, E. Sild, and G. Selldén (1998). Growth stage dependence of the grain yield response to ozone in spring wheat (*Triticum asetivum* L.). *Agriculture, Ecosystems and Environment* 70: 61-68.

Pleijel, H., J. Gelang, E. Sild, H. Danielsson, S. Younis, P.-K. Karlsson, G. Wallin, L. Skärby, and G. Selldén (2000). Effects of elevated carbon dioxide, ozone and water availability on spring wheat growth and yield. *Physiologia Plantarum* 108: 61-70.

Pleijel, H., L. Mortensen, J. Fuhrer, K. Ojanperä, and H. Danielsson (1999). Grain protein accumulation in relation to grain yield of spring wheat (*Triticum aestivum* L.) grown in open-top chambers with different concentrations of ozone, carbon dioxide and water availability. *Agriculture, Ecosystems and Environment* 72: 265-270.

Pleijel, H., L. Skärby, G. Wallin, and G. Selldén (1991). Yield and grain quality of spring wheat (*Triticum aestivum* L., cv. Drabant) exposed to different concentrations of ozone in open-top chambers. *Environmental Pollution* 69: 151-168.

Poljakoff-Mayber, A. and H.R. Lerner (1999). Plants in saline environments. In *Handbook of Plant and Crop Stress*, ed. M. Pessarakli. New York: Marcel Dekker, pp. 125-152.

Polle, A. (1996). Protection from oxidative stress in trees as affected by elevated CO_2 and environmental stress. In *Terrestrial Ecosystem Response to Elevated Carbon Dioxide*, eds. G.W. Koch and H.A. Mooney. San Diego: Academic Press, pp. 299-315.

Polle, A., K. Chakrabarti, W. Schuermann, and H. Rennenberg (1990). Composition and properties of hydrogen peroxide decomposing systems in extracellular and total extracts from needles of Norway spruce (*Picea abies* L., Karst.). *Plant Physiology* 94: 312-319.

Polle, A., T. Pfirrmann, S. Chakrabarti, and H. Rennenberg (1993). The effects of enhanced ozone and enhanced carbon dioxide concentrations on biomass, pigments and antioxidative enzymes in spruce needles (*Picea abies* L.). *Plant, Cell and Environment* 16: 311-316.

Polley, H.W. (2002). Implications of atmospheric and climatic change for crop yield and water use efficiency. *Crop Science* 42: 131-140.

Pollock, C.J. (1986a). Environmental effects on sucrose and fructan metabolism. *Current Topics in Plant Biochemistry and Physiology* 5: 32-46.

Pollock, C.J. (1986b). Fructans and the metabolism of sucrose in vascular plants. *New Phytologist* 104: 1-24.

Poorter, H. and O. Nagel (2000). The role of biomass allocation in the growth response of plants to different levels of light, CO_2, nutrients and water: A quantitative review. *Australian Journal of Plant Physiology* 27: 595-607.

Poorter, H., C. Roumet, and B.D. Campbell (1996). Interspecific variation in the growth response of plants to elevated CO_2: A search for functional types. In *Carbon Dioxide, Populations, and Communities,* eds. C. Körner and F.A. Bazzaz. San Diego: Academic Press, pp. 375-412.

Poorter, H., Y. Van Berkel, R. Baxter, J. Den Hertog, P. Dijkstra, R.M. Gifford, K.L. Griffin, C. Roumet, J. Roy, and S.C. Wong (1997). The effect of elevated

CO_2 on the chemical composition and construction costs of leaves of 27 C_3 species. *Plant, Cell and Environment* 20: 472-482.

Poorter, H., A. van der Werf, O.K. Atkin, and H. Lambers (1991). Respiratory energy requirements of roots vary with the potential growth rate of a plant species. *Physiologia Plantarum* 83: 469-475.

Postel, S.L. (1990). Water for food production: Will there be enough in 2025? *BioScience* 48: 629-637.

Prasad, P.V.V., K.J. Boote, L.H. Allen Jr., and J.M.G. Thomas (2002). Effects of elevated temperature and carbon dioxide on seed-set and yield of kidney bean (*Phaseolus vulgaris* L.). *Global Change Biology* 8: 710-721.

Prather, M., D. Ehhalt, F. Dentener, R. Derwent, E. Dlugokencky, E. Holland, I. Isaksen, J. Katima, V. Kirchhoff, P. Matson, P. Midgley, M. Wang, et al. (2001). Atmospheric chemistry and greenhouse gases. In *Climate Change 2001: The Scientific Basis,* eds. J.T. Houghton, Y. Ding, D.J. Griggs, M. Noguer, P.J. van der Linden, X. Dai, K. Maskell, and C.A. Johnson. Cambridge, UK: Cambridge University Press, pp. 239-287.

Prentice, I.C., G.D. Farquhar, M.J.R. Fasham, M.L. Goulden, M. Heimann, V.J. Jaramillo, H.S. Kheshgi, C. Le Quéré, R.J. Scholes, D.W.R. Wallace, et al. (2001). The carbon cycle and atmospheric carbon dioxide. In *Climate Change 2001: The Scientific Basis,* eds. J.T. Houghton, Y. Ding, D.J. Griggs, M. Noguer, P.J. van der Linden, X. Dai, K. Maskell, and C.A. Johnson. Cambridge, UK: Cambridge University Press, pp. 183-237.

Prior, S.A., H.H. Rogers, G.B. Runion, and J.R. Mauney (1994). Effects of free-air CO_2 enrichment on cotton root growth. *Agricultural and Forrest Meteorology* 70: 69-86.

Prior, S.A., H.A. Torbert, G.B. Runion, G.L. Mullins, H.H. Rogers, and J.R. Mauney (1998). Effects of CO_2 enrichment on cotton nutrient dynamics. *Journal of Plant Nutrition* 21: 1407-1426.

Pritchard, J. (1994). The control of cell expansion in roots. *New Phytologist* 127: 3-26.

Pritchard, S.G. and H.H. Rogers (2000). Spatial and temporal deployment of crop roots in CO_2-enriched environments. *New Phytologist* 147: 55-71.

Pritchard, S.G., C. Mosjidis, C.M. Peterson, G.B. Runion, and H.H. Rogers (1998). Anatomical and morphological alterations in longleaf pine needles resulting from growth in elevated CO_2: Interactions with soil resource availability. *International Journal of Plant Sciences* 159: 1002-1009.

Pritchard, S.G., H.H. Rogers, S.A. Prior, and C.M. Peterson (1999). Elevated CO_2 and plant structure: A review. *Global Change Biology* 5: 807-837.

Pritchard, S.G., Z. Ju, E. van Santen, J. Qiu, D.B. Weaver, S.A. Prior, and H.H. Rogers (2000). The influence of elevated CO_2 on the activities of antioxidative enzymes in two soybean genotypes. *Australian Journal of Plant Physiology* 27: 1061-1068.

Quisenberry, J.E., G.B. Cartwright, and B.L. McMichael (1984). Genetic relationship between turgor maintenance and growth in cotton germplasm. *Crop Science* 24: 479-482.

Radin, J.W. (1990). Response of transpiration and hydraulic conductance to root temperature in nitrogen- and phosphorus-deficient cotton seedlings. *Plant Physiology* 92: 855-857.

Radoglou, K.M. and P.G. Jarvis (1990). Effects of CO_2 enrichment on four poplar clones. I. Growth and leaf anatomy. *Annals of Botany* 65: 617-626.

Ranasinghe, S. and G. Taylor (1996). Mechanism for increased leaf growth in elevated CO_2. *Journal of Experimental Botany* 47: 349-358.

Rao, G.G. and G.R. Rao (1982). Anatomical changes in the leaves and their role in adaptation to salinity in pigeon pea (*Cajanus indicus* Spreng) and gingelley (*Sesamum indicum* L.). *Proceedings of the Indian National Society of Academics* B48: 774-778.

Ravikovich, S. (1973). Effects of brackish irrigation water and fertilizers on millet and corn. *Experimental Agriculture* 9: 181-188.

Rawson, H.M. (1995). Yield responses of two wheat genotypes to carbon dioxide and temperature in field studies using temperature gradient tunnels. *Australian Journal of Plant Physiology* 22: 23-32.

Rawson, H.M., R.M. Gifford, and B.N. Condon (1995). Temperature gradient chambers for research on global environment change. I. Portable chambers for research on short-stature vegetation. *Plant, Cell and Environment* 18: 1048-1054.

Rawson, H.M. and R. Munns (1984). Leaf expansion in sunflower as influenced by salinity and short-term changes in carbon fixation. *Plant, Cell and Environment* 7: 207-217.

Raychaudhuri, S.S. (2000). The role of superoxide dismutase in combating oxidative stress in higher plants. *The Botanical Review* 66: 89-98.

Rayle, D.L., P.M. Haughton, and R. Cleland (1970). An in vitro system that simulates plant cell extension growth. *Proceedings of the National Academy of Sciences* USA 67: 1814-1817.

Read, J.J. and J.A. Morgan (1996). Growth and partitioning in *Pascopyrum smithii* (C_3) and *Bouteloua gracilis* (C_4) as influenced by carbon dioxide and temperature. *Annals of Botany* 77: 487-496.

Reddy, C.P.P., B. Acock, and M.C. Acock (1989). Seasonal carbon and nitrogen accumulation in relation to net carbon dioxide exchange in a carbon dioxide-enriched soybean canopy. *Agronomy Journal* 81: 78-83.

Reddy, K.R., H.F. Hodges, and B.A. Kimball (2000). Crop ecosystem responses to climatic change: Cotton. In *Climate Change and Global Crop Productivity,* eds. K.R. Reddy and H.F. Hodges. Wallingford, UK: CABI Publishing, pp. 161-187.

Reddy, P.R. and J.A. Goss (1971). Effect of salinity on pollen. I. Pollen viability as altered by increasing osmotic pressure with NaCl, $MgCl_2$, and $CaCl_2$. *American Journal of Botany* 58: 721-725.

Reekie, E.G. (1996). The effect of elevated CO_2 on developmental processes and its implications for plant-plant interactions. In *Carbon Dioxide, Populations, and Communities,* eds. C. Körner and F.A. Bazzaz. San Diego: Academic Press, pp. 333-346.

Reeves, D.W., H.H. Rogers, S.A. Prior, C.W. Wood, and G.B. Runion (1994). Elevated atmospheric carbon dioxide effects on sorghum and soybean nutrient status. *Journal of Plant Nutrition* 17: 1939-1954.

Reich, P.B., A.W. Schoettle, and R.G. Amundson (1985). Effects of low concentrations of O_3, leaf age and water stress on leaf diffusive conductance and water use efficiency in soybean. *Physiologia Plantarum* 63: 58-64.

Reid, R.J. (1999). Kinetics of nutrient uptake by plant cells. In *Mineral Nutrition of Crops,* ed. Z. Rengel. Binghamton, NY: Food Products Press, pp. 41-66.

Reinhardt, D.H. and T.L. Rost (1995). Salinity accelerates endodermal development and induces an exodermis in cotton seedling roots. *Environmental and Experimental Botany* 35: 563-574.

Renaudin, J.P., J. Colasanti, H. Rime, Z. Yuan, and V. Sundaresan (1994). Cloning of four cyclins from maize indicates that higher plants have three structurally distinct groups of mitotic cyclins. *Proceedings of the National Academy of Science, USA* 91: 7375-7379.

Rennenberg, H., C. Herschbach, and A. Polle (1996). Consequences of air pollution on shoot-root interactions. *Journal of Plant Physiology* 148: 296-301.

Reuveni, J. and B. Bugbee (1997). Very high CO_2 reduces photosynthesis, dark respiration and yield in wheat. *Annals of Botany* 80: 539-546.

Reuveni, J. and J. Gale (1985). The effect of high levels of carbon dioxide on dark respiration and growth of plants. *Plant, Cell and Environment* 8: 623-628.

Reuveni, J., J. Gale, and A.M. Mayer (1993). Reduction of respiration by high ambient CO_2 and the resulting error in measurements of respiration made with O_2 electrodes. *Annals of Botany* 72: 129-131.

Reuveni, J., J. Gale, and M. Zeroni (1997). Differentiating day from night effects of high ambient [CO_2] on the gas exchange and growth of *Xanthium strumarium* L. exposed to salinity stress. *Annals of Botany* 79: 191-196.

Reynolds, M.P., M. Balota, M.I.B. Delgado, I. Amani, and R.A. Fischer (1994). Physiological and morphological traits associated with spring wheat yield under hot, irrigated conditions. *Australian Journal of Plant Physiology* 21: 717-730.

Richards, J.F., J.S. Olson, and R.M. Rotty (1983). *Development of a Data Base for Carbon Dioxide Releases Resulting from Conversion of Land to Agricultural Uses.* ORAU/IEA-82-10(M) ORNL/TM-8801. Oak Ridge, TN: Oak Ridge National Laboratory, U.S. Department of Energy.

Richards, R.A. (2000). Selectable traits to increase crop photosynthesis and yield of grain crops. *Journal of Experimental Botany* 51: 447-458.

Rillig, M.C., S.F. Wright, B.A. Kimball, P.J. Pinter, G.W. Wall, M.J. Ottman, and S.W. Leavitt (2001). Elevated carbon dioxide and irrigation effects on water stable aggregates in a sorghum field: A possible role for arbuscular mycorrhizal fungi. *Global Change Biology* 7: 333-337.

Ritchie, J.T. and D.S. NeSmith (1991). Temperature and crop development. In *Modeling Plant and Soil Systems,* eds. R.J. Hanks and J.T. Ritchie. Agronomy Monograph 31. Madison, WI: ASA Press, pp. 5-29.

Roberts, E.H. (1988). Temperature and seed germination. In *Plants and Temperature*, eds. S.P. Long and F.I. Woodward. Cambridge, UK: Society for Experimental Biology, pp. 109-132.

Robertson, C.J. (1956). The expansion of the arable area. *Scottish Geographical Magazine* 72: 1-20.

Robinson, J.M. and S.J. Britz (2000). Tolerance of a field grown soybean cultivar to elevated ozone level is concurrent with higher leaflet ascorbic acid level, higher ascorbate-dehydroascorbate redox status, and long term photosynthetic productivity. *Photosynthesis Research* 64: 77-87.

Robinson, M.F., J. Heath, and T.A. Mansfield (1998). Disturbances in stomatal behaviour caused by air pollutants. *Journal of Experimental Botany* 49: 461-469.

Rogers, H.H. and R.C. Dahlman (1993). Crop responses to CO_2 enrichment. *Vegetatio* 104/105: 117-131.

Rogers, H.H., W.W. Heck, and A.S. Heagle (1983). A field technique for the study of plant responses to elevated carbon dioxide concentrations. *Journal of the Air Pollution Control Association* 33: 42-44.

Rogers, H.H., S.A. Prior, G.B. Runion, and R.J. Mitchell (1992). Cotton root and rhizosphere responses to free-air CO_2 enrichment. *Critical Reviews in Plant Sciences* 11: 251-263.

Rogers, H.H., S.A. Prior, G.B. Runion, and R.J. Mitchell (1996). Root to shoot ratio of crops as influenced by CO_2. *Plant and Soil* 187: 229-248.

Rogers, H.H., G.B. Runion, and S.V. Krupa (1994). Plant responses to atmospheric CO_2 enrichment with emphasis on roots and the rhizosphere. *Environmental Pollution* 83: 155-189.

Rogers, H.H., G.B. Runion, S.V. Krupa, and S.A. Prior (1997). Plant responses to atmospheric carbon dioxide enrichment: Implications in root-soil-microbe interactions. In *Advances in Carbon Dioxide Effects Research,* eds. L.H. Allen, M.B. Kirkman, D.M. Olyszyk, and C.M. Whitman. Madison, WI: American Society of Agronomy, pp. 1-31.

Rogers, H.H., G.B. Runion, S.A. Prior, and H.A. Torbert (1999). Response of plants to elevated atmospheric CO_2: Root growth, mineral nutrition, and soil carbon. In *Carbon Dioxide and Environmental Stress,* eds. Y. Luo and H.A. Mooney. San Diego: Academic Press, pp. 215-244.

Rogers, H.H., J.F. Thomas, and G.E. Bingham (1983). Response of agronomic and forest species to elevated atmospheric carbon dioxide. *Science* 220: 428-429.

Roitsch, T., R. Ehness, M. Goetz, B. Hause, M. Hofmann, and A.K. Sinha (2000). Regulation and function of extracellular invertase from higher plants in relation to assimilate partitioning, stress responses and sugar signaling. *Australian Journal of Plant Physiology* 27: 815-825.

Rolland, F., B. Moore, and J. Sheen (2002). Sugar sensing and signaling in plants. *Plant Cell* supplement: S185-S205.

Rosenzweig, C. and D. Hillel (1998). *Climate Change and the Global Harvest: Potential Impacts of the Greenhouse Effect on Agriculture.* New York: Oxford University Press.

Rosenzweig, C., F.N. Tubiello, R. Goldberg, E. Mills, and J. Bloomfield (2002). Increased crop damage in the US from excess precipitation under climate change. *Global Environmental Change* 12: 197-202.

Rowland-Bamford, A.J., J.T. Baker, L.H. Allen Jr., and G. Bowes (1996). Interactions of CO_2 enrichment and temperature on carbohydrate accumulation and partitioning in rice. *Environmental and Experimental Botany* 36: 111-124.

Roy, M. and B. Ghosh (1996). Polyamines, both common and uncommon, under heat stress in rice *(Oryza sativa)* callus. *Physiologia Plantarum* 98: 196-200.

Ruckenbauer, P. (1975). Photosynthetic and translocation pattern in contrasting winter wheat varieties. *Annals of Applied Biology* 79: 351-359.

Rudd, J.J. and V.E. Franklin-Tong (2001). Unraveling response-specificity in Ca^{2+} signaling pathways in plant cells. *New Phytologist* 151: 7-33.

Rudorff, B.F.T., C.L. Mulchi, C.S.T. Daughtry, and E.H. Lee (1996). Growth, radiation use efficiency, and canopy reflectance of wheat and corn grown under elevated ozone and carbon dioxide atmospheres. *Remote Sensing of Environment* 55: 163-173.

Rufty, T.W., S.C. Huber, and P.S. Kerr (1985). Association between sucrose-phosphate synthase activity in leaves and plant growth rate in response to altered aerial temperature. *Plant Science* 39: 7-12.

Runeckles, V.C. and B.I. Chevone (1992). Crop responses to ozone. In *Surface Level Ozone Exposures and Their Effects on Vegetation,* ed. A.S. Lefohn. Chelsea, MI: Lewis Publishers, pp. 189-270.

Runion, G.B., E.A. Curl, H.H. Rogers, P.A. Backman, R. Rodriguez-Kabana, and B.E. Helms (1994). Effects of free-air CO_2 enrichment on microbial populations in the rhizosphere and phylosphere of cotton. *Agricultural and Forestry Meteorology* 70: 117-130.

Russell, E.J. and D.J. Watson (1940). *The Rothamsted Field Experiments on the Growth of Wheat*. Technical Communication No. 40. Harpenden, England: Imperial Bureau of Soil Science.

Ruuska, S., T.J. Andrews, M.R. Badger, G.S. Hudson, A. Laisk, G.D. Price, and S. von Caemmerer (1998). The interplay between limiting processes in C_3 photosynthesis studied by rapid-response gas exchange using transgenic tobacco impaired in photosynthesis. *Australian Journal of Plant Physiology* 25: 859-870.

Ryle, G.J.A., C.E. Powell, and V. Tewson (1992). Effect of elevated CO_2 on the photosynthesis, respiration and growth of perennial ryegrass. *Journal of Experimental Botany* 43: 811-818.

Ryle, G.J.A., J. Woledge, V. Tewson, and C.E. Powell (1992). Influence of elevated CO_2 and temperature on the photosynthesis and respiration of white clover dependent on N_2 fixation. *Annals of Botany* 70: 213-220.

Saadalla, M.M., J.S. Quick, and J.F. Shanahan (1990). Heat tolerance in winter wheat: II. Membrane thermostability and field performance. *Crop Science* 30: 1248-1251.

Sachs, R.M. (1965). Stem elongation. *Annual Review of Plant Physiology* 16: 73-96.

Sadras, V.O. and N. Trápani (1999). Leaf expansion and phenological development: Key determinants of sunflower plasticity, growth and yield. In *Crop Yield: Physiology and Processes,* eds. D.L. Smith and C. Hamel. Berlin: Springer-Verlag, pp. 205-234.

Saftner, R.A., J. Daie, and R.E. Wyse (1983). Sucrose uptake and compartmentation in sugar beet taproot tissue. *Plant Physiology* 72: 1-6.

Sage, R.F. (1995). Was low atmospheric CO_2 during the Pleistocene a limiting factor for the origin of agriculture? *Global Change Biology* 1: 93-106.

Saini, H.S. and D. Aspinalli (1982). Abnormal sporogenesis in wheat (*Triticum aestivum* L.) induced by short periods of high temperature. *Annals of Botany* 49: 835-846.

Sakai, H., K. Yagi, K. Kobayashi, and S. Kawashima (2001). Rice carbon balance under elevated CO_2. *New Phytologist* 150: 241-249.

Sakaki, T., K. Tanaka, and M. Yamada (1994). General metabolic changes in leaf lipids in response to ozone. *Plant and Cell Physiology* 35: 53-62.

Salehuzzaman, S.N.I.M., E. Jacobsen, and R.G.F. Visser (1994). Expression patterns of two starch biosynthetic genes in in vitro cultured cassava plants and their induction by sugars. *Plant Science* 98: 53-62.

Salim, M. and M. Akbar (1995). Salinity-induced changes in rice: Effects on plant-insect interactions. In *Handbook of Plant and Crop Physiology,* ed. M. Pessarakli. New York: Marcel Dekker Inc., pp. 661-678.

Sandermann, H., Jr. (2000). Ozone/disease interactions: Molecular biomarkers as a new experimental tool. *Environmental Pollution* 108: 327-332.

Sanders, G.E., J.J. Colls, A.G. Clark, S. Galaup, J. Bonte, and J. Cantuel (1992). *Phaseolus vulgaris* and ozone: Results from open top chamber experiments in France and England. *Agriculture, Ecosystems and Environment* 38: 31-40.

Santa-Cruz, A., M.T. Estañ, A. Rus, M.C. Bolarin, and M. Acosta (1997). Effects of NaCl and mannitol iso-osmotic stresses on the free polyamine levels in discs of tomato species differing in salt tolerance. *Journal of Plant Physiology* 151: 754-758.

Sasek, T.W. and B.R. Strain (1989). Effects of carbon dioxide enrichment on the expansion and size of kudzu *(Pueraria lobata)* leaves. *Weed Science* 37: 23-28.

Sato, S., M.M. Peet, and J.F. Thomas (2000). Physiological factors limit fruit set in tomato (*Lysopersicon esculentum* Mill.) under chronic mild heat stress. *Plant, Cell and Environment* 23: 719-726.

Sattelmacher, B., H. Marschner, and R. Kühne (1990). Effects of temperature of the rooting zone on the growth and development of roots of potato *(Solanum tuberosum). Annals of Botany* 65: 27-36.

Savin, R., P.J. Stone, and M.E. Nicolas (1996). Responses of grain growth and malting quality of barley to short periods of high temperature in field studies using portable chambers. *Australian Journal of Agricultural Research* 47: 465-477.

Sawada, S., M. Kuninaka, K. Watanabe, A. Sato, H. Kawamura, K. Komine, T. Sakamoto, and M. Kasai (2001). The mechanism to suppress photosynthesis through end-product inhibition in single-rooted soybean leaves during acclimation to CO_2 enrichment. *Plant Cell Physiology* 42: 1093-1102.

Schäffner, A.R. (1998). Aquaporin function, structure, and expression: Are there more surprises to surface in water relations? *Planta* 204: 131-139.

Schapendonk, A.H.C.M., M. van Oijen, P. Dijkstra, C. S. Pot, W.J.R.M. Jordi, and G.M. Stoopen (2000). Effects of elevated CO_2 concentration on photosynthetic acclimation and productivity of two potato cultivars grown in open-top chambers. *Australian Journal of Plant Physiology* 27:1119-1130.

Schiefelbein, J.W., J.D. Masucci, and H. Wang (1997). Building a root: The control of patterning and morphogenesis during root development. *Plant Cell* 9: 1089-1098.

Schneider, M., P. Schweizer, P. Meuwly, and J.P. Mètraux (1996). Systemic acquired resistance in plants. *International Review of Cytology* 168: 303-340.

Schöffl, F., F. Prändl, and A. Reindl (1998). Regulation of the heat-shock response. *Plant Physiology* 117: 1135-1141.

Schortemeyer, M., U. Hartwig. G. Hendry, and M.J. Sadowsky (1996). Microbial community changes in the rhizospheres of white clover and perennial ryegrass exposed to free air carbon dioxide enrichment (FACE). *Soil Biology and Biochemistry* 28: 1717-1724.

Schrier, A.A., G. Hoffmann-Thoma, and A.J.E. van Bel (2000). Temperature effects on symplastic and apoplasmic phloem loading and loading-associated carbohydrate processing. *Australian Journal of Plant Physiology* 27: 769-778.

Schulz, A. (1998). Phloem, structure related to function. *Progress in Botany* 59: 429-475.

Schütz, M. and A. Fangmeier (2001). Growth and yield responses of spring wheat (*Triticum aestivum* L. cv. Minaret) to elevated CO_2 and water limitation. *Environmental Pollution* 114: 187-194.

Schwanz, P. and A. Polle (1998). Antioxidative systems, pigment and protein contents in leaves of adult Mediterranean oak species (*Quercus pubescens* and *Q. ilex*) with lifetime exposure to elevated CO_2. *New Phytologist* 140: 411-423.

Schwarz, M. and J. Gale (1984). Growth response to salinity at high levels of carbon dioxide. *Journal of Experimental Botany* 35: 193-196.

Seeman, J.R. and T.D. Sharkey (1986). Salinity and nitrogen effects on photosynthesis, ribulose-1, 5-bisphosphate carboxylase and metabolite pool sizes in *Phaseolus vulgaris* L. *Plant Physiology* 82: 555-560.

Seligman, N.G., R.S. Loomis, J. Burke, and A. Abshahi (1983). Nitrogen nutrition and canopy temperature in field-grown spring wheat. *Journal of Agricultural Science, Cambridge* 101: 691-697.

Serraj, R. and J.-J. Drevon (1998). Effects of salinity and nitrogen source on growth and nitrogen fixation in alfalfa. *Journal of Plant Nutrition* 21: 1805-1818.

Serraj, R. and T.R. Sinclair (2002). Osmolyte accumulation: Can it really help increase crop yield under drought conditions? *Plant, Cell and Environment* 25: 333-341.

Serraj, R., T.R. Sinclair, and L.H. Allen (1998). Soybean nodulation and N_2 fixation response to drought under carbon dioxide enrichment. *Plant, Cell and Environment* 21: 491-500.

Shabala, S. (2000). Ionic and osmotic components of salt stress specifically modulate net ion fluxes from bean leaf mesophyll. *Plant, Cell and Environment* 23: 825-837.

Shafer, S.R. and M.M. Schoeneberger (1994). Air pollution and ecosystem health: The mycorrhizal connection. In *Mycorrhizae and Plant Health,* eds. F.L. Pfleger and R.G. Linderman. St. Paul, MN: APS Press, pp. 153-188.

Sheen, J. (1996). Ca^{2+}-dependent protein kinases and stress signal transduction in plants. *Science* 274: 1900-1902.

Shennan, C., S.R. Grattan, D.M. May, C.J. Hillhouse, D.P. Schachtman, M. Wander, B. Roberts, S. Tafoya, R.G. Burau, C. McNeish, and L. Zelinski (1995). Feasability of cyclic reuse of saline drainage in a tomato-cotton rotation. *Journal of Environmental Quality* 24: 476-488.

Siddiqi, M.Y., A.R. Memon, and A.D.M. Glass (1984). Regulation of K^+ influx in barley: Effects of low temperature. *Plant Physiology* 74: 730-733.

Signora, L., N. Galtier, L. Skot, H. Lucas, and C.H. Foyer (1998). Over-expression of sucrose phosphate synthase in *Arabidopsis thaliana* results in increased foliar sucrose/starch ratios and favours decreased foliar carbohydrate accumulation in plants after prolonged growth with CO_2 enrichment. *Journal of Experimental Botany* 49: 669-680.

Simko, I. (1994). Sucrose application causes hormonal changes associated with potato-tuber induction. *Journal of Plant Growth Regulators* 13: 73-77.

Sims, D.A., Y. Luo, and J.R. Seemann (1998). Importance of leaf versus whole plant CO_2 environment for photosynthetic acclimation. *Plant, Cell and Environment* 21: 1189-1196.

Sinclair, T.R. (1992). Mineral nutrition and plant growth response to climate change. *Journal of Experimental Botany* 43: 1141-1146.

Sinclair, T.R. (1994). Limits to crop yield? In *Physiology and Determination of Crop Yield,* eds. K.J. Boote, J.M. Bennett, T.R. Sinclair, and G.M. Paulsen. Madison, WI: American Society of Agronomy, pp. 509-532.

Sinclair, T.R., P.J. Pinter, B.A. Kimball, F.J. Adamsen, R.L. LaMorte, G.W. Wall, D.J. Hunsaker, N.R. Adam, T.J. Brooks, R.L. Garcia, T.L. Thompson, S.W. Leavitt, and A.D. Matthias (2000). Leaf nitrogen concentration of wheat subjected to elevated [CO_2] and either water or nitrogen deficits. *Agriculture, Ecosystems, and Environment* 79: 53-60.

Singh, N.K., A.K. Handa, P.M. Hasegawa, and R.A. Bressan (1985). Proteins associated with adaptation of cultured tobacco lines to NaCl. *Plant Physiology* 79: 126-137.

Sionit, N., H. Hellmers, and B.R. Strain (1980). Growth and yield of wheat under CO_2 enrichment and water stress. *Crop Science* 20: 687-690.

Sionit, N., B.R. Strain, and E.P. Flint (1987). Interaction of temperature and CO_2 enrichment on soybean: Growth and dry matter partitioning. *Canadian Journal of Plant Sciences* 67: 59-67.

Skirvin, D.J., J.N. Perry, and R. Harrington (1997). The effect of climate change on an aphid-coccinellid interaction. *Global Climate Change Biology* 3: 1-11.

Slaughter, L.H., C.L. Mulchi, and E.H. Lee (1993). Wheat-kernel growth characteristics during exposure to chronic ozone pollution. *Environmental Pollution* 81: 73-79.

Smertenko, A., P. Dráber, V. Viklicky, and Z. Opatrny (1997). Heat stress affects the organization of microtubules and cell division in *Nicotiana tabacum* cells. *Plant, Cell and Environment* 20: 1534-1542.

Smith, B.D. (1995). *The Emergence of Agriculture*. New York: Scientific American Library.

Smith, B.D. (2001). Documenting plant domestication: The consilience of biological and archaeological approaches. *Proceedings of the National Academy of Sciences, USA* 98: 1324-1326.

Smith, C.J. and J.R. Gallon (2001). Living in the real world: How plants perceive their environment. *New Phytologist* 151: 1-6.

Smucker, A.J.M. (1993). Soil environmental modifications of root dynamics and measurement. *Annual Review of Phytopathology* 31: 191-216.

Song, J., D. Nada, and S. Tachibana (1999). Ameliorative effect of polyamines on the high temperature inhibition of in vitro pollen germination in tomato (*Lysopersicon esculentum* Mill.). *Scientia Horticulturae* 80: 203-212.

Soni, R., J.P. Carmichael, Z.H. Shah, and J.A.H. Murray (1995). A family of cyclin D homologues from plants differently controlled by growth regulators and containing the conserved retinoblastoma protein interaction motif. *Plant Cell* 7: 85-103.

Spector, W.S. (Ed.) (1956). *Handbook of Biological Data*. Philadelphia and London: W.B. Saunders Company.

Squire, G.R., B. Marshall, A.C. Terry, and J.L. Monteith (1984). Response to temperature in a stand of pearl millet. VI. Light interception and dry matter production. *Journal of Experimental Botany* 35: 599-610.

Sreenivasulu, N., R. Grimm, U. Wobus, and W. Weschke (2000). Differential response of antioxidant compounds to salinity stress in salt-tolerant and salt-sensitive seedlings in foxtail millet *(Setaria italica)*. *Physiologia Plantarum* 109: 435-442.

Stanhill, G. (1977). Quantifying weather–crop relations. In *Environmental Effects on Crop Physiology,* eds. J.J. Landsberg and C.V. Cutting. London: Academic Press, pp. 23-37.

Stephen, J., R. Woodfin, J.E. Corlett, N.D. Paul, H.G. Jones, and P.G. Ayres (1999). Response of barley and pea crops to supplementary UV-B radiation. *Journal of Agricultural Science* 132: 253-261.

Steudle, E. and C.A. Peterson (1998). How does water get through roots? *Journal of Experimental Botany* 49: 775-788.

Stewart, C.A., V.J. Black, C.R. Black, and J.A. Roberts (1996). Direct effects of ozone on the reproductive development of *Brassica* species. *Journal of Plant Physiology* 148: 172-178.

Stitt, M. (1991). Rising CO_2 levels and their potential significance for carbon flow in photosynthetic cells. *Plant, Cell and Environment* 14: 741-762.

Stitt, M. and A. Krapp (1999). The interaction between elevated carbon dioxide and nitrogen nutrition: The physiological and molecular background. *Plant, Cell and Environment* 22: 583-621.

Stitt, M., and W. Scheible (1998). Understanding allocation to shoot and root growth will require molecular information about which compounds act as signals for the plant nutrient status, and how meristem activity and cellular growth are regulated. *Plant and Soil* 201: 259-263.

Stone, J.A. and H.M. Taylor (1983). Temperature and development of the taproot and lateral roots of four indeterminate soybean cultivars. *Agronomy Journal* 75: 613-618.

Stone, P. (2001). The effect of heat stress on cereal yield and quality. In *Crop Responses and Adaptations to Temperature Stress,* ed. A.S. Basra. Binghamton, NY: Food Products Press, pp. 243-291.

Stone, P.J. and M.E. Nicolas (1995). Effect of timing of heat stress during grain filling on two wheat varieties differing in heat tolerance. I. Grain growth. *Australian Journal of Plant Physiology* 22: 927-934.

Stoy, V. (1965). Photosynthesis, respiration, and carbohydrate accumulation in spring wheat in relation to yield. *Physiologia Planatrum Supplementum IV.*

St. Pierre, B. and N. Brisson (1995). Induction of the plastidic starch-phosphorylase gene in potato storage sink tissue. *Planta* 195: 339-344.

Stulen I. and J. den Hertog (1993). Root growth and functioning under atmospheric CO_2 enrichment. *Vegetatio* 104/105: 99-115.

Surano, K., P. Daley, J. Houpis, J. Shinn, J. Helms, R. Pallasou, and M. Costella (1986). Growth and physiological responses of *Pinus ponderosa* to long-term elevated CO_2 concentrations. *Tree Physiology* 2: 243-259.

Sweetlove, L.J. and S.A. Hill (2000). Source metabolism dominates the control of source to sink carbon flux in tuberizing potato plants throughout the diurnal cycle and under a range of environmental conditions. *Plant, Cell and Environment* 23: 523-529.

Tanaka, A. and M. Osaki (1983). Growth and behaviour of photosynthesized ^{14}C in various crops in relation to productivity. *Soil Science and Plant Nutrition* 29: 147-158.

Tanaka, A. and J. Yamaguchi (1968). The growth efficiency in relation to the growth of the rice plant. *Soil Science and Plant Nutrition* 14: 110-116.

Tardieu, F. and W.J. Davies (1993). Integration of hydraulic and chemical signaling in the control of stomatal conductance and water status of droughted plants. *Plant, Cell and Environment* 16: 341-349.

Tardieu, F. and S. Pellerin (1991). Influence of soil temperature during root appearance on the trajectory of nodal roots of field grown maize. *Plant and Soil* 131: 207-214.

Tashiro, T. and I.F. Wardlaw (1989). A comparison of the effect of high temperature on grain development in wheat and rice. *Annals of Botany* 64: 59-65.

Tashiro, T. and I.F. Wardlaw (1991). The effect of high temperature on kernel dimensions and the type and occurrence of kernel damage in rice. *Australian Journal of Agricultural Research* 42: 485-496.

Taylor, G. and R. Ferris (1996). Influence of air pollution on root physiology and growth. In *Plant Responses to Air Pollution,* eds. M. Yunus and M. Iqbal. London: John Wiley and Sons, pp. 375-393.

Taylor, G., S. Ranasinghe, C. Bosac, T.J. Flowers, M. Crookshanks, and L. Dolan (1994). Elevated CO_2 and plant growth: Cellular mechanisms and responses of whole plants. *Journal of Experimental Botany* 45: 1761-1774.

Temple, P.J. (1990). Growth form and yield responses of four cotton cultivars to ozone. *Agronomy Journal* 82: 1045-1050.

Temple, P.J. (1991). Variations in responses of dry bean *(Phaseolus vulgaris)* cultivars to ozone. *Agriculture, Ecosystems, and Environment* 36: 1-11.

Temple, P.J. and S. Bisessar (1979). Response of white bean to bacterial blight, ozone, and antioxidant protection in the field. *Phytopathology* 69: 101-103.

Tenhaken, R., A. Levine, L.F. Brisson, R.A. Dixon, and C. Lamb (1995). Function of the oxidative burst in hypersensitive disease resistance. *Proceedings of the National Academy of Sciences USA* 92: 4158-4163.

Tester, R.F., W.R. Morrison, R.H. Ellis, J.R. Piggot, G.R. Batts, T.R. Wheeler, J.I.L. Morison, P. Hadley, and D.A. Ledward (1995). Effects of elevated growth temperature and carbon dioxide levels on some physicochemical properties of wheat starch. *Journal of Cereal Science* 22: 63-71.

Thomas, J.F. and C.H. Harvey (1983). Leaf anatomy of four species grown under continuous CO_2 enrichment. *Botanical Gazette* 144: 303-309.

Thomas, M.D. and G.R. Hill (1949). Photosynthesis under field conditions. In *Photosynthesis in Plants,* eds. J. Franck and W.E. Loomis. Ames, IA: Iowa State College Press, pp. 19-52.

Thomas, R.B. and K.L. Griffin (1994). Direct and indirect effects of atmospheric carbon dioxide enrichment on leaf respiration of *Glycine max* (L.) Merr. *Plant Physiology* 104: 355-361.

Thomas, R.B., C.D. Reid, R. Ybema, and B.R. Strain (1993). Growth and maintenance components of leaf respiration of cotton grown in elevated carbon dioxide partial pressure. *Plant, Cell and Environment* 16: 539-546.

Thompson, G.B., J.K. Brown, and F.I. Woodward (1993). The effects of host carbon dioxide, nitrogen and water supply on the infection of wheat by powdery mildew and aphids. *Plant, Cell and Environment* 16: 687-694.

Thompson, G.B. and F.I. Woodward (1994). Some influences of CO_2 enrichment, nitrogen nutrition and competition on grain yield and quality in spring wheat and barley. *Journal of Experimental Botany* 45: 937-942.

Thompson, J.F., J.T. Madison, and A.-M.E. Muenster (1977). In vitro culture of immature cotyledons of soya bean (*Glycine max* L. Merr.). *Annals of Botany* 41: 29-39.

Thompson, L.M. (1970). Weather and technology in the production of soybeans in the central United States. *Agronomy Journal* 62: 232-236.

Thompson, L.M. (1986). Climatic change, weather variability, and corn production. *Agronomy Journal* 78: 649-653.

Thorne, G.N. and D.W. Wood (1987). Effects of radiation and temperature on tiller survival, grain number and grain yield in winter wheat. *Annals of Botany* 59: 413-426.

Tiburcio, A.F., T. Altabella, A. Borrell, and C. Masgrau (1997). Polyamine metabolism and its regulation. *Physiologia Plantarum* 100: 664-674.

Tiedemann, A.v. and K.H. Firsching (2000). Interactive effects of elevated ozone and carbon dioxide on growth and yield of leaf rust-infected versus non-infected wheat. *Environmental Pollution* 108: 357-363.

Tingey, D.T. and U. Blum (1973). Effects of ozone on soybean nodules. *Journal of Environmental Quality* 2: 341-342.

Tingey, D.T., K.D. Rodecap, E.H. Lee, W.E. Hogsett, and J.W. Gregg (2002). Pod development increases the ozone sensitivity of *Phaseolus vulgaris. Water, Air and Soil Pollution* 139: 325-341.

Tipping, C. and D.R. Murray (1999). Effect of atmospheric CO_2 concentration on leaf anatomy and morphology in *Panicum* species representing different photosynthetic modes. *International Journal of Plant Sciences* 160: 1063-1073.

Tjoelker, M.G., P.B. Reich, and J. Oleksyn (1999). Changes in leaf nitrogen and carbohydrate underlie temperature and CO_2 acclimation of dark respiration in five boreal tree species. *Plant, Cell and Environment* 22: 767-778.

Tollenaar, M. (1999). Yield improvement in temperate maize is attributable to greater stress tolerance. *Crop Science* 39: 1597-1604.

Tollenaar, M. and L.M. Dwyer (1999). Physiology of maize. In *Crop Yield: Physiology and Processes,* eds. D.L. Smith and C. Hamel. Berlin: Springer-Verlag, pp. 169-204.

Tremmel, D.C. and D.T. Patterson (1993). Responses of soybean and five weeds to CO_2 enrichment under two temperature regimes. *Canadian Journal of Plant Science* 73: 1249-1260.

Tremmel, D.C. and D.T. Patterson (1994). Effects of elevated CO_2 and temperature on development in soybean and five weeds. *Canadian Journal of Plant Science* 74: 43-50.

Treseder, K.K. and M.F. Allen (2000). Mycorrhizal fungi have a potential role in soil carbon storage under elevated CO_2 and nitrogen deposition. *New Phytologist* 147: 189-200.

Trotman, A.P. and R.W. Weaver (1995). Tolerance of clover rhizobia to heat and dessication stresses in soil. *Soil Science Society of America Journal* 59: 466-470.

Turcsányi, E., T. Lyons, M. Plöchl, and J. Barnes (2000). Does ascorbate in the mesophyll cell walls form the first line of defence against ozone? Testing the concept using broad bean (*Vicia faba* L.). *Journal of Experimental Botany* 51: 901-910.

Turgeon, R. (2000). Plasmodesmata and solute exchange in the phloem. *Australian Journal of Plant Physiology* 27: 521-529.

Tyagi, R.K. and N.S. Rangaswamy (1993). Screening of pollen grains vis-à-vis whole plants of oilseed brassicas for tolerance to salt. *Theoretical and Applied Genetics* 87: 343-346.

Tyerman, S.D., H.J. Bohnert, C. Maruel, E. Steudle, and J.A.C. Smith (1999). Plant aquaporins: Their molecular biology, biophysics and significance for plant water relations. Journal of Experimental Botany 50: 1055-1071.

Urao, T., B. Yakubov, R. Satoh, K. Yamaguchi-Shinozaki, M. Seki, T. Hirayama, and K. Shinozaki (1999). A transmembrane hybrid-type histidine kinase in *Arabidopsis* functions as an osmosensor. *The Plant Cell* 11: 1743-1754.

U.S. Department of Agriculture, National Agricultural Statistics Service (2002). *Agricultural Statistics 2002.* Washington, DC: United States Government Printing Office.

Vandermeiren, K., L. De Temmerman, A. Staquet, and H. Baeten (1992). Effects of air filtration on spring wheat grown in open-top field chambers at a rural site. II. Effects on mineral partitioning, sulphur, and nitrogen metabolism and on grain quality. *Environmental Pollution* 77: 7-14.

van der Werf, A. and O.W. Nagel (1996). Carbon allocation to shoots and roots in relation to nitrogen supply is mediated by cytokinins and sucrose. *Plant and Soil* 185: 21-32.

van Hoorn, J.W., N. Katerji, A. Hamdy, and M. Mastrorilli (2001). Effect of salinity on yield and nitrogen uptake of four grain legumes and on biological nitrogen contributions from the soil. *Agricultural and Water Management* 51: 87-98.

Van Oijen, M., A.H.C.M. Schapendonk, M.J.H. Jansen, C.S. Pot, and R. Maciorowski (1999). Do open-top chambers overestimate the effects of rising CO_2 on plants? An analysis using spring wheat. *Global Change Biology* 5: 411-421.

Van Oosten, J.-J. and R.T. Besford (1994). Sugar feeding mimics effect of acclimation to high CO_2-rapid down regulation of RuBisCO small subunit transcripts but not of the large subunit transcripts. *Journal of Plant Physiology* 143: 306-312.

Van Oosten, J.J., D. Wilkens, and R.T. Besford (1995). Acclimation of tomato to different carbon dioxide concentrations. Relationships between biochemistry and gas exchange during leaf development. *New Phytologist* 130: 357-367.

Van Vuuren, M.M.I., D. Robinson, A.H. Fitter, S.D. Chasalow, L. Williamson, and J.A. Raven (1997). Effects of elevated atmospheric CO_2 and soil water availability on root biomass, root length and N, P, and K uptake by wheat. *New Phytologist* 135: 455-466.

Vierling, E. (1991). The roles of heat shock proteins in plants. *Annual Review of Plant Physiology and Molecular Biology* 42: 579-620.

Volkmar, K.M., Y. Hu, and H. Steppuhn (1998). Physiological responses of plants to salinity: A review. *Canadian Journal of Plant Science* 78: 19-27.

Vu, C.V, L.H. Allen, K.J. Boote, and G. Bowes (1997). Effects of elevated CO_2 and temperature on photosynthesis and rubisco in rice and soybean. *Plant, Cell and Environment* 20: 68-76.

Vu, C.V., L.H. Allen, and G. Bowes (1983). Effects of light and elevated atmospheric CO_2 on the ribulose bisphosphate carboxylase activity and ribulose bisphosphate level of soybean leaves. *Plant Physiology* 73: 729-734.

Wagenet, R.J., R.R. Rodriguez, W.F. Campbell, and D.L. Turner (1983). Fertilizer and salty water effects on *Phaseolus. Agronomy Journal* 75: 161-166.

Waggoner, P.E. (1997). How much land can American farmers spare? In *Crop and Livestock Technologies: RCA III Symposium,* eds. B.C. English, R.L. White, and L.-H. Chuang. Ames, IA: Iowa State University Press, pp. 23-44.

Wagner, J., A. Lüscher, C. Hillebrand, B. Kobald, N. Spitaler, and W. Larcher (2001). Sexual reproduction of *Lolium perenne* L. and *Trifolium repens* L. under free air CO_2 enrichment (FACE) at two levels of nitrogen application. *Plant, Cell and Environment* 24: 957-965.

Wagner, W., F. Keller, and A. Wiemken (1983). Fructan metabolism in cereals: Induction in leaves and compartmentation in protoplasts and vacuoles. *Zeitschreft für Pflanzenphysiologie* 112: 359-372.

Wahid, A., E. Rasul, and A. Rao (1999). Germination of seeds and propagules under salt stress. In *Handbook of Plant and Crop Stress,* ed. M. Pessarakli. New York: Marcel Dekker, pp. 153-167.

Walker, J.M. (1969). One degree increments in soil temperature affect seedling behavior. *Soil Science Society of America Proceedings* 33: 729-736.

Wallace, H.A. (1920). Mathematical enquiry into the effect of weather on corn yields. *Monthly Weather Review* 48: 439-446.

Walmsley, L., M.R. Ashmore, and J.N.B. Bell (1980). Adaptation of radish *Raphanus sativus* L. in response to continuous exposure to ozone. *Environmental Pollution* 23: 165-177.

Wang, H., Q. Qi, P. Schorr, A.J. Cutler, W. Crosby, and L.C. Fowke (1998). ICK1, a cyclin-dependent protein kinase inhibitor from *Arabidopsis thaliana* interacts with both Cdc2a and CycD3, and its expression is induced by abscisic acid. *Plant Journal* 15: 501-510.

Wardlaw, I.F. (1990). The control of carbon partitioning in plants. *New Phytologist* 116: 341-381.

Wardlaw, I.F. and C.W. Wrigley (1994). Heat tolerance in temperate cereals: An overview. *Australian Journal of Plant Physiology* 21: 695-703.

Warrag, M.O.A. and A.E. Hall (1984). Reproductive responses of cowpea (*Vigna unguiculata* (L.) Walp.) to heat stress. II. Responses to night air temperature. *Field Crops Research* 8: 17-33.

Warren Wilson, J. (1967). Ecological data on dry-matter production by plants and plant communities. In *The Collection and Processing of Field Data,* eds E.F. Bradley and O.T. Denmead. New York: John Wiley and Sons Ltd., pp. 77-123.

Watling, J.R., M.C. Press, and W.P. Quick (2000). Elevated CO_2 induces biochemical and ultrastructural changes in leaves of C_4 cereal sorghum. *Plant Physiology* 123: 1143-1152.

Watson, D.J. (1963). Climate, weather, and plant yield. In *Environmental Control of Plant Growth,* ed. L.T. Evans. New York: Academic Press, pp. 337-349.

Watson, R., J. Houghton, and D. Yihui (Eds.) (2001). *Climate Change 2001: The Scientific Basis. Summary for Policymakers and Technical Summary of the Working Group I Report.* Cambridge, UK: Cambridge University Press.

Watt, B.K. and A.L. Merrill (1963). *Composition of Foods.* Washington, DC: U.S. Department of Agriculture.

Wechsung, G., F. Wechsung, G.W. Wall, F.J. Adamsen, B.A. Kimball, P.J. Pinter, R.L. LaMorte Jr., R.L. Garcia, and T. Kartschall (1999). The effects of free-air CO_2 enrichment and soil water availability on spatial and seasonal patterns of wheat root growth. *Global Change Biology* 5: 519-529.

Weigand, C.L. and J.A. Cuellar (1981). Duration of grain filling and kernel weight of wheat as affected by temperature. *Crop Science* 21: 95-101.

Went, F.W. (1959). Effects of environment of parent and grandparent generations on tuber production in potatoes. *American Journal of Botany* 46: 277-278.

Wheeler, T.R., P.Q. Craufurd, R.H. Ellis, J.R. Porter, and P.V. Vara Prasad (2000). Temperature variability and the yield of annual crops. *Agriculture, Ecosystems and Environment* 82: 159-167.

Wheeler, T.R., T.D. Hong, R.H. Ellis, G.R. Batts, J.I.L. Morison, and P. Hadley (1996). The duration and rate of grain growth, and harvest index, of wheat (*Triticum aestivum* L.) in response to temperature and CO_2. *Journal of Experimental Botany* 47: 623-630.

White, P.R. (1937). Seasonal fluctuations in growth rates of excised tomato root tips. *Plant Physiology* 12: 183-190.

Whittaker, J.B. (2001). Insects and plants in a changing atmosphere. *Journal of Ecology* 89: 507-518.

Whittaker, J.B. and N.P. Tribe (1996). An altitudinal transect as an indicator of responses of insects to climate change. *European Journal of Entomology* 93: 319-324.

Wignarajah, K. (1995). Mineral nutrition of plants. In *Handbook of Plant and Crop Physiology,* ed. M. Pessarakli. New York: Marcel Dekker Inc., pp. 193-221.

Wilhelm, E.P., R.E. Mullen, P.L. Keeling, and G.W. Singletary (1999). Heat stress during grain filling in maize: Effects on kernel growth and metabolism. *Crop Science* 39: 1733-1741.

Williams, J.H.H., A.L. Winters, and J.F. Farrar (1992). Sucrose: A novel plant growth regulator. In *Molecular, Biochemical and Physiological Aspects of Plant Respiration,* eds. H. Lambers and L.H.W. van der Plas. The Hague: SPB Academic, pp. 463-469.

Williams, L.E., T.M. DeJong, and D.A. Phillips (1981). Carbon and nitrogen limitations on soybean seedling development. *Plant Physiology* 68: 1206-1209.

Williams, M., P.R. Shewry, D.W. Lawlor, and J.L. Harwood (1995). The effects of elevated temperature and atmospheric carbon dioxide concentration on the quality of grain lipids in wheat (*Triticum aestivum* L.) grown at two levels of nitrogen application. *Plant, Cell and Environment* 18: 999-1009.

Wilson, A.T. (1978). Pioneer agriculture explosion and CO_2 levels in the atmosphere. *Science* 273: 40-41.

Wink, M. (1997). Special nitrogen metabolism. In *Plant Biochemistry,* eds. P.M. Dey and J.B. Harborne. San Diego: Academic Press, pp. 439-486.

Wolters, J.H.B. and M.J.M. Martens (1987). Effects of air pollutants on pollen. *Botanical Review* 53: 372-414.

World Meteorological Organization (1998). *Scientific Assessment of Ozone Depletion: 1998, Executive Summary.* Geneva: World Meteorological Organization. <www.al.noaa.gov/WWWHD/pubdocs/Assessment98/ExecSum98.pdf>.

Wright, S.R., H.D. Coble, C.D. Raper, and T.W. Rufty (1999). Comparative responses of soybean *(Glycine max),* sicklepod *(Senna obtusifolia),* and Palmer amaranth *(Amaranthus palmeri)* to root zone and aerial temperatures. *Weed Science* 47: 167-174.

Xiong, L. and J.-K. Zhu (2002). Molecular and genetic aspects of plant responses to osmotic stress. *Plant, Cell and Environment* 25: 131-139.

Xu, J., G.H. Pemberton, E.C. Almira, D.R. McCarty, and K.E. Koch (1995). The Ivr1 gene for invertase in maize. *Plant Physiology* 108: 1293-1294.

Yamaguchi, J. (1978). Respiration and the growth efficiency in relation to crop productivity. *Journal of the Faculty of Agriculture, Hokkaido University* 59: 59-129.

Yamamura, K. and K. Kiritani (1998). A simple method to estimate the potential increase in the number of generations under global warming in temperate zones. *Applied Entomology and Zoology* 33: 289-298.

Yan, F., X. Liang, and X. Zhu (1999). The role of DIMBOA on the feeding of Asian corn borer, *Ostrinia furnacalis* (Guenee) (Lep., Pyralidae). *Journal of Applied Entomology* 123: 49-53.

Yang, Y., H.B. Kwon, H.P. Peng, and M.C. Shih (1993). Stress responses and metabolic regulation of glyceraldehyde-3-phosphate dehydrogenase genes in *Arabidopsis. Plant Physiology* 101: 209-216.

Yegappan, T.M., D.M. Paton, and C.T. Gates (1982). Water stress in sunflower (*Helianthus annuus* L.) 2. Effects on leaf cells and leaf area. *Annals of Botany* 49: 63-68.

Yeh, K.-W., T.-L. Jinn, C.-H. Yeh, Y.-M. Chen, and C.-Y. Lin (1994). Plant low-molecular-mass heat-shock proteins: Their relationship to the acquisition of thermotolerance in plants. *Biotechnology and Applied Biochemistry* 19: 41-49.

Yeo, A. (1983). Salinity resistance: Physiologies and prices. *Physiologia Plantarum* 58: 214-222.

Yokota, H. (1986). Alleviation of NaCl-induced inhibition of pollen tube growth by monovalent and divalent cations. *Soil Science and Plant Nutrition* 32: 343-350.

Yokoyama, Y., K. Lambeck, P. De Deckker, P. Johnston, and L.K. Fifield (2000). Timing of the Last Glacial Maximum from observed sea-level minima. *Nature* 406: 713-716.

Yong, W.H.J., S.C. Chin, S.D. Latham, C.H. Hocart, and G.D. Farquhar (2000). Effects of elevated [CO_2] and nitrogen nutrition on cytokinins in the xylem sap and leaves of cotton. *Plant Physiology* 124: 767-779.

Young, K.J. and S.P. Long (2000). Crop ecosystem responses to climatic change: Maize and sorghum. In *Climate Change and Global Crop Productivity,* eds. K.R. Reddy and H.F. Hodges. Wallingford, UK: CABI Publishing, pp. 107-132.

Younglove, T., P.M. McCool, R.C. Musselman, and M.E. Kahl (1994). Growth-stage dependent crop yield response to ozone exposure. *Environmental Pollution* 86: 287-295.

Yuan, L., Z. Yanqun, C. Haiyan, C. Jianjun, Y. Jilong, and H. Zhide (2000). Intraspecific responses in crop growth and yield of 20 wheat cultivars to enhanced ultraviolet-B radiation under field conditions. *Field Crops Research* 67: 25-33.

Zadoks, J.C., T.T. Chang, and C.F. Konzak (1974). A decimal code for the growth stages of cereals. *Weed Research* 14: 415-421.

Zanetti, S., U.A. Hartwig, and J. Nösberger (1998). Elevated atmospheric CO_2 does not affect per se the preference for symbiotic nitrogen as opposed to mineral nitrogen of *Trifolium repens* L. *Plant, Cell and Environment* 21: 623-630.

Zeng, L. and M.C. Shannon (2000). Salinity effects on seedling growth and yield components of rice. *Crop Science* 40: 996-1003.

Zeng, L., M.C. Shannon, and S.M. Lesch (2001). Timing of salinity stress affects rice growth and yield components. *Agricultural Water Management* 48: 191-206.

Zerihun, A., B.A. McKenzie, and J.D. Morton (1998). Photosynthate costs associated with the utilization of different nitrogen-forms: Influence on the carbon balance of plants and shoot-root biomass partitioning. *New Phytologist* 138: 1-11.

Zhang, J., J.E. Specht, G.L. Graef, and B.E. Johnson (1992). Pubescence density effects on soybean seed yield and other agronomic traits. *Crop Science* 32: 641-648.

Zhao, K., R. Munns, and R.W. King (1991). Abscisic acid levels in NaCl-treated barley, cotton and saltbush. *Australian Journal of Plant Physiology* 18: 17-24.

Zheng, Y., H. Shimizu, and J.D. Barnes (2002). Limitations to CO_2 assimilation in ozone-exposed leaves of *Plangato Major. New Phytologist* 155: 67-78.

Zhong, H. and A. Läuchli (1993). Changes in cell wall composition and polymer size in primary roots of cotton seedlings under high salinity. *Journal of Experimental Botany* 44: 773-778.

Zhou, X., R. Harrington, I.P. Woiwod, J.N. Perry, J.S. Bale, and S.J. Clark (1995). Effects of temperature on aphid phenology. *Global Change Biology* 1: 303-313.

Zhu, J.-K. (2000). Genetic analysis of plant salt tolerance using *Arabidopsis. Plant Physiology* 124: 941-948.

Zhu, J.-K. (2001). Plant salt tolerance. *Trends in Plant Science* 6: 66-71.

Zhu, J.-K., P.M. Hasegawa, and R.A. Bressan (1997). Molecular aspects of osmotic stress in plants. *Critical Reviews in Plant Sciences* 16: 253-277.

Ziska, L.H. (2000). The impact of elevated CO_2 on yield loss from a C_3 and C_4 weed in field-grown soybean. *Global Change Biology* 6: 899-905.

Ziska, L.H. (2001). Changes in competitive ability between a C_4 crop and a C_3 weed with elevated carbon dioxide. *Weed Science* 49: 622-627.

Ziska, L.H. (2003). Evaluation of the growth response of six invasive species to past, present and future atmospheric carbon dioxide. *Journal of Experimental Botany* 54: 395-404.

Ziska, L.H. and J.A. Bunce (1994). Direct and indirect inhibition of single leaf respiration by elevated CO_2 concentrations: Interaction with temperature. *Physiologia Plantarum* 90: 130-138.

Ziska, L.H., J.A. Bunce, and F.A. Caulfield (2001). Rising atmospheric carbon dioxide and seed yield of soybean genotypes. *Crop Science* 41: 385-391.

Ziska, L.H., S. Faulkner, and J. Lydon (2004). Changes in biomass and root:shoot ratio of field-grown Canada thistle *(Cirsium arvense),* a noxious, invasive weed, with elevated CO_2: Implications for control with glyphosate. *Weed Science* 52: 584-588.

Ziska, L.H., P.A. Manalo, and R.A. Ordonez (1996). Intraspecific variation in the response of rice (*Oryza sativa* L.) to increased CO_2 and temperature: Growth and yield response of 17 cultivars. *Journal of Experimental Botany* 47: 1353-1359.

Ziska, L.H. and J.R. Teasdale (2000). Sustained growth and increased tolerance to glyphosate observed in a C_3 perennial weed, quackgrass *(Elytrigia repens),* grown at elevated carbon dioxide. *Australian Journal of Plant Physiology* 27: 159-166.

Ziska, L.H., J.R. Teasdale, and J.A. Bunce (1999). Future atmospheric carbon dioxide may increase tolerance to glyphosate. *Weed Science* 47: 608-615.

Index

Page numbers followed by the letter "f" indicate figures; those followed by the letter "t" indicate tables; and those followed by the letter "b" indicate boxes.